(x)Fx = Everything is F (For all x, x, is F)
(∃x)Fx = Something is F (For some x, x, is F)

reverse squiggle rules	
~(∃x)Fx ∴ (x)~Fx	~(x)Fx ∴ (∃x)~Fx

dropping existentials	
(∃x)Fx ∴ Fa	(use a NEW constant for "a")

Drop (∃x)'s before (x)'s!

dropping universals	
(x)Fx ∴ Fa	(use ANY constant for "a")

Fa a = b ∴ Fb	equals may substitute for equals

self-identity
∴ a = a

MODAL LOGIC

□A = A is necessary (true in all possible worlds)
◇A = A is possible (true in some possible world)

reverse squiggle rules	
~◇A ∴ □~A	~□A ∴ ◇~A

dropping diamonds	
◇A W ∴ A	(use a NEW string of W's)

Drop ◇'s before □'s!

dropping boxes	
□A W ∴ A	(use ANY string of W's — or no W's at all)

SYMBOLIC LOGIC

CLASSICAL AND ADVANCED SYSTEMS

HARRY J. GENSLER
Loyola University of Chicago

PRENTICE HALL, Englewood Cliffs, New Jersey 07632

Library of Congress Cataloging-in-Publication Data

Gensler, Harry J.
    Symbolic logic: classical and advanced systems / Harry J.
  Gensler.
        p. cm.
    Includes bibliographical references.
    ISBN   0-13-879941-5
    1. Logic, Symbolic and mathematical.   I. Title.
  BC135.G39   1990   89-48288
  160—dc20

Editorial/production supervision: *Edith Riker/Jeanne Sillay Jacobson*
Cover design: *Bruce Kenselaar*
Manufacturing buyer: *Carol Bystrom*

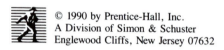
Printed in the United States of America
10   9   8   7   6   5   4   3   2   1

ISBN   0-13-879941-5

Prentice-Hall International (UK) Limited, *London*
Prentice-Hall of Australia Pty. Limited, *Sydney*
Prentice-Hall Canada Inc., *Toronto*
Prentice-Hall Hispanoamericana, S.A., *Mexico*
Prentice-Hall of India Private Limited, *New Delhi*
Prentice-Hall of Japan, Inc., *Tokyo*
Simon & Schuster Asia Pte. Ltd., *Singapore*
Editora Prentice-Hall do Brasil, Ltda., *Rio de Janeiro*

Contents

Preface

This introduction to symbolic logic doesn't presume any previous study of the subject. It covers these topics:

- the classical systems: propositional and quantificational logic (Chapters 2 to 5)
- some advanced philosophical systems: modal, deontic, and belief logic (Chapters 6 to 8)
- the formalization of an ethical theory (Chapter 9)

I've stressed matters that relate closely to other areas of philosophy. Chapters 8 and 9 contain original systems and each chapter builds on the previous ones.

The various systems use a common proof method that I've developed. The method is easy to learn and has three advantages over the standard approach: It uses a smaller and more natural set of rules; it uses an automatic proof strategy instead of guesswork; and it leads to a refutation of invalid arguments.

The book is filled with important and interesting arguments, many of which are taken from philosophy. The rigorous tools of logic are used to clarify reasoning on real issues.

The text is user-friendly. Explanations are clear and direct. I avoid technical jargon as much as possible. And the many exercises are carefully planned to lead the student by small steps to master the material.

I've developed a software program, *LogiCola*, to supplement classroom activity. LogiCola is an interactive computer program to help students learn logic. It randomly generates problems, gives feedback on answers, provides explanations, and keeps homework records. Most of the sections in this book with exercises have corresponding LogiCola exercises. The program runs on IBM (or IBM-compatible) personal computers and is available from Prentice Hall.

This book overlaps with my more basic text, *Logic: Analyzing and Appraising Arguments*. Chapters 3–6 of that book (with minor improvements and shorter exercise sets) have become Chapters 2–5 of this present book. At first I was going to include only a brief version of this material. But most of my students (including those with a previous logic course) preferred the fuller explanations.

I wish to thank various people who helped in this project. My teachers stimulated my interest in logic—especially Arthur Burks (of Michigan, where I did my doctorate) and Hector-Neri Castañeda (who strongly influenced my deontic material). My students at Loyola pushed me to develop my teaching methods. Paul Moser and Tom Carson, my colleagues at Loyola, encouraged me on the project. Alvin Plantinga gave me some hints on the modal logic section. Patricia Haggard, a Loyola grad student, helped me improve my explanations. Keith Cooper, who wrote the study guide for my more basic logic book, gave me ideas on how to improve a couple of my explanations. Joseph

Heider and Caroline Carney, the Prentice Hall philosophy editors, did a great job of guiding the project. Loyola University gave me a sabbatical semester to work on the book. Finally, I give special thanks to a fellow Jesuit and former logic student of mine, Tim Clancy. Without Tim's impetus, this project wouldn't have started.

If you have suggestions for improving this text (or interesting arguments that I might add), please send them to me. I thank you in advance.

Harry J. Gensler
Philosophy Department
Loyola University
6525 North Sheridan
Chicago, IL 60626

1

Introduction to Logic

1.1 PRELIMINARIES

Logic is the analysis and appraisal of arguments. Logic is something you *do*. You *do* logic when you try to clarify reasoning and separate good from bad reasoning. You'll do a lot of logic as you work through this book.

You'll examine philosophical reasoning on things like free will and determinism, the existence of God, and the nature of morality. You'll also study reasoning on backpacking, football, and the Bible. Logic isn't an irrelevant game with funny symbols. Rather, it's a useful means to clarify and evaluate our reasoning, whether on everyday topics or on life's deeper questions. Many people study logic to improve their analytical powers. I hope this is one of your goals. You'll mostly learn by doing; this book contains a lot of exercises.

We'll start simply and won't presume any previous study of logic. We'll cover the classical systems of symbolic logic and then move into some advanced philosophical systems.

1.2 VALID ARGUMENTS

I begin my lower-level logic course with a multiple-choice test. The test has ten problems. Each presents some information (premises) and asks what conclusion follows logically. The problems are easy, but most of my students get only about half of them right.

I take two problems from the clever fifth-grade logic textbook, *Harry Stottle-meier's Discovery*.* ("Harry Stottlemeier" sounds like "Aristotle," the ancient Greek philosopher who invented logic.) Here's one problem:

* Matthew Lipman, *Harry Stottlemeier's Discovery* (Caldwell, NJ: Universal Diversified Services, 1974).

If you overslept, you'll be late.
You aren't late.

THEREFORE: ???

[Correct answer: You didn't oversleep.]

Almost everyone picks the correct conclusion. In the book, Harry's mother keeps saying "If you oversleep, you'll be late for school." Assume that she's right. Assume further that Harry *isn't* late. Then we can conclude that he *didn't* oversleep.

Here's another problem:

If you overslept, you'll be late.
You didn't oversleep.

THEREFORE: ???

Many students conclude: "You aren't late." But this is wrong. The first premise says only what happens if you oversleep. It doesn't say what happens if you *don't* oversleep. You might be late for other reasons. In the story, Harry sometimes gets up on time but yet is delayed on the way to school. So he might get up on time but still be late. The correct answer is that none of the listed conclusions follow.

Untrained intuitions about what follows from what are often unreliable. But logical intuitions can be developed. Yours will improve as you work through this book. You'll also learn special techniques for testing arguments.

Here are two important definitions:

> *Logic* is the analysis and appraisal of arguments.

> An *argument* is a set of statements
> consisting of premises and a conclusion.

An *argument* here isn't a quarrel or a fight. Rather, it's the verbal expression of a reasoning process.

Consider our first argument again (here "∴" is short for *therefore*):

If you overslept, you'll be late.
You aren't late.
∴ You didn't oversleep.

This argument is *valid*. *Valid* is the most important technical term in this book.

> A *valid* argument is one in which
> it would be contradictory for
> the premises to be true
> but the conclusion false.

In calling an argument *valid*, we aren't saying whether the premises are true. We're just saying that the conclusion *follows from* the premises—that *if* the premises were true, *then* the conclusion would also have to be true. In saying this, we implicitly assume that there's no shift in the meaning or reference of the terms. We must use "overslept," "late," and "you" in exactly the same way throughout the argument.

Let's use "1" to stand for "You oversleep" and "2" for "You're late." This way of writing the argument better reveals its *logical form* or *structure*:

If you overslept, you'll be late.	If 1 then 2
You aren't late.	Not-2
∴ You didn't oversleep.	∴ Not-1

Our argument is valid because its form is valid. If we take another argument of the same form (but substituting other ideas for "1" and "2"), this second argument will also be valid. Here's an example:

If we're close to the top of Forester Pass, then we're in a treeless tundra.	If 1 then 2
We *aren't* in a treeless tundra.	Not-2
∴ We *aren't* close to the top of Forester Pass.	∴ Not-1

Logic studies *forms* of reasoning. The *content* might deal with anything—backpacking, mathematics, cooking, nuclear physics, ethics, or whatever. When you learn logic you're learning tools of reasoning that can be applied to any subject.

Consider again our example of an invalid argument:

If you overslept, you'll be late.	If 1 then 2
You didn't oversleep.	Not-1
∴ You aren't late.	∴ Not-2

Here the second premise denies *part-1* of the if-then. This is enough to make the logical form defective. In our valid example, we denied *part-2*.

Here's another argument with the same invalid form:

If you're on top of Forester Pass, then you have a magnificent view.	If 1 then 2
You *aren't* on top of Forester Pass.	Not-1
∴ You *don't* have a magnificent view.	∴ Not-2

This is invalid. The premises could be true while the conclusion is false. Maybe you have a magnificent view of Lake Michigan!

We don't have to visit Forester Pass to tell whether an argument about it is valid. We need only appeal to abstract matters like logical form. But to tell whether *premises* about the pass are true, we (or someone) may have to climb the John Muir Trail in California to check it out.

1.3 SOUND ARGUMENTS

Logicians distinguish *valid arguments* from *sound arguments*. I define these two terms as follows:

> A *valid* argument is one in which
> it would be contradictory for
> the premises to be true
> but the conclusion false.

> A *sound* argument is one that
> is valid and has true premises.

Calling an argument "valid" says nothing about whether its premises are true. But calling it "sound" says that it's valid *and* has true premises. We could also define a sound argument as one in which (1) the premises are all true and (2) the conclusion follows logically from the premises.

Here's an example of a sound argument:

If you're reading this, you aren't illiterate. You're reading this. ∴ You aren't illiterate.	TRUE premises. VALID argument: the conclusion follows. SOUND argument.

The conclusion of a sound argument must be true.

An argument could be unsound in one of two ways. It might have a false premise. Or it might be invalid. The conclusion of an unsound argument could be either true or false.

This following argument is unsound (and defective) because it has a false premise:

Gensler owns General Motors. If Gensler own General Motors, then he's a millionaire. ∴ Gensler is a millionaire.	FALSE first premise. VALID argument: the conclusion follows from the premises. UNSOUND argument.

Since this argument is *unsound*, it doesn't prove its conclusion. But the conclusion might happen to be true anyway.

The next argument is unsound (and defective) because it's invalid:

Gensler was born in America. If Gensler was born in Chicago, then he was born in America. ∴ Gensler was born in Chicago.	TRUE premises. INVALID argument: the conclusion doesn't follow. UNSOUND argument.

Since this is *unsound*, it doesn't prove its conclusion. But the conclusion, again, might just happen to be true.

These examples illustrate this pair of principles:

> The conclusion of a sound argument
> (a valid argument with true premises)
> is always true.
>
> The conclusion of an unsound argument
> (an argument that is invalid
> or has false premises)
> might be either true or false.

The first principle is important when we try to prove a conclusion. Then we must make sure of two things: (1) that our premises are true, and (2) that our conclusion follows from our premises. If we know these two things, then we can know that our conclusion has to be true.

The second principle is crucial when we try to refute an opponent's argument by showing it to be *unsound*. There are two strategies we might use. We might try to show that one of the premises is false. Or we might try to show that the argument is invalid (that the conclusion doesn't follow from the premises). If the argument has a false premise or is invalid, then our opponent *hasn't proved* the conclusion. But the conclusion still might be true—and our opponent might later discover a better argument to prove it. To show a view to be false, we must do more than just refute an argument for it. We have to invent an argument of our own that shows the view to be false.

Besides asking whether premises are true, we could ask *how certain* they are (to ourselves or to others). We'd like our premises to be utterly certain and obvious to everyone. We usually have to settle for less than this. Our premises are often educated guesses or personal convictions. Our arguments are only as strong as their premises. This suggests a third strategy for criticizing an argument: We could try to show that one or more of the premises are very uncertain.

Arguments can go wrong in other ways. They might be circular (a premise might be a mere rephrasing of the conclusion). Or they might be ambiguous or unclear or needlessly complex. Or they might be irrelevant to the issue at hand.

Often, especially in philosophy, one argument leads to further arguments. Consider this objection to belief in God stated by St. Thomas Aquinas*:

All beliefs unnecessary to explain our experience ought to be rejected.

The belief that there is a God is unnecessary to explain our experience.

∴ The belief that there is a God ought to be rejected.

The argument is *valid*—the conclusion follows from the premises. Are the premises *true*? Aquinas might have clarified and then rejected the first premise as self-refuting. (Isn't the first premise itself unnecessary to explain our experience?) Instead he rejected the second premise. He gave further arguments to show that belief in God *is* necessary to explain our experience (of motion, causality, and so on). Are Aquinas's further arguments sound? We must leave debate on this to philosophy of religion. But logic

---

* In this book, I often say that an argument is from a given philosopher. I mean that the person's writings contain or suggest the ideas in the argument; the phrasing is usually mine. In the present case, for example, Aquinas worded the argument differently.

can clarify the discussion. It can help us express reasoning clearly, determine whether a conclusion follows from the premises, and focus on key premises to defend or criticize.

In a few cases, an argument is valid because of its *content*, not because of its *form*. Here's an example:

This is green.   x is G
∴ This is not red.   ∴ x is not R

This is valid, since it would be impossible for the premise to be true but the conclusion false. The validity is due, not to the form, but to the logical connections between "green" and "red." We can make the argument *formally valid* by adding a premise expressing these logical connections—namely, "It's impossible for something to be both green and red." This book will focus on formal validity.

Logicians classify arguments as "deductive" or "inductive." A *deductive argument* is one in which the truth of the premises is supposed to *necessitate* the truth of the conclusion. All the examples I've given so far have been deductive. An *inductive argument* is one in which the truth of the premises is only supposed to make the conclusion *probably* true. Here's an inductive argument:

Most people living in Chicago were born in Illinois.
Gensler lives in Chicago.
This is all we know about the matter.
∴ Probably Gensler was born in Illinois.

Suppose the premises are true. Then it's *likely* that Gensler was born in Illinois. But the premises don't guarantee this. Inductive reasoning is a reasoned form of guessing. This book will focus on deductive arguments.

I have two final points on terminology. We'll call statements *true* or *false* (not valid or invalid). And we'll call arguments *valid* or *invalid* (not true or false). This is conventional usage. It pains a logician's ears to hear "invalid statement" or "false argument."

1.4 THE PLAN OF THIS BOOK

This book is about *deductive* or *formal logic*. It deals with the question, "Is this argument valid; does the conclusion follow from the premises?" This validity question is very objective. With few exceptions (and neglecting calculation errors), all who have mastered the logical tools should, despite philosophical differences, agree on whether an argument is valid or invalid.

We'll set up symbolic languages (called logical calculuses) to help us test arguments. We'll see the propositional calculus in the next chapter. This language reflects in a simplified way the logical functions of words like *if-then* and *not*. A calculus isn't supposed to reflect every subtlety of English usage. Rather it's a tool for testing arguments. An appropriate symbolism makes argument-testing easier.

A *logical calculus* (or formal system) is more precisely defined as an artificial language with notational grammar rules and notational rules for determining validity.

The chapters of this book present a series of logical systems. Each is expressed as a logical calculus and builds on previous systems.

We won't cover informal logic. *Informal logic* deals with aspects of logic other than validity testing—for example, probability, inductive reasoning, definitions, and informal fallacies. I cover informal logic in the second half of my more basic text, *Logic*: *Analyzing and Appraising Arguments*.

2

Basic Propositional Logic

2.1 PROPOSITIONAL FORMULAS

Our *propositional calculus* (PC) is a language for testing those arguments whose validity depends on one or more of these ideas: *not*, *and*, *or*, *if-then*, and *if and only if*. Logicians have developed PC over the last hundred years or so. PC is the foundational system of modern symbolic logic. Our further systems all build on PC.

PC uses capital letters (including capitals with primes) to represent true or false *statements*. It uses parentheses to group things together. And it uses five special symbols:

SYMBOL	MEANING	NAME
~	not	squiggle
·	and	dot
∨	or	vee
⊃	if-then	horseshoe
≡	if and only if	three bar

Grammatical sentences of PC are called *wffs*, or well-formed formulas. PC wffs are typographical strings constructable using the following rules:

1. Any capital letter (by itself or with primes) is a wff.
2. The result of prefixing any wff with ''~'' is a wff.
3. The result of joining any two wffs by ''·'' or ''∨'' or ''⊃'' or ''≡'' and enclosing the result in parentheses is a wff.

Rule 1 gives us wffs like these:

P   ["I've been to Paris"]
Q   ["I've been to Quebec"]

The capital letters here represent true or false statements. Applying rule 2 to these gives us more wffs:

~P   Not-P   ["I *haven't* been to Paris"]
~Q   Not-Q   ["I *haven't* been to Quebec"]

We could also read "P" as "It's *true* that P," and "~P" as "It's *false* that P." Repeating rule 2 gives us longer and longer wffs:

~P   Not-P
~~P   Not-not-P   [= P]
~~~P   Not-not-not-P   [= Not-P]
~~~~P   Not-not-not-not-P   [= P]
              . . . and so forth.

There's no longest PC wff.
     Rule 3 gives us wffs like these:

(P · Q)   P and Q
(P ∨ Q)   P or Q
(P ⊃ Q)   If P then Q
(P ≡ Q)   P if and only if Q

We'll sometimes read "(P · Q)" as "*Both* P *and* Q," and "(P ∨ Q)" as "*Either* P *or* Q." Applying rules 2 and 3 on wffs already constructed gives us further wffs like these:

((P · Q) ⊃ P)   If P and Q, then P
(~Q ⊃ ~P)   If not-Q, then not-P
~(P ∨ Q)   Not either P or Q

     Our rules require a pair of parentheses for each "·," "∨," "⊃," or "≡." So this isn't a wff:

P · Q ⊃ P

This ambiguous form could mean either of these:

((P · Q) ⊃ P)   If P-and-Q, then P
(P · (Q ⊃ P))   P, and if-Q-then-P

We don't add parentheses for "~." So these aren't wffs:

~(P)   (~P)

The correct form is simply "~P." Since "~P" isn't ambiguous, parentheses here serve no purpose.

This isn't a wff:

~P · Q

This form could mean either of these (the usual convention is that it means the second):

~(P · Q)   (~P · Q)

Don't read them both in the same way, as "not P and Q." The two differ in what is negated. The first negates "(P · Q)" while the second just negates "P." Reading "both" for "(" brings out the difference:

Read "~(P · Q)" as "Not *both* P and Q."
Read "(~P · Q)" as "*Both* not-P and Q."

The first says "Not both are true; in other words, at least one is false." The second says "P is false and Q is true." We could also use pauses to mark the difference:

Read "~(P · Q)" as "Not (pause) P and Q."
Read "(~P · Q)" as "Not-P (pause) and (pause) Q."

When you read formulas, try to group the parts correctly.

Similarly, don't read these two formulas in the same way, as "not P or Q":

~(P ∨ Q)   (~P ∨ Q)

Rather, read "either" in place of "(":

Read "~(P ∨ Q)" as "Not *either* P or Q."
Read "(~P ∨ Q)" as "*Either* not-P or Q."

The first says "Not either is true; in other words, both are false." The second says "Either P is false or Q is true." Or use pauses:

Read "~(P ∨ Q)" as "Not (pause) P or Q."
Read "(~P ∨ Q)" as "Not-P (pause) or (pause) Q."

Read "if" in place of the "(" that goes with "⊃":

Read "~(P ⊃ Q)" as "Not *if* P then Q."
Read "(~P ⊃ Q)" as "*If* not-P then Q."

The first says "P doesn't imply Q." The second says "Not-P implies Q." Or, with pauses:

> Read "~(P ⊃ Q)" as "Not (pause) P implies Q."
> Read "(~P ⊃ Q)" as "Not-P (pause) implies (pause) Q."

We'll sometimes read "⊃" as "implies" instead of "if-then." The two are similar in English, except that we use quotation marks or "that" clauses with "implies":

> *If* it's snowing, *then* it's cold.

> "It's snowing" *implies* "It's cold."
> That it's snowing *implies* that it's cold.

2.2 TRANSLATIONS

Our first rule for translating from English to PC mirrors how we suggested reading the wffs:

> "Both," "either," and "if" generally translate as a left-hand parenthesis.

For *"not either"* write "~("; and for *"either not"* write "(~":

It *won't either* rain or snow.	*Either* it *won't* rain or it will snow.
= *Not either* R or S.	= *Either not* R or S.
= ~(R ∨ S)	= (~R ∨ S)

Similarly, for *"if both"* write "((" —as in the first of these two examples:

If it *both* snows and rains then I'll get wet.	It snows and *if* it rains then I'll get wet.
= If *both* S and R then W.	= S and *if* R then W.
= ((S · R) ⊃ W)	= (S · (R ⊃ W))

Translate *"not both not"* as "~(~":

> *Not both not* A and not B
> = ~(~A · ~B)

Here the "(" keeps the negations from canceling out.

Our boxed rule has exceptions. For example, sometimes it's silly to translate "both" as "(":

> I saw them *both*.
> = S([wrong!]

Rules for translating from English to logic are rough guidelines that we should apply with common sense.

Here's another rule for translating from English to PC:

> A comma often indicates a formula's logical middle (so that the parts on either side glob together as units).

Here are two examples:

If A, then B and C   If A then B, and C
= (A ⊃ (B · C))   = ((A ⊃ B) · C)

If a sentence has no comma, ask yourself where the comma would naturally go. Consider this sentence:

If it snows then I'll go outside and I'll ski.

The sentence likely means this:

If it snows, then I'll go outside and I'll ski.
= If S, then G and K.
= (S ⊃ (G · K))

If a sentence has more than one comma, decide where the comma would go if the sentence were to use only *one* comma:

If it snows, then, if my cold is better, I'll ski.
= If it snows, then if my cold is better I'll ski.
= If S, then if B then K.
= (S ⊃ (B ⊃ K))

Make sure that your capital letters stand for whole statements. This translation is correct:

Tom and Judy were there.
= (T · J)

This means "Tom was there and Judy was there." "And" here connects whole statements. But this translation is wrong:

Tom and Judy got married to each other.
= (T · J) [wrong!]

This doesn't mean "Tom got married and Judy got married." Nor does it mean "Tom got married to each other and Judy got married to each other"! "And" here doesn't connect whole statements. Here's the proper translation:

Tom and Judy got married to each other.

= M

Long English sentences can be intimidating. Suppose you want to translate this sentence:

> If attempts to prove "God exists" fail in the same way as proofs for "There are other conscious beings besides myself," then belief in God is reasonable if and only if belief in other conscious beings is reasonable.

If the sentence confuses you, try underlining the logical words (*not*, *both*, *and*, *either*, *or*, *if*, *then*, and *if-and-only-if*) and commas:

> <u>If</u> attempts to prove "God exists" fail in the same way as proofs for "There are other conscious beings besides myself<u>,</u>" <u>then</u> belief in God is reasonable <u>if and only if</u> belief in other conscious beings is reasonab<u>le</u>.

Then replace the not-underlined parts with dots (or letters) and translate into PC:

If . . . , then . . . if and only if . . . .

= If F, then G if and only if O.

= $(F \supset (G \equiv O))$

Don't let complex wording intimidate you.

2.2A Exercises

Translate each of these into a PC wff.*

0. Both not A and B.   ANSWER: $(\sim A \cdot B)$

1. Not both A and B.
2. Both A and either B or C.
3. Either not A or B.
4. Either A or B, and C.
5. Either A, or B and C.
6. If A then not either not A or B.
7. If A, then B or C.
8. If A then B, or C.
9. Not either A or B.
10. Either both A and B or C.
11. If there's knowledge, then either there's an endless reasoning chain proving every premise by a prior proof or else some things are known without proof.
12. The statue is either by Cellini or by Michelangelo.
13. The statue isn't by either Cellini or Michelangelo.

---

* Each exercise section in this book has a problem 0 with the answer given. The "Answers to Selected Problems" section in the back of the book gives answers for problems 1, 3, 5, 10, 15, 20, 25, 30, and so forth.

14. If I don't have either 90¢ in exact change or a bus pass, then I won't ride the bus today.
15. If it's true that if the Bears change their defensive strategy then their defense will be mediocre, then if the Bears change their defensive strategy then they won't return to the Super Bowl.
16. Either you drove through both Cincinnati and Dayton, or you drove through Louisville.
17. She had hamburgers and french fries and a soft drink.
18. I'm going to Atlanta or Boston and you're going to Chicago.
19. I'm going to Atlanta or you're going to Boston and Chicago.
20. Everyone is male or female.

2.3 SIMPLE TRUTH TABLES

Capital letters represent true or false statements. Let "P" represent "I've been to Paris" and "Q" represent "I've been to Quebec." Each could be *true* or *false* (the two *truth values*)—represented by "1" and "0." There are four possible truth-value combinations for P and Q:

P	Q	
0	0	P and Q are both false
0	1	P is false but Q true
1	0	P is true but Q false
1	1	P and Q are both true

In the first case, I've been to neither Paris nor Quebec. In the second, I haven't been to Paris but I've been to Quebec. And so on.

A *truth table* gives a "logical diagram" for a wff. It lists all possible truth-value combinations for the letters and shows whether the wff is true or false on each combination. The truth table for "·" ("and") is very simple:

P	Q	(P · Q)	
0	0	0	"I've been to Paris *and* I've been to Quebec."
0	1	0	"(P · Q)" is a *conjunction*; P and Q are its *conjuncts*.
1	0	0	
1	1	1	

The table conveniently summarizes these facts:

- If P and Q are both false, then "(P · Q)" is false.
- If P is false but Q true, then "(P · Q)" is false.
- If P is true but Q false, then "(P · Q)" is false.
- If P and Q are both true, then "(P · Q)" is true.

Our "·" claims that *both* parts are true. So "(P · Q)" is true if both letters are true but is false if at least one is false. "I've been to Paris *and* Quebec" is true if "I've been to Paris" and "I've been to Quebec" are both true. Otherwise it's false.

These *truth equivalences* give the same information:

(0 · 0) = 0	(false · false) = false
(0 · 1) = 0	(false · true) = false
(1 · 0) = 0	(true · false) = false
(1 · 1) = 1	(true · true) = true

Here "(0 · 0) = 0," for example, says that a conjunction with both parts false is itself false.

Translations of "·" as "and" are only approximate. Our English word "and" suggests that the two parts have some *relevance* to each other. So this sentence is strange:

> 2 + 2 = 4 and I like Pepsi.

But the wff "(T · L)" is perfectly fine and is true just if both parts are true. Also, "and" often means "*and then*," as in these two quite different sentences:

> Suzy got married and had a baby.
> Suzy had a baby and got married.

Our "·" says that both parts are true, but it doesn't say which became true first. "·" abstracts from relevance and temporal sequence. "·" is simpler and more abstract than our English "and." Likewise, the other propositional symbols are simple abstract notions not completely identical in meaning with the English words used to translate them.

Here are the truth table and equivalences for "$\lor$" ("or"):

P	Q	(P $\lor$ Q)
0	0	0
0	1	1
1	0	1
1	1	1

$(0 \lor 0) = 0$
$(0 \lor 1) = 1$
$(1 \lor 0) = 1$
$(1 \lor 1) = 1$

"I've been to Paris *or* I've been to Quebec."

"(P $\lor$ Q)" is a *disjunction*. P and Q are its *disjuncts*.

Our "$\lor$" claims that *at least one* part is true. So "(P $\lor$ Q)" is true if one or both letters are true, but is false if both letters are false. "I've been to Paris *or* Quebec" is true if "I've been to Paris" or "I've been to Quebec" (or both) are true. Otherwise it's false.

Our English "or" has two main senses. "You get mashed potatoes or french fries" could mean either of these, depending on whether you're at a buffet or at an ordinary restaurant:

> "You get mashed potatoes or french fries *or both*" (inclusive "or").

> "You get mashed potatoes or french fries *but not both*" (exclusive "or").

Our "$\vee$" represents the *inclusive* sense: "one or the other *or both*" (sometimes put as "and/or"). This chart compares the two sense of "or" and shows how to translate each:

INCLUSIVE "OR"	EXCLUSIVE "OR"
P or Q or both	P or Q but not both
$(P \vee Q)$	$((P \vee Q) \cdot {\sim}(P \cdot Q))$

We'll generally take "or" to mean "$\vee$" (inclusive sense).

Here are the truth table and equivalences for "$\supset$" ("if-then" or "implies"):

P	Q	$(P \supset Q)$	
0	0	1	$(0 \supset 0) = 1$
0	1	1	$(0 \supset 1) = 1$
1	0	0	$(1 \supset 0) = 0$
1	1	1	$(1 \supset 1) = 1$

"*If* I've been to Paris *then* I've been to Quebec."

"$(P \supset Q)$" is a *conditional*. P is its *antecedent* and Q its *consequent*.

Our "$\supset$" claims that what you *don't* have is the first part true and the second part false. So "$(P \supset Q)$" is always true except when P is true and Q is false.

The truth table for "$\supset$" is difficult. Suppose you say:

> "*If* I've been to Paris, *then* I've been to Quebec."

Our truth table says that you're speaking truthfully if you've been to neither place, or to both places, or to Quebec but not Paris. But you're speaking falsely if you've been to Paris but *not* to Quebec. Does this seem correct to you? Most people think it does, but some have doubts.

The truth table for "$\supset$" can produce strange results. Take this example:

> *If* I had eggs for breakfast *then* the world will $(E \supset W)$
> end at noon.

Suppose I *didn't* have eggs for breakfast (so E = 0). Then our truth table says that the conditional is true [since if E = 0 then $(E \supset W) = 1$]. This is strange. We don't normally think that my *not* having eggs would make this if-then sentence true. Rather we view the sentence as false *regardless* of whether or not I had eggs. We take the sentence to suggest or assert a causal connection that doesn't exist:

> My having eggs for breakfast would *cause* the world to end
> at noon.

And we take the falsity of this causal connection to entail the falsity of the conditional. So translating "if-then" as "$\supset$" doesn't seem entirely satisfactory. Something fishy is going on when we translate "if-then" as "$\supset$." What's happening here?

I mentioned that "$\cdot$" abstracts from some elements of the English "and" (for example, relevance and temporal sequence). Similarly, "$\supset$" abstracts from some elements of the English "if-then" (for example, causal connections). The truth-table meaning of "⊃" is very elementary. "(P ⊃ Q)" simply asserts that what we *don't* have is P-true-and-Q-false:

$$(P \supset Q) \qquad\qquad\qquad \sim(P \cdot \sim Q)$$

If P is true, then Q is true.	↔	What we don't have is P-true-and-Q-false.

In a very simple sense of "if-then," the two boxes are saying the same thing—and we could infer the truth of either box from the truth of the other. This simple sense of "if-then" is sometimes called the *material conditional*. An English "If P then Q" always asserts at least this much. But it may also add other elements. For example, it might suggest or assert that P *causes* Q. Our "⊃" captures only *part* of what such an "if-then" suggests or asserts. But the *part* it captures is what normally determines whether an argument is valid. It usually doesn't hurt to simplify things and translate "if-then" as "⊃." In the cases in which it does hurt, we can translate "if-then" using a more advanced system (for example, modal logic; see Chapter 6).

The truth table and equivalences for "⊃" are difficult to remember. These three slogans may help:

Truth does not imply falsity. In other words, true-implies-false is false.	$(1 \supset 0) = 0$
Anything implies truth. In other words, true-implies-true is true, and false-implies-true is true.	$(\ \supset 1) = 1$ $(1 \supset 1) = 1$ $(0 \supset 1) = 1$
Falsity implies anything. In other words, false-implies-true is true, and false-implies-false is true.	$(0 \supset \ ) = 1$ $(0 \supset 1) = 1$ $(0 \supset 0) = 1$

Note that "$(1 \supset 0) = 0$" but "$(0 \supset 1) = 1$." People often get these two mixed up.

Here are the truth table and equivalences for "≡" ("if-and-only-if" or "mutually implies"):

P	Q	(P ≡ Q)		
0	0	1	$(0 \equiv 0) = 1$	"I've been to Paris *if and only if* I've also been to Quebec."
0	1	0	$(0 \equiv 1) = 0$	"(P ≡ Q)" is a *biconditional*.
1	0	0	$(1 \equiv 0) = 0$	
1	1	1	$(1 \equiv 1) = 1$	

Our "≡" claims that both parts have the *same* truth value. So "(P ≡ Q)" is true if the letters are both true or both false, but is false if one letter is true but the other false. "I've been to Paris *if and only if* I've also been to Quebec" is true if I've been to *both* places or to *neither* place. It's false if I've been to one place but not the other. "≡" is much like "equals."

Here are the truth table and equivalences for "~" ("not"):

P   ~P   "I *haven't* been to Paris."

0   1   ~0 = 1   "~P" is a *negation*.
1   0   ~1 = 0

"~P" has the opposite value of "P." If "P" is true then "~P" is false, and if "P" is false then "~P" is true.

Almost everything else in this book presupposes these truth equivalences. Try to master them right away. Most of the equivalences are easy, but those for "⊃" can be confusing.

2.4 TRUTH EVALUATIONS

We can calculate the truth values of a wff if we are given the truth values for its letters. Consider this problem:

> Suppose P = 1, Q = 0, and R = 0. Then what is the truth value of "((P ⊃ Q) ≡ ~R)"?

We first substitute "1" and "0" for the respective letters. We substitute "1" for "P," and "0" for "Q" and "R":

((P ⊃ Q) ≡ ~R)
((1 ⊃ 0) ≡ ~0)

Then we simplify from the inside out, using our truth equivalences, until we get "1" or "0." We first substitute "1" for "~0" (using our "~0 = 1" equivalence):

((1 ⊃ 0) ≡ 1)

Next we use our "(1 ⊃ 0) = 0" equivalence and substitute "0" for "(1 ⊃ 0)":

(0 ≡ 1)

Finally we use our "(0 ≡ 1) = 0" equivalence to get:

0

So "((P ⊃ Q) ≡ ~R)" is false.

Some like to work out truth values vertically:

```
              0
        ┌─────┴─────┐
        0           1
    ┌───┴───┐      ┌─┴─┐
  ((1  ⊃  0)  ≡   ~0)
```

Some like to work them out horizontally:

$$((1 \supset 0) \equiv \sim 0) = (0 \equiv 1) = 0$$

Still others like to work them out in their heads.

Simplify starting from the *inside*. If you have a wff of the form "~( )," first work out the part in parentheses to get 1 or 0. Then apply "~" to the result. Study these two examples (which I've done both vertically and horizontally):

Right: $\sim(1 \lor 0) = \sim 1 = 0$

Wrong: $\sim(1 \lor 0) = (\sim 1 \lor \sim 0) = (0 \lor 1) = 1$

Don't distribute "~" the way the "wrong" example does it.

2.4A Exercises

You're given that A = 1 and B = 1 (A and B are both true) while X = 0 and Y = 0 (X and Y are both false). Calculate the truth value of each wff below.

0. $(\sim A \lor \sim X)$

ANSWER: The formula is *true*. I'll work it out both vertically and horizontally:

$$(\sim A \lor \sim X) = (\sim 1 \lor \sim 0) = (0 \lor 1) = 1$$

1. $\sim(A \cdot X)$	**2.** $(\sim A \cdot \sim X)$
3. $\sim(\sim A \cdot \sim X)$	**4.** $(A \supset X)$
5. $(\sim X \equiv Y)$	**6.** $(\sim B \supset A)$
7. $\sim(A \supset X)$	**8.** $(B \cdot (X \lor A))$
9. $(\sim(X \cdot A) \lor X)$	**10.** $(\sim A \lor \sim(X \supset Y))$
11. $(B \supset (X \equiv \sim Y))$	**12.** $(X \lor (A \cdot X))$
13. $(\sim(X \cdot A) \supset (A \lor X))$	**14.** $\sim(X \equiv (X \equiv X))$

15. $((A \cdot \sim X) \supset \sim B)$ **16.** $(\sim A \supset (X \vee \sim B))$
17. $(\sim X \vee \sim (\sim A \equiv B))$ **18.** $(\sim Y \supset (A \cdot X))$
19. $((A \supset X) \supset (B \supset Y))$ **20.** $(A \vee \sim A)$

2.5 EVALUATIONS WITH AN UNKNOWN

Suppose we know that Texas won and Florida lost, but we don't know whether the University of Southern California (USC) won:

> T = 1   ("Texas won" is True)
> F = 0   ("Florida won" is False)
> U = ?   ("USC won" is Unknown)

Suppose you bet $10 that *Texas or USC will win*. The victory of Texas ensures that you win the $10. USC doesn't matter. More formally, you bet that "$(T \vee U)$" is true. We can plug in "1" and "?":

> $(T \vee U)$
> $(1 \vee ?)$

We might just see that the formula is true, since a disjunction with one part true is itself true. Or we might try it both ways. The "?" could turn out to be either "1" or "0." So we write "1" above the "?" and "0" below it. We evaluate the formula for each case. It comes out true either way:

$$\begin{matrix} & 1 & & 1 \\ (1 & \vee & ?) & = & 1 \\ & 0 & & 1 \end{matrix}$$

So the formula is true. You win $10!

Suppose you had bet that *if USC wins then Florida will win*—"$(U \supset F)$." Again we plug in "?" and "0":

> $(U \supset F)$
> $(? \supset 0)$

We work it out both ways:

$$\begin{matrix} & 1 & & & 0 \\ (? & \supset & 0) & = & ? \\ & 0 & & & 1 \end{matrix}$$

The formula could be true or false, so its value is unknown. You must wait for the USC score to find out if you won $10.

2.5A Exercises

You're given that T = 1 (T is true), F = 0 (F is false), and U = ? (U is unknown). Calculate the truth value of each of the following wffs:

0. (~T · U)

ANSWER: The formula is *false*. It works out as follows:

(~T · U) = (~1 · ?) = (0 · ?) = 0

1. (U ⊃ ~T)	**2.** (U ⊃ ~F)
3. (U ∨ ~F)	**4.** (U · T)
5. (U · F)	**6.** (F ⊃ U)
7. (T · U)	**8.** (U ∨ ~T)
9. (~T ⊃ U)	**10.** (~F · U)
11. (T ∨ U)	**12.** (~T ∨ U)
13. (T ⊃ U)	**14.** (U · ~T)
15. (U ⊃ F)	**16.** (~F ⊃ U)
17. (~F ∨ U)	**18.** (F · U)
19. (U · ~F)	**20.** (U ⊃ T)

2.6 COMPLEX TRUTH TABLES

A *truth table* for a PC wff is a chart listing all possible truth-value combinations for the letters and showing whether the wff is true or false in each case. We've done some simple tables already. Now we'll work on more difficult ones.

A wff using n distinct letters has 2^n possible truth-value combinations:

WITH 1 LETTER WE GET 2^1, OR 2 COMBINATIONS:	WITH 2 LETTERS WE GET 2^2, OR 4 COMBINATIONS:	WITH 3 LETTERS WE GET 2^3, OR 8 COMBINATIONS:
A	A B	A B C
0	0 0	0 0 0
1	0 1	0 0 1
	1 0	0 1 0
	1 1	0 1 1
		1 0 0
		1 0 1
		1 1 0
		1 1 1

One letter gives two possible combinations: The letter might be false or true. Two letters give four combinations: Both letters might be false, just the second might be true, just the first might be true, or both might be true. Three letters give eight combinations, four letters give sixteen, and so forth.

Here's an easy way to get every combination. Alternate 0's and 1's for the last letter the required number of times. Then alternate 0's and 1's for each previous letter at half again the rate: by twos, and then fours, and so forth. This numbers each row in ascending order in base 2.

A truth table for "$\sim$(A $\lor$ $\sim$B)" begins like this:

A	B	$\sim$(A $\lor$ $\sim$B)
0	0	
0	1	
1	0	
1	1	

The formula goes on the right. On the left we put each letter used in the formula. We write each letter just *once*. Then we write every possible truth-value combination for the letters. (With two letters we have 2^2, or 4, combinations.) Finally, we figure out the value of the formula for each truth combination.

The first combination has A = 0 and B = 0. We work out the value of the formula on paper (vertically or horizontally) or in our heads:

$$\sim(A \lor \sim B) = \sim(0 \lor \sim 0) = \sim(0 \lor 1) = \sim 1 = 0$$

So on the first combination our formula is false:

A	B	$\sim$(A $\lor$ $\sim$B)
0	0	0
0	1	
1	0	
1	1	

Then we work out the other lines in a similar way. Eventually we finish the truth table for the formula:

A	B	$\sim$(A $\lor$ $\sim$B)
0	0	0
0	1	1
1	0	0
1	1	0

The table shows that "~(A $\lor$ ~B)" is true only when A is false and B is true. The simpler wff "(~A · B)" says the same thing (in other words, is true in the same cases).

"~(A $\lor$ ~B)" could be either true or false. It's true in one case and false in three others. A PC wff that could be either true or false is called a *contingent* wff.

Here's the truth table for "(P $\lor$ ~P)" (the *law of the excluded middle*):

P	(P $\lor$ ~P)	"I've been to Paris, or I haven't been to Paris."
0	1	
1	1	

"(P $\lor$ ~P)" is *true* in every possible case. Such wffs are called *tautologies*, or *logical truths*.

It's doubtful that the law of the excluded middle holds universally. Is "It's raining" true or false if there's a slight drizzle? Is "My shirt is white" true or false if it's a light cream color? Our claims are sometimes too vague to be either true or false. The law of the excluded middle is idealized when applied to ordinary language. Since I want this law to hold in PC, I've *stipulated* that capital letters in PC represent true or false statements. So statements too vague to be either true or false shouldn't be translated into PC.

Here's the truth table for "(P · ~P)":

P	(P · ~P)	"I've been to Paris, and I haven't been to Paris."
0	0	
1	0	

"(P · ~P)" is *false* in every possible case. Such wffs are called *contradictions,* or *logical falsehoods*.

Let's think about "(P · ~P)" for a moment. Consider this sentence:

> "I've been to Paris, and I haven't been to Paris."

You could use this sentence to say something *true*, so long as you shift the meaning of "been to Paris." You might explain your sentence as follows:

> "I landed once at the Paris airport (and in that sense I've been to Paris). But I didn't spend any significant time there (and in that sense I haven't been to Paris)."

Thus understood, the English sentence has the form "(P · ~Q)." This could be true, and so isn't a contradiction. What "P" affirms differs from what "~Q" denies. When we say that "(P · ~P)" is a contradiction, we mean that it's a contradiction to affirm and deny the *exact same thing*.

The denial of "(P · ~P)" is the *law of noncontradiction*:

> ~(P · ~P)

"Not both P and not-P."

"A statement can't be both *true* and (taking the statement in the exact same sense) also *not true*."

"~(P · ~P)" is true in every possible case. So it's a *tautology*, or *logical truth*.

2.6A Exercises

Do a truth table for each wff.

0. (P ⊃ (Q ∨ R))   ANSWER:

P	Q	R	(P ⊃ (Q ∨ R))
0	0	0	1
0	0	1	1
0	1	0	1
0	1	1	1
1	0	0	0
1	0	1	1
1	1	0	1
1	1	1	1

1. (~P ⊃ Q)   **2.** (P ⊃ ~P)
3. (~P · Q)   **4.** ((P ≡ Q) ⊃ Q)
5. (~Q ⊃ ~P)   **6.** ((~P · Q) ⊃ R)
7. (P ∨ ~(Q · R))   **8.** ((P ⊃ Q) ≡ ~ R)
9. (P ∨ (Q · ~R))   **10.** ((P ∨ ~Q) ⊃ R)

2.7 THE TRUTH-TABLE TEST

Suppose we want to test this argument:

> If I overslept, then I'll be late.   (S ⊃ L)
> I didn't oversleep.   ~S
> ∴ I won't be late.   ∴ ~L

We first do a truth table for the premises and conclusion. We start as follows:

S	L	(S ⊃ L), ~S ∴ ~ L
0	0	
0	1	
1	0	
1	1	

Then we evaluate the three wffs on each truth combination. The first combination has S = 0 and L = 0. The three wffs are true on this combination:

$$(S \supset L) \qquad \sim S \qquad \sim L$$
$$= (0 \supset 0) \qquad = \sim 0 \qquad = \sim 0$$
$$= 1 \qquad \qquad = 1 \qquad = 1$$

So the first line of our truth table looks like this:

S	L	(S ⊃ L),	~S	∴ ~L
0	0	1	1	1

We then work out the other three lines, as follows:

S	L	(S ⊃ L),	~S	∴ ~L	INVALID
0	0	1	1	1	
0	1	1	1	0	
1	0	0	0	1	
1	1	1	0	0	

So we *can* have the premises all true and the conclusion false. The second line has 110. The argument is invalid:

INVALID = You *can* have premises 1 and conclusion 0.

It might happen that you didn't oversleep (S = you overslept = 0) but you'll still be late (L = you'll be late = 1) for other reasons. Then the premises would be true and the conclusion false. So the argument is invalid.

Consider a second example:

If I overslept, then I'll be late. (S ⊃ L)
I won't be late. ~L
∴ I didn't oversleep. ∴ ~S

Again we do a truth table for the premises and the conclusion:

S	L	(S ⊃ L),	~L	∴ ~S	VALID
0	0	1	1	1	
0	1	1	0	1	
1	0	0	1	0	
1	1	1	0	0	

In no possible case are all the premises true but the conclusion false. We never have 110. So the argument is valid:

> VALID: You *can't* have premises 1 and conclusion 0.

Let's compare the two arguments we just tested:

> INVALID: (S ⊃ L), ~S ∴ ~L
> [The second premise denies the *antecedent*.]

> VALID: (S ⊃ L), ~L ∴ ~S
> [The second premise denies the *consequent*.]

The difference may seem small. But it's enough to make one argument invalid and the other valid.

There's a "short cut" form of the truth-table test. Remember that all we're looking for is 110 (premises all true and conclusion false). The argument is invalid if 110 sometimes occurs; otherwise, it's valid. To save time, we can first evaluate an easy wff and then cross out any lines that couldn't be 110. We might work out "~S" first:

S	L	(S ⊃ L), ~L ∴ ~S
0	0	--------------------1--
0	1	--------------------1--
1	0	0
1	1	0

The first two lines can't be 110 (since the last digit is 1). So we cross them out and ignore the remaining values. The last two lines could turn out to be 110, so we work them out further. Next we evaluate "~L":

S	L	(S ⊃ L), ~L ∴ ~S
0	0	--------------------1--
0	1	--------------------1--
1	0	1 0
1	1	-----------0-------0--

The last line can't be 110 (since the second digit is 0). So we cross it out. We have only one line to evaluate when we get to the complex wff "(S ⊃ L)." On this line "(S ⊃ L)" comes out false. We never get 110. So the argument is valid:

S	L	(S ⊃ L), ~L ∴ ~S	VALID
0	0	--------------------1--	
0	1	--------------------1--	
1	0	0 1 0	
1	1	-----------0-------0--	

With a two-premise argument, look for 110. With three premises, look for 1110. Whatever the number of premises, look for a case in which the premises are all true but the conclusion false.

Suppose an argument tests as invalid on the truth-table test. All this shows is that the argument is *either* invalid *or else* valid on grounds that go beyond propositional logic. Consider this argument:

| Everything is green. | E |
| ∴ This is green. | ∴ T |

This argument is valid but tests as invalid on the truth-table test. We need a stronger system (the system of Chapter 4) to express its logical form adequately and prove it valid. We must understand our invalidity results here and throughout the book as qualified in a certain way. Suppose we use a system to "show that an argument is invalid." What we've really shown is that the argument is *either* invalid *or else* valid on grounds that go beyond the system in question.

2.7A Exercises

For each argument, first evaluate intuitively and then translate into PC (using the letters provided) and test using the truth-table method:

0. The keys are in my left pocket or my right pocket.
   The keys aren't in my left pocket.
∴ The keys are in my right pocket.
   [Use R and L.]

	L	R		(L ∨ R),	~L	∴ R	
ANSWER:							VALID (since we never get true premises and false conclusion)
	0	0		0	1	0	
	0	1		1	1	1	
	1	0		1	0	0	
	1	1		1	0	1	

1. If you're in Chicago, then you're in Illinois.
   You're in Illinois.
∴ You're in Chicago.
   [Use C and I.]
2. If you're in Chicago, then you're in Illinois.
   You're in Chicago.
∴ You're in Illinois.
   [Use C and I.]
3. If television is always right, then Anacin is better than Bayer.
   If television is always right, then Anacin *isn't* better than Bayer.
∴ Television isn't always right.
   [Use T and A. Can you always believe television ads?]
4. Michigan will win a Bowl game if and only if it learns how to pass.
   Michigan won't learn how to pass.

∴ Michigan won't win a Bowl game.
[Use W and P. My alma mater has had a poor Bowl record.]

5. If I get Grand Canyon reservations and get a group together, then I'll explore canyons during spring break.
I've got a group together.
I can't get Grand Canyon reservations.
∴ I won't explore canyons during spring break.
[Use R, T, and E.]

6. If there is knowledge, then either there is an endless reasoning chain proving every premise by a prior proof or else some things are known without proof.
There is no endless reasoning chain proving every premise by a prior proof.
There is knowledge.
∴ Some things are known without proof.
[Use K, E, and S. From Aristotle.]

7. If it rains and your tent leaks, then your down sleeping bag will become like mush.
Your tent won't leak.
∴ Your down sleeping bag won't become like mush.
[Use R, T, and D.]

8. There's a moral law.
If there's a moral law, then someone gave the moral law.
If someone gave the moral law, then there is a God.
∴ There is a God.
[Use M, S, and G.]

9. Matter always existed.
If there is a God, then God created the universe.
If God created the universe, then matter didn't always exist.
∴ There is no God.
[Use M, G, and C.]

10. If you modified your computer or you didn't send in the registration card, then the warranty is void.
You didn't modify your computer.
You sent in the registration card.
∴ The warranty isn't void.
[Use M, S, and V.]

2.8 THE TRUTH-ASSIGNMENT TEST

The truth-table test is a calculation method that always gives the right answer (unless we miscalculate). It provides a "mechanical decision procedure" for PC. But the test is tedious, especially for longer arguments. An argument with 6 letters requires a table with 64 lines. An argument with 10 letters requires 1,024 lines! Fortunately, easier tests are available. Now we'll learn the truth-assignment test for simple arguments. Later we'll learn the formal-proof test for complex arguments.

The *truth-assignment test* is based on the meanings of *valid* and *invalid*.

> VALID = You *can't* have premises 1 and conclusion 0.
> INVALID = You *can* have premises 1 and conclusion 0.

In this test, we first assign 1 to the premises and 0 to the conclusion. Then we see whether this involves making contradictory assignments. Take this clearly valid argument:

It's in the left or the right hand. $(L \lor R)$

It's not in the left hand. $\sim L$

∴ It's in the right hand.   ∴ R

We begin by setting each premise equal to 1 and the conclusion equal to 0 (just to see if this is possible):

$(L \lor R) = 1$

$\sim L = 1$

∴ $R = 0$

Then we figure out the values of as many letters as we can. The conclusion has $R = 0$. So R is false. We show this by giving *each* R an 0-superscript:

$(L \lor R^0) = 1$

$\sim L = 1$

∴ $R^0 = 0$

The second premise has $\sim L = 1$. Since $\sim L$ is true, L is false. We show this by giving *each* L an 0-superscript:

$(L^0 \lor R^0) = 1$

$\sim L^0 = 1$

∴ $R^0 = 0$

The superscript in the second premise indicates, *not* that $\sim L$ is 0, but that L is 0. Then the first premise has to be 0, since a disjunction with both parts 0 is 0:

$$\overset{0}{\underset{}{\overbrace{(L^0 \lor R^0)}}} \neq 1 \qquad \text{VALID}$$

$\sim L^0 = 1$

∴ $R^0 = 0$

Since 0 is not 1, we slash the " = ." It's impossible to have true premises and a false conclusion, since this would make the first premise both 1 and 0. So the argument is valid.

In doing the test, assign 1 to the premises and 0 to the conclusion, just to see if this is possible. Figure out the truth values for the letters. Then figure out the truth values for the larger formulas. If you have to cross something out, then the assignment *isn't* possible, so the argument is valid.

Consider this clearly invalid argument:

It's in the left or the right hand. $(L \lor R)$
It's not in the left hand. $\sim L$
∴ It's not in the right hand.   ∴ $\sim R$

If we work this out, we find that we can make the premises true and the conclusion false. So the argument is invalid:

$$\overset{\overset{1}{\downarrow}}{(L^0 \lor R^1)} = 1 \qquad \text{INVALID}$$
$$\sim L^0 = 1$$
$$\therefore \sim R^1 = 0$$

The truth-*table* test gives us true premises and a false conclusion when $L = 0$ and $R = 1$:

L	R	$(L \lor R)$,	$\sim L$	∴	$\sim R$	INVALID
0	1	1	1		0	

The truth-*assignment* test gives us this result more directly.

Take another clearly invalid argument:

It's in the left or the right hand. $(L \lor R)$
∴ It's in the right hand.   ∴ R

If we work this out, we get a 0 for R:

$$(L \lor R^0) = 1$$
$$\therefore R^0 = 0$$

We can get a value for L using the first premise. L has to be 1 if the disjunction is to be 1 and yet the second disjunct is 0. With L as 1, it's possible for the premises to be true and the conclusion false. So the argument is invalid:

$$\overset{\overset{1}{\downarrow}}{(L^1 \lor R^0)} = 1 \qquad \text{INVALID}$$
$$\therefore R^0 = 0$$

Alternatively, we might try both values for L and see whether either gives true premises and a false conclusion. If there's *any* possible way to get true premises and a false conclusion, then the argument is invalid.

2.8A Symbolic Argument Exercises

Test these arguments for validity using the truth-assignment test:

0. $((A \lor \sim B) \supset (\sim C \cdot D))$
   C
$\therefore$ B

ANSWER: This argument is valid. It's impossible to make all the premises true but the conclusion false. If you try it, then you have to cross something out:

$$((A \lor \sim B^0) \supset (\sim C^1 \cdot D)) \neq 1 \qquad \text{VALID}$$
$$C^1 = 1$$
$$\therefore B^0 = 0$$

Here we get no values for A and D. But the first premise comes out false regardless of the values for A and D.

1. $((T \lor \sim M) \supset O)$
   $\sim M$
$\therefore$ O

2. $\sim(N \equiv H)$
   N
$\therefore \sim H$

3. $((J \cdot \sim D) \supset Z)$
   $\sim Z$
   D
$\therefore \sim J$

4. $((L \cdot F) \supset S)$
   S
   F
$\therefore$ L

5. $((W \cdot C) \supset Z)$
   $\sim Z$
$\therefore \sim C$

6. $((\sim Q \lor N) \supset \sim U)$
   U
$\therefore$ Q

7. $(P \supset (S \supset L))$
   L
   S
$\therefore$ P

8. $(Q \supset (\sim R \supset K))$
   $\sim R$
$\therefore$ K

9. $((B \cdot V) \supset C)$
   $\sim C$
   B
$\therefore \sim V$

10. K
   $((K \cdot E) \supset \sim A)$
   $\sim A$
$\therefore$ E

11. $((A \cdot U) \supset \sim B)$ **12.** $(E \lor (Y \cdot X))$
    B $\sim E$
    A   Y
$\therefore$ U $\therefore$ X

13. P   **14.** $\sim P$
$\therefore (P \lor Q)$ $\therefore (Q \supset P)$

15. $(\sim T \supset (P \supset J))$ **16.** $((\sim M \cdot \sim G) \supset \sim R)$
    P   R
    $\sim J$ $\sim G$
$\therefore$ T $\therefore$ M

17. P   **18.** Q
$\therefore \sim(Q \cdot P)$ $\therefore (P \supset Q)$

19. $\sim(O \equiv I)$ **20.** A
    $\sim O$ $\sim A$
$\therefore$ I $\therefore$ B

2.8B English Argument Exercises

In each case, first evaluate the argument intuitively. Then translate it into PC and test for validity using the truth-assignment test.

  0. If "good" means "socially approved" and it's virtuous to try to do what is good, then it's virtuous to try to do what is socially approved.
     It's virtuous to try to do what is good.
     It isn't virtuous to try to do what is socially approved.
$\therefore$ "Good" doesn't mean "socially approved."
     [An attack on cultural relativism.]

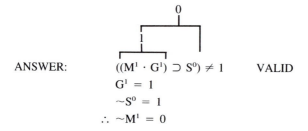

ANSWER: $((M^1 \cdot G^1) \supset S^0) \neq 1$ VALID
                 $G^1 = 1$
                 $\sim S^0 = 1$
           $\therefore \sim M^1 = 0$

  1. If there's no first cause, then either there's an endless chain of causes-and-effects or some being caused itself.
     There's no endless chain of causes-and-effects.
     No being caused itself.
$\therefore$ There's a first cause.
     [A "first cause" (often identified with God) is a being that causes other things to exist but yet nothing caused it to exist.]

  2. If you pass and it's intercepted, then the other side gets the ball.
     You pass.
     It isn't intercepted.

∴ The other side doesn't get the ball.
3. If existence is a perfection and God by definition has all perfections, then God by definition must exist.
God by definition has all perfections.
Existence is a perfection.
∴ God by definition must exist.
[From René Descartes.]
4. If the United States will be weak, then there will be war.
The United States won't be weak.
∴ There won't be war.
5. If maximizing human enjoyment is always good and the sadist's dog-torturing maximizes human enjoyment, then the sadist's act is good.
The sadist's act isn't good.
The sadist's dog-torturing maximizes human enjoyment.
∴ Maximizing human enjoyment isn't always good.
[An attack on one form of utilitarianism.]
6. If "good" is definable in experimental terms, then ethical judgments are scientifically provable and ethics has a rational basis.
Ethical judgments aren't scientifically provable.
∴ Ethics doesn't have a rational basis.
7. If Newton's gravitational theory is correct and there's no undiscovered planet near Uranus, then the orbit of Uranus would be such-and-such.
Newton's gravitational theory is correct.
The orbit of Uranus isn't such-and-such.
∴ There's an undiscovered planet near Uranus.
[This argument led scientists to search for and discover the planet Neptune. If their search proved vain, they would have questioned one or more of the premises.]
8. If attempts to prove "God exists" fail in the same way as proofs for "There are other conscious beings besides myself," then belief in God is reasonable if and only if belief in other conscious beings is reasonable.
Attempts to prove "God exists" fail in the same way as proofs for "There are other conscious beings besides myself."
Belief in other conscious beings is reasonable.
∴ Belief in God is reasonable.
[From Alvin Plantinga, who argued extensively for the second premise.]
9. If knowledge is sensation, then pigs have knowledge.
Pigs don't have knowledge.
∴ Knowledge isn't sensation.
[From Plato.]
10. If you pack intelligently, then either this teddy bear will be useful on the backpack trip or you won't pack it.
This teddy bear won't be useful on the backpack trip.
You won't pack it.
∴ You pack intelligently.
[My young nephew Keith insisted that it *would* be useful to bring his teddy bear on our backpack trip! I exerted my authority and said *no*.]
11. If "X is good" means "Hurrah for X!" and it makes sense to say "If X is good," then it makes sense to say "If hurrah for X!"

It makes sense to say "If X is good."
It doesn't make sense to say "If hurrah for X!"
∴ "X is good" doesn't mean "Hurrah for X!"
[Use M for " 'X is good' means 'Hurrah for X!,' " G for "It makes sense to say 'If X is good,' " and H for "It makes sense to say 'If hurrah for X!' " This argument from Hector-Neri Castañeda attacks a form of ethical emotivism—that "This is good" means "Hurrah for this!"]

12. If there are moral authorities, then there's a way to justify moral beliefs that doesn't appeal to individual judgment or else some people have superior judgment.
There's no way to justify moral beliefs that doesn't appeal to individual judgment.
Some people have superior judgment.
∴ There are moral authorities.

13. If we have an idea of substance, then "substance" denotes a simple sensation or a complex constructed out of simple sensations.
"Substance" doesn't denote a simple sensation.
∴ We don't have an idea of substance.
[From David Hume.]

14. If belief in God were a purely intellectual matter, then either all smart people would be believers or all smart people would be nonbelievers.
Not all smart people are believers.
Not all smart people are nonbelievers.
∴ Belief in God isn't a purely intellectual matter.
[From the Jesuit theologian John Powell.]

15. If we have an idea of "substance" and the idea of "substance" doesn't derive from sensations, then "substance" is a thought category of pure reason.
The idea of "substance" doesn't derive from sensations.
We do have an idea of "substance."
∴ "Substance" is a thought category of pure reason.
[From Immanuel Kant and against Hume.]

16. Either the United States has a strong military, or it'll be unable to defend its allies and the Soviets will foment revolution around the globe.
The United States has a strong military.
The United States is able to defend its allies.
∴ The Soviets won't foment revolution around the globe.

17. If capital punishment is justified and justice doesn't demand a vindication for past injustices, then either capital punishment reforms the offender or else capital punishment effectively deters crime.
Capital punishment doesn't reform the offender.
Capital punishment doesn't effectively deter crime.
∴ Capital punishment isn't justified.

18. If you're lost, then you should call for help or head toward the river.
You're lost.
∴ You should call for help.

19. If we have ethical knowledge and any justification for an ethical judgment would appeal to more-basic ethical knowledge, then either our justifications could go on forever or else we have some basic ethical knowledge that we can't justify by anything further.
We do have ethical knowledge.

Any justification for an ethical judgment would appeal to more-basic ethical knowledge.
Our justifications couldn't go on forever.
∴ We have some basic ethical knowledge that we can't justify by anything further.
[An argument for ethical intuitionism.]

20. [Here T is the claim "All truths are knowable through experimental science." Use T for "T is true" and K for "T is knowable through experimental science."]
If T is true, then T is knowable through experimental science.
T isn't knowable through experimental science.
∴ T isn't true.
[T is a self-refuting claim. Other such claims include "*Every* rule has exceptions" (including this rule?), "*No* statements are true" (including this one?), and "I *know* that knowledge is impossible."]

2.9 IDIOMS

Now we'll learn about translating from idiomatic English to PC. But first let's recall two rules from Section 2.2:

> "Both," "either," and "if" generally translate as a left-hand parenthesis.

Not either A or B.   Either not A or B.
= ~(A ∨ B)   = (~A ∨ B)

> A comma often indicates a formula's logical middle (so that the parts on either side glob together as units).

If A, then B and C.   If A then B, and C.
= (A ⊃ (B · C))   = ((A ⊃ B) · C)

Here are three new rules:

> Translate "but" ("yet," "however," "although," and so on) as "and."

Michigan played *but* they lost.
= P *and* L.
= (P · L)

The element of contrast (or surprise) is lost in the translation, but this doesn't affect validity.

> Translate "unless" as "or."

Unless you breathe you'll die.
= B *or* D.
= (B ∨ D)

You'll die *unless* you breathe.
= D *or* B.
= (D $\lor$ B)

"(B $\lor$ D)" and "(D $\lor$ B)" mean the same; the order of disjuncts doesn't matter. "Unless" is also equivalent to "if not":

= If you don't breathe then you'll die.
= If not B, then D.
= ($\sim$B $\supset$ D)

Translate "just if" and "iff" (a logician word) as "if and only if."

I'll agree *just if* you pay me $10,000.
= I'll agree *iff* you pay me $10,000.
= A *if and only if* P.
= (A $\equiv$ P)

Our next two rules are tricky to apply. The first one governs most conditional words:

The part after "if" ("provided that," "assuming that," " on the condition that," and so on) is the antecedent (the "if" part, the part before the horseshoe).

If you're in Chicago, then you're in Illinois.
= *If* C then I.
= (C $\supset$ I)

You're in Illinois *if* you're in Chicago.
= I *if* C.
= *If* C then I.
= (C $\supset$ I)

Provided that you're in Chicago, you're in Illinois.
= *Provided that* C, I.
= *If* C then I.
= (C $\supset$ I)

You're in Illinois, *provided that* you're in Chicago.
= I, *provided that* C.
= *If* C then I.
= (C $\supset$ I)

"Only if" is different and follows its own rule:

> The part after "only if" is the consequent (the "then" part, the part after the horseshoe).

You're in Chicago *only if* you're in Illinois.
= C *only if* I.
= If C *then* I.
= (C ⊃ I)

Only if you're in Illinois are you in Chicago.
= *Only if* I, C.
= If C *then* I.
= (C ⊃ I)

Here's another way to put the same point:

> Write "⊃" where you see "then" or "only if."

If A then B = (A ⊃ B)
A only if B = (A ⊃ B)

With this rule, "Only if A, B" would become "⊃ A, B"—which we would put as "(B ⊃ A)."

Sometimes our rules for "only if" produce a correct but awkward translation. Take this example:

You'll pass *only if* you take the exam.
= P *only if* E.
= If P *then* E.
= If you'll pass, then you take the exam.
= (P ⊃ E)

This rephrasing means the same, but sounds better:

= If not E then not P.
= If you don't take the exam then you won't pass.
= (~E ⊃ ~P)

Here "(P ⊃ E)" and "(~E ⊃ ~P)" are contrapositives. To form the *contrapositive* of a conditional, negate antecedent and consequent and switch the two around. Contrapositives have the same meaning. But one may sound better than the other.

The terms "sufficient" and "necessary" translate this way:

"A is sufficient for B" means "If A then B."
"A is necessary for B" means "If not A then not B."
"A is necessary and sufficient for B" means "A if and only if B."

Taking the exam is *sufficient* for getting an A.
= If you take the exam then you get an A.
= (E ⊃ A)

Taking the exam is *necessary* for getting an A.
= If you don't take the exam then you don't get an A.
= (~E ⊃ ~A)

Taking the exam is *necessary and sufficient* for getting an A.
= You take the exam if and only if you get an A.
= You get an A if and only if you take the exam. [This means the same but sounds more natural.]
= (E ≡ A) = (A ≡ E) = (A ⊃ E)

These translation rules are rough guidelines and don't always work. Sometimes you have to puzzle out the meaning on your own.

2.9A Exercises

Translate each of these into a PC wff.

0. A, presuming that B.   ANSWER: (B ⊃ A)

1. A is a sufficient condition for B.
2. A, unless B.
3. A is a necessary and sufficient condition for B.
4. Assuming that A, if B and C then either D or E.
5. A is necessary for B.
6. You're a man just if you're a rational animal.
7. Unless you have faith, you'll die.
8. She neither asserted it nor hinted at it.
9. Only if you exercise are you fully alive.
10. Assuming that your belief is false, you don't know.
11. I'll go, assuming that you go.
12. Having a true belief is a necessary condition for having knowledge.
13. You're alive only if you have oxygen.
14. You get mashed potatoes or french fries, but not both.
15. You're wrong if you say that.

2.10 TWISTED ARGUMENTS

So far, the arguments in this chapter have been neatly phrased and have had a clear premise-conclusion format. This makes them easier to understand.

Unfortunately, arguments in real life are often twisted. The ideas may be expressed in idioms like those we studied in the last section. There may be extraneous material.

Important parts of the argument may be omitted or hinted at. And it may be difficult to pick out the premises and conclusion. It often takes hard work to reconstruct a clearly stated argument from a passage.

In reconstructing an argument, first pick out the premises and the conclusion. Premises often come before the conclusion. But sometimes the conclusion comes first or in the middle. Words like "hence," "thus," "so," and "therefore" often indicate the conclusion. And words like "because," "for," and "since" often indicate premises. When you don't have this help, ask yourself what is being argued *for* (this is the conclusion) and what is being argued *from* (these are the premises).

After you pick out the premises and conclusion, translate them into PC. To make a plausible argument, you may need to add premises that are unstated but implicit. The "principle of charity" should prevail: Interpret unclear reasoning in the way that gives the best argument. Finally, test for validity.

Here's an easy example:

> "The gun must have been shot recently! It's still hot."

First we pick out the premises and the conclusion:

The gun is still hot.	*Premise*
∴ The gun was shot recently.	*Conclusion*

We add the implicit premise "If the gun is still hot then it was shot recently," translate into PC, and test for validity:

If the gun is still hot, then it was shot recently.	(H ⊃ S)	VALID
The gun is still hot.	H	
∴ The gun was shot recently.	∴ S	

2.10A Exercises

Pick out the conclusions of the arguments in Section 2.10B.

0. The conclusion in example 0 is "Knowledge is good in itself." "So" here indicates the conclusion.

2.10B Exercises

For each argument, first evaluate intuitively, and then translate into PC and test for validity using the truth-assignment test. Supply implicit premises if needed.

0. Knowledge is good in itself only if it's capable of being desired for its own sake. So knowledge is good in itself, since it's capable of being desired for its own sake.

$$
\begin{array}{ll}
\text{ANSWER:} & (G^0 \supset C^1) = 1 \qquad \text{INVALID} \\
& C^1 = 1 \\
& \therefore G^0 = 0
\end{array}
$$

The first premise means "If knowledge is good in itself, then it's capable of being desired for its own sake" ["(G ⊃ C)"]. The argument is invalid. It would be valid if we used this premise: "If knowledge is capable of being desired for its own sake, then it's good in itself" ["(C ⊃ G)"].

1. Knowledge can't be sensation. If it were, then we couldn't know something that we aren't presently sensing. [From Plato.]
2. Presuming that we followed the map, then unless the map is wrong there's a pair of lakes just over the pass. We followed the map. There's no pair of lakes just over the pass. Hence the map is wrong.
3. If they blitz but don't get to our quarterback, then our wide receiver will be open. But our wide receiver won't be open, as shown by the fact that they won't blitz.
4. Unless you give me a raise, I'll quit. Therefore I'm quitting!
5. Taking the exam is a necessary condition for getting an A. You didn't take the exam. Consequently you don't get an A.
6. My true love will marry me *only if* I buy her a Rolls Royce. It follows that she'll marry me, since I'll buy her a Rolls Royce.
7. That your views are logically consistent is a necessary condition for your views to be sensible. Your ideas are logically consistent. Consequently your views are sensible.
8. We have tea bags in the house *only if* your sister Carol drinks tea. This shows that we don't have tea bags.
9. We have no ethical knowledge. This is proved by the fact that if we did have such knowledge then basic moral principles would have to be either provable or else self-evident. [An argument for ethical skepticism.]
10. Sam must have stolen the money. After all, he knew where the money was.
11. If Ohio State wins but Alabama doesn't, then the Buckeyes will be national champions. It looks like the Buckeyes won't be national champs, because Alabama clearly is going to win.
12. The filter capacitor can't be blown. This is indicated by the following facts. You'd hear a hum, presuming that the silicon diodes work but the filter capacitor is blown. But you don't hear a hum. And the silicon diodes work. [A person fixing a radio often reasons like this.]
13. We can't still be on the right trail. We'd see the white Appalachian Trail blazes on the trees if we were still on the right trail.
14. My true love will marry me *if* I buy him a Rolls Royce. I won't buy him a Rolls Royce. This implies that he won't marry me.
15. There will be a fire! My reason for saying this is that only if there's oxygen present will there be a fire. Of course there's oxygen present.
16. Isabel couldn't have committed the murder in Boston on July 4th. She was with me in Miami all day on July 4th.
17. The basic principles of ethics can't be self-evident truths, since if they were then they'd be equally obvious to everyone.
18. It must be a touchdown, because the ball broke the plane of the end zone.
19. Assuming that it wasn't an inside job, then the lock was forced unless the thief stole the key. The thief didn't steal the key. We may infer that the robbery was an inside job, inasmuch as the lock wasn't forced.
20. Taking the exam is a sufficient condition for getting an A. You didn't take the exam. This means you don't get an A.

2.11 THE S-RULES

We'll now study some simple inference rules: the S-rules (which we use to *simplify* a *single* premise) and the I-rules (which we use to *infer* a conclusion from *two* premises). These rules form the building blocks of formal proofs, which we'll start in the next chapter. Formal proofs reduce a complex argument to a series of simple steps. The S- and I-rules provide the simple steps. These rules are also important in their own right, since they represent common forms of reasoning. The accuracy of these rules should be obvious. But you can check them using the truth-table or truth-assignment tests, if you desire.

We'll now take the three most important S-rules. Our first S-rule deals with "and" and is very simple. Here it is in English and symbolic forms:

This and that.	$(P \cdot Q)$
∴ This.	
∴ That.	P, Q

"It's cold and windy; therefore it's cold, and therefore it's windy." Given a conjunction, we may infer each conjunct. The same holds if one or both conjuncts are negative:

It isn't cold and it isn't windy. $(\sim C \cdot \sim W)$

∴ It isn't cold.

∴ It isn't windy. $\sim C, \sim W$

But if the conjunction as a whole is negative (the negation is outside the parentheses), then we can infer nothing about the truth or falsity of each part:

You aren't both in Atlanta $\sim(A \cdot B)$
and also in Boston.

∴ no conclusion   no conclusion

You can't be in both cities at the same time. But this fact doesn't tell us where you are; you might be in Atlanta, or in Boston, or in some third place. From "$\sim(A \cdot B)$" we can't infer the truth value for A or for B. We just know that *not both* are true (that is, at least one is false); but we can't tell which one is false.

Our second S-rule deals with "or":

Not either this or that.	$\sim(P \lor Q)$
∴ Not this.	
∴ Not that.	$\sim P, \sim Q$

"It isn't either cold or windy; therefore it isn't cold, and therefore it isn't windy." Given the denial of a disjunction, we may infer the denial of each disjunct. The same holds if one or both of the two disjuncts are negative. Here again we infer the *opposite* of the disjuncts (the opposite of "$\sim A$" and "$\sim B$" being "A" and "B"):

Not either not-A or not-B.	$\sim(\sim A \lor \sim B)$
$\therefore$ A.	
$\therefore$ B.	A, B

So if not either letter is false then both are true. But from a positive disjunction we can infer nothing about the truth or falsity of each part:

You're in either Atlanta	$(A \lor B)$
or Boston.	
$\therefore$ no conclusion	no conclusion

You might be in Atlanta (and not Boston), or in Boston (and not Atlanta). From "$(A \lor B)$" we can't tell the truth value for A or for B. We know only that at least one of the two is true; but we can't tell which is the true one.

Our third S-rule deals with "if-then":

False *if-then*.	$\sim(P \supset Q)$
$\therefore$ First part true.	
$\therefore$ Second part false.	P, $\sim$Q

This holds because, according to the truth tables, a conditional is false in just the *one case* in which the antecedent is true and the consequent false:

$$(1 \supset 0) = 0$$

So if we know that the conditional is false, we can then infer that the first part is true and the second false. Alternatively, you might recall that we explained "$(P \supset Q)$" as meaning "What we *don't* have is P-true-and-Q-false"; so "$\sim(P \supset Q)$" gives us "What we *do* have is P-true-and-Q-false."

This "false if-then" rule isn't very intuitive. I suggest memorizing it instead of appealing to logical intuitions or concrete examples. You'll use this rule so much in doing formal proofs that it'll become second nature.

If one or both parts of the false if-then are themselves negative, we again infer the truth of the first part and the falsity of the second. We bring down the first part just as it is. And we bring down the *opposite* of the second part:

$\sim(\sim A \supset B)$	$\sim(A \supset \sim B)$	$\sim(\sim A \supset \sim B)$
$\sim A, \sim B$	A, B	$\sim A$, B

This diagram might help you follow what is going on here:

$\sim($ first part	$\supset$	second part	)
write the		write the opposite	
first part		of the second part	

If the conditional itself is positive (there's no negation outside the parentheses), then we can infer nothing about the truth or falsity of each part. So from "(A ⊃ B)" by itself, we can infer nothing about A or about B.

Let me summarize the forms that we can/can't simplify:

CAN SIMPLIFY	
(·)	*Both* are true.
~(∨)	*Not either* is true.
~(⊃)	*False if-then.*

CAN'T SIMPLIFY	
~(·)	*Not both* are true.
(∨)	*At least one* is true.
(⊃)	*If-then.*

If you're not sure whether something follows, you could check it using the truth-table or truth-assignment tests. But try to learn these rules so well that you can apply them immediately.

2.11A Exercises

Draw whatever simple conclusions (a letter or its negation) follow from these premises. If nothing follows, leave blank.

0. (C · ~R)   ANSWER: (C · ~R)

C, ~R

[For example: "It's cold and not raining; therefore it's cold, and therefore it's not raining."]

1. ~(I ∨ ~V)   **2.** ~(H ⊃ ~I)   **3.** ~(~P · U)

4. (~O ∨ ~X)   **5.** (~Q · B)   **6.** (F ⊃ ~G)

7. ~(~A · ~J)   **8.** (~T ⊃ ~H)   **9.** ~(~N ∨ ~E)

10. (A · P)   **11.** (E ⊃ O)   **12.** ~(~K ∨ R)

13. ~(F ⊃ M)   **14.** (M ∨ ~W)   **15.** (~N ⊃ S)

16. ~(J · ~V)   **17.** ~(D ∨ S)   **18.** (~D · ~Z)

2.12 THE I-RULES

We'll use the I-rules to INFER a conclusion from *two* premises. There are six I-rules— two each for "·," "∨," and "⊃."

Our first pair of I-rules deals with "and":

| Not both are true. | $\sim(A \cdot B)$ | $\sim(A \cdot B)$ |
This one is true.	A	B
$\therefore$ The other is false.	$\sim B$	$\sim A$

Deny conjunction.
Affirm one conjunct.
$\therefore$ *Deny* other conjunct.

Here are two examples:

You can't be simultaneously in Atlanta and also in Boston.	You can't be simultaneously in Atlanta and also in Boston.
You're in Atlanta.	You're in Boston.
$\therefore$ You're not in Boston.	$\therefore$ You're not in Atlanta.

The inference also holds if one or both of the conjuncts are negative. If we *affirm* one conjunct, we can *deny* the other:

| $\sim(\sim A \cdot \sim B)$ | $\sim(A \cdot \sim B)$ | $\sim(A \cdot \sim B)$ |
$\sim A$	A	$\sim B$
B	B	$\sim A$

Each second premise *affirms* (says the *same* as) one conjunct. And each conclusion *denies* (says the *opposite* from) the other conjunct.

If we *deny* one conjunct, then we can't draw any conclusion about the other conjunct:

| $\sim(A \cdot B)$ | Not both are true. |
$\sim A$	The first is false.
nil	$\therefore$ no conclusion

People want to conclude "B must be true"; but maybe A and B are both false. Here's a concrete example:

> You can't simultaneously be in Atlanta and also in Boston.
> You aren't in Atlanta.
> $\therefore$ no conclusion

We can't conclude "You must be in Boston"; maybe you're in neither place. If we deny a conjunction then, to derive a conclusion, we must *affirm* one conjunct.

Our second pair of I-rules deals with "or":

At least one is true.	(L $\lor$ R)	(L $\lor$ R)
This one is false.	~L	~R
∴ The other is true.	R	L

> *Affirm* disjunction.
> *Deny* disjunct.
> ∴ *Affirm* other disjunct.

To give two examples:

At least one hand (left or right) has M&Ms.	At least one hand (left or right) has M&Ms.
The left hand doesn't have M&Ms.	The right hand doesn't have M&Ms.
∴ The right hand has M&Ms.	∴ The left hand has M&Ms.

The inference also holds if one or both of the disjuncts are negative. If we deny one disjunct, then we can affirm the other:

(~P $\lor$ ~Q)	(P $\lor$ ~Q)	(P $\lor$ ~Q)
P	~P	Q
~Q	~Q	P

Each second premise *denies* (says the *opposite* from) one disjunct. And each conclusion *affirms* (says the *same* as) the other disjunct.

If we *affirm* one disjunct, then we can't draw any conclusion about the other disjunct:

(A $\lor$ B)	At least one is true.
A	This one is true.
nil	∴ no conclusion

People want to conclude "B must be false"; but maybe A and B are both true. Here's a concrete example:

At least one hand has M&Ms.
The left hand has M&Ms.
∴ no conclusion

We can't conclude "The right hand can't have M&Ms"; maybe both hands have M&Ms. If we affirm a disjunction then, to derive a conclusion, we must *deny* one disjunct.

Let me summarize the valid/invalid forms of these inferences:

VALID	INVALID

Not both are true.
 This one is true.
   → The other is false.
	Not both are true.
	 This one is false.
	   → *No conclusion.*

At least one is true.
 This one is false.
   → The other is true.
	At least one is true.
	 This one is true.
	   → *No conclusion.*

Our last two I-rules are *modus ponens* (affirming mode) and *modus tollens* (denying mode). Both deal with "if-then":

If-then. Affirm first.	(P ⊃ Q) P
∴ Affirm second.	Q

If-then. Deny second.	(P ⊃ Q) ~Q
∴ Deny first.	~P

Here are examples of these common forms:

If you're in Chicago, then you're in Illinois. You're in Chicago.	(C ⊃ I) C
∴ You're in Illinois.	I

If you're in Chicago, then you're in Illinois. You aren't in Illinois.	(C ⊃ I) ~I
∴ You aren't in Chicago	~C

Both are clearly valid. The same forms hold if one or both parts are negative. If we affirm the antecedent, then we can affirm the consequent:

(~P ⊃ ~Q) ~P	(P ⊃ ~Q) P	(~P ⊃ Q) ~P
~Q	~Q	Q

And if we deny the consequent, then we can deny the antecedent:

(~P ⊃ ~Q) Q	(P ⊃ ~Q) Q	(~P ⊃ Q) ~Q
P	~P	P

If we *deny* the *antecedent*, then we can't draw any conclusion about the consequent:

If you're in Chicago, then you're in Illinois.	(C ⊃ I)
You aren't in Chicago.	~C
∴ no conclusion	no conclusion

People want to conclude "You aren't in Illinois"; but you might be some place in Illinois outside of Chicago. Remember that an if-then tells us what follows if the first part is true—but *not* what follows if the first part is false.

Similarly, if we *affirm* the *consequent*, then we can't draw any conclusion about the antecedent:

If you're in Chicago, then you're in Illinois.	(C ⊃ I)
You're in Illinois.	I
∴ no conclusion	no conclusion

People want to conclude "You're in Chicago"; but again you might be some place in Illinois outside of Chicago.

Let me summarize the valid/invalid forms of if-then inferences:

VALID	INVALID
Affirm first. → Affirm second.	Affirm second. → *No conclusion.*
Deny second. → Deny first.	Deny first. → *No conclusion.*

So to infer with an if-then, we need to have either the *first part true* or the *second part false*:

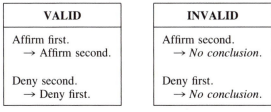

$$(+ \supset -)$$

The invalid forms are called the fallacies of *denying the antecedent* and *affirming the consequent*.

Try to master the S- and I-rules before starting formal proofs in the next chapter.

2.12A Exercises

Draw whatever simple conclusions (a letter or its negation) follow from these premises. If nothing follows, leave blank.

0. (~Q ∨ ~M)	ANSWER: (~Q ∨ ~M)
Q	Q
	~M

[At least one is true. It isn't the first, so it must be the second.]

1. ~(W · T)   **2.** (~Y ⊃ K)   **3.** (S ∨ ~L)
    W   Y   ~S
    ――――   ――――   ――――

4. (X ⊃ E)   **5.** (~M ∨ ~B)   **6.** ~(B · S)
    E   ~M   ~S
    ――――   ――――   ――――

7. (U ⊃ G)   **8.** ~(~F · ~Q)   **9.** (C ⊃ ~V)
    U   ~F   ~C
    ――――   ――――   ――――

10. (U ∨ J)   **11.** (H ⊃ ~B)   **12.** (~N ∨ ~A)
    ~U   H   A
    ――――   ――――   ――――

13. (~P ⊃ ~N)   **14.** (H ⊃ ~J)   **15.** ~(~B · C)
    P   J   C
    ――――   ――――   ――――

16. (~T ∨ D)   **17.** ~(C · ~P)   **18.** (Z ⊃ ~X)
    ~T   P   ~X
    ――――   ――――   ――――

19. (F ∨ H)   **20.** (~E ⊃ ~O)   **21.** ~(~P · ~G)
    ~H   ~E   G
    ――――   ――――   ――――

22. (~Z ⊃ D)   **23.** (~Z ∨ ~N)   **24.** (~I ⊃ Y)
    D   ~N   ~I
    ――――   ――――   ――――

2.12B Exercises

This exercise mixes the S- and the I-rules. Use the S-rules on the problems with *one* premise and the I-rules on those with *two* premises. In either case, draw whatever simple conclusions (a letter or its negation) follow from these premises. If nothing follows, leave blank.

0. (A ⊃ ~B)   ANSWER: Leave blank (*no conclusion*).
    ~A
    ――――

[This has two premises, so we use the I-rules. Here nothing follows. To infer with an if-then, we need either the first part true or the second part false.]

1. ~(~D · ~O)   **2.** ~(~D · ~O)   **3.** (~E ∨ L)
    ――――   D   ~L
                                  ――――   ――――

4. ~(D $\lor$ ~J)   **5.** (~R $\cdot$ A)   **6.** (W $\supset$ L)
_____   _____   ~L
                                                           _____

7. (~J $\lor$ K)   **8.** (~X $\supset$ F)   **9.** ~(~Z $\cdot$ Q)
J   _____   Z
_____   _____

10. (G $\lor$ V)   **11.** (~R $\lor$ ~G)   **12.** ~(L $\supset$ ~T)
G   _____   _____
_____

13. ~(W $\cdot$ ~X)   **14.** ~(L $\cdot$ M)   **15.** ~(~B $\lor$ C)
~W   _____   _____
_____

16. (D $\lor$ ~J)   **17.** ~(~C $\supset$ D)   **18.** ~(~R $\cdot$ A)
D   _____   ~R
_____   _____

2.13 EXTENDED INFERENCES

Consider our first S-rule:

(A $\cdot$ B)
_____
A, B

So far we've used this rule on little formulas. But the rule also works on big formulas. Consider this premise:

((C $\equiv$ D) $\cdot$ (E $\supset$ F))

Try to visualize this as a conjunction with two parts:

((C $\equiv$ D) $\cdot$ (E $\supset$ F))
($$$$$$ $\cdot$ ####)

The S-rules allow us to infer each part of a conjunction:

((C $\equiv$ D) $\cdot$ (E $\supset$ F))   ($$$$$$ $\cdot$ ####)
_____   _____
(C $\equiv$ D), (E $\supset$ F) $$$$$$, ####

The inference is just like the simpler ones we are used to. But now the parts are more complicated.
    Consider this premise:

~(C $\supset$ (D $\supset$ F))

Try to visualize this as a false if-then with two parts:

~(C ⊃ (D ⊃ F))
~($ ⊃ ####)

The S-rules allow us to infer from a false if-then that the first part is true and the second is false:

~(C ⊃ (D ⊃ F))	~($ ⊃ ####)
C, ~(D ⊃ F)	$, ~####

We could simplify ''~(D ⊃ F)'' further into ''D'' and ''~F.'' But now we are just doing one-step inferences with the S- and I-rules.

Consider this premise:

((C ≡ D) ⋁ (E · F))

This is a disjunction with two parts:

((C ≡ D) ⋁ (E · F))
($$$$$$$ ⋁ ####)

A disjunction says ''At least one part is true.'' If that's all we know, we can't infer anything about the truth or falsity of the parts. But suppose we add a second premise:

((C ≡ D) ⋁ (E · F))
~(C ≡ D)

Given these *two* premises, we can infer a conclusion:

| ((C ≡ D) ⋁ (E · F)) | ($$$$$$$ ⋁ ####) |
~(C ≡ D)	~$$$$$$$
(E · F)	####

At least one is true; it isn't the first, so it must be the second. This uses an I-rule.

Consider this long formula:

((C · D) ⊃ (E ⊃ F))

This is an if-then with two parts:

((C · D) ⊃ (E ⊃ F))
($$$$$$$ ⊃ ####)

We could infer with this formula using the I-rules if we had one of two further premises. Given a premise affirming the *first* part, we could draw a conclusion affirming the *second* part:

$((C \cdot D) \supset (E \supset F))$	$(\$\$\$\$\$\$\$ \supset ####)$
$(C \cdot D)$	$\$\$\$\$\$\$\$$
$(E \supset F)$	$####$

Given a premise denying the *second* part, we could draw a conclusion denying the *first* part:

$((C \cdot D) \supset (E \supset F))$	$(\$\$\$\$\$\$\$ \supset ####)$
$\sim(E \supset F)$	$\sim####$
$\sim(C \cdot D)$	$\sim\$\$\$\$\$\$\$$

These are the *only correct* S- and I-rule inferences using this if-then. But people are tempted to make several *incorrect* inferences. Let me mention four of these. This first one is the fallacy of denying the antecedent:

$((C \cdot D) \supset (E \supset F))$	$(\$\$\$\$\$\$\$ \supset ####)$
$\sim(C \cdot D)$	$\sim\$\$\$\$\$\$\$$
$\sim(E \supset F)$ [wrong!]	$\sim####$ [wrong!]

The next is the fallacy of affirming the consequent:

$((C \cdot D) \supset (E \supset F))$	$(\$\$\$\$\$\$\$ \supset ####)$
$(E \supset F)$	$####$
$(C \cdot D)$ [wrong!]	$\$\$\$\$\$\$\$$ [wrong!]

And this next one presumes that we are given that the first part of the if-then is true (whereas all we are given is that *if* the first part is true *then* the second part is true):

$((C \cdot D) \supset (E \supset F))$
C, D [wrong!]

And this last one presumes that we are given that the second part of the if-then is true (whereas all we are given is that *if* the first part is true *then* the second part is true):

$((C \cdot D) \supset (E \supset F))$
E
F [wrong!]

2.13A Exercises

Draw whatever conclusions follow from these premises by a single application of the S- or I-rules. If nothing follows in this way, then leave blank.

0. $\sim(\sim A \lor (B \cdot C))$ ANSWER: $\sim(\sim A \lor (B \cdot C))$

                                                        A, $\sim(B \cdot C)$

[Not either part is true, and so each part is false.]

1. $\sim((A \cdot B) \supset \sim C)$
   _____

2. $((A \cdot B) \supset \sim C)$
   $\sim(A \cdot B)$
   _____

3. $\sim((G \vee H) \cdot (I \vee J))$
   _____

4. $\sim((G \vee H) \cdot (I \vee J))$
   $(G \vee H)$
   _____

5. $((A \cdot B) \vee (C \supset D))$
   _____

6. $((A \cdot B) \vee (C \supset D))$
   C
   _____

7. $\sim((A \supset B) \supset C)$
   _____

8. $((A \supset B) \supset C)$
   $(A \supset B)$
   _____

9. $((G \equiv H) \supset \sim(I \cdot J))$
   $\sim(I \cdot J)$
   _____

10. $((G \equiv H) \supset \sim(I \cdot J))$
    I
    _____

2.14 COMPUTER APPLICATIONS

Logic was an important force in moving us into the computer age. One of my logic teachers at the University of Michigan, Arthur Burks, was part of the team that invented the first large-scale electronic computer, the ENIAC, in the 1940s. Computers make heavy use of logic.

In a computer, instead of truth and falsity we have different physical states. Suppose that "1" represents a positive voltage and "0" a zero voltage. An *and-gate* would then be a device with two inputs and one output, where the *output* has a positive voltage if and only if both *inputs* have positive voltages:

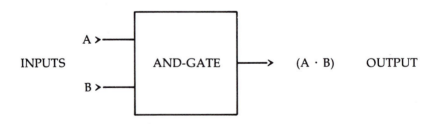

Our truth table for "and" describes the input/output function of this device. Similarly, we could construct a *not-gate* that has a positive *output* if and only if the *input* is zero. We could hook gates together, as in this example:

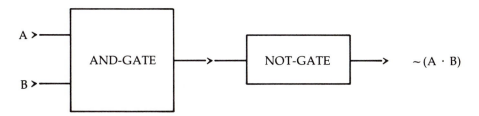

This gives a positive output if and only if not both inputs are positive. It's fairly easy to simulate any PC wff in an electronic *logic switch* constructed out of transistors. Modern computers contain millions of such logic switches. You might picture a computer as containing millions of little humanoids doing truth tables.

A computer codes information using 1's and 0's. A 1 or 0 is called a *bit* (for "binary digit"). When you type on a computer, each letter is encoded as a *byte*, which is a series of eight bits; for example, "A" encodes as "01000001." This book takes up about a million bytes—about eight million 1's and 0's! Electronic logic switches manipulate the 1's and 0's and help convert them back to letters on the screen or on a piece of paper.

I wrote the computer program *LogiCola* to go with this book. LogiCola is a set of computer instructions encoded as magnetic 1's and 0's on a floppy disk. You put the floppy disk into a personal computer. LogiCola then generates logic problems (for example, truth-table problems), tells you if you're doing them right, and gives you help when you need it. LogiCola uses a programming language called PILOT (invented by John Starkweather).

Computer languages such as PILOT often use logical operations like "or" or "if-then." PILOT uses the following logical symbols:

\ is for *not*
! is for *or*
& is for *and*
= is for *if and only if*
< > is for *exclusive or*

Unlike BASIC and PASCAL, PILOT has no symbols for if-then. Instead, PILOT places a formula between a command symbol and a semicolon to formulate a conditional command. *If* the formula is true *then* the command is executed. LogiCola uses commands like this one:

t(x = 7!y = 0):Terrific job!!

This types "Terrific job!!" on the screen if variable x equals 7 or variable y equals 0. The values for the variables are set by further parts of the program or by your responses. PILOT can also calculate whether a logical formula is true or false (1 or 0) on the basis of knowing the truth value of its parts. Such logical notions are used in writing programs of all sorts, not just programs for logic instruction.

3

Propositional Proofs

3.1 THE BUTLER DID IT

From now on, we'll use formal proofs as our *sole* method for testing arguments. Formal proofs provide a convenient test for complex PC arguments and can be extended to cover more advanced systems. They're also useful in developing reasoning ability. Since proofs presume a mastery of the S- and I-rules (Sections 2.11 to 2.13), you might want to review these rules from time to time.

Let's start with a simple proof in English. Suppose we want to prove that the butler did it. We know the following premises:

1. If the only people in the mansion at the time of the killing were the butler and the maid, then either the butler or the maid did it.
2. The only people in the mansion at the time of the killing were the butler and the maid.
3. If the maid did it, then the maid had a motive for the killing.
4. The maid didn't have a motive for the killing.

Using an indirect proof strategy, we first assume (line 5) that the butler *didn't* do it. Then we show that this is impossible—because it leads to a contradiction. We derive the contradiction from our assumption (line 5) and our premises (lines 1–4):

5. Assume: The butler didn't do it.
6. ∴ Either the butler or the maid did it. [from lines 1 and 2]
7. ∴ The maid did it. [from lines 5 and 6]
8. ∴ The maid had a motive for the killing. [from lines 3 and 7, contradicting 4]

Given our assumption, it follows that the maid both did (line 8) and didn't (line 4) have a motive for the killing. Our assumption that the butler *didn't* do it thus leads to an impossibility. We conclude that, *given our data* (and our argument is only as strong as its premises), the butler *must* have done it.

This symbolic version of the argument is a "formal proof":

* **1.**   (T ⊃ (B ∨ M))   VALID
  2.   T
* **3.**   (M ⊃ H)
  4.   ~H
          [∴ B
  5.   ⎡ asm: ~B
* **6.**   │ ∴ (B ∨ M) (from 1 and 2]
  7.   │ ∴ M [from 5 and 6]
  8.   ⎣ ∴ H [from 3 and 7, contradicting 4]
  9.   ∴ B [from 5, 4, and 8]

This mirrors our intuitive reasoning. First we block off the conclusion (showing that we can't use it in deriving further lines) and in line 5 assume its opposite. Then we derive further lines using the S- and I-rules until we get a contradiction. We get line 6 from lines 1 and 2 by the I-rule that goes "If-then, affirm first, so affirm second." Here the "first" is "T" and the "second" is "(B ∨ M)." We get 7 from 5 and 6 by the I-rule that goes "At least one is true; this one is false, so the other is true." And 8 follows from 3 and 7 by the I-rule that goes "If-then, affirm first, so affirm second." But 4 and 8 contradict. So 9 follows, using RAA, the new *reductio ad absurdum* ("reduction to absurdity") rule. RAA says that an assumption that leads to a contradiction must be false. As we apply RAA, we block off lines 5 through 8 to show that we can't use them in deriving further lines. Thus we prove the argument valid.

There are often various ways to do a proof. Instead of deriving "H" in line 8, we might have used 3 and 4 to derive "~M," thus contradicting 7. Inferences and contradictions can use any lines at all (premises, assumptions, and derived steps), except ones that are already blocked off.

In our proof, we starred lines 1, 3, and 6 when we used them. A *star* means "We can henceforth ignore this line." Here are the starring rules—and two examples:

Star any line that you simplify using an S-rule.

When you use an I-rule to derive a line from previous lines, star the *longer* previous line used.

---

* (A · B)
  ―――――――
  ∴ A
  ∴ B

* (A ⊃ B)
  A
  ―――――――
  ∴ B

(The *shorter* wff used in an I-rule inference might have to be used again.) A starred line is redundant since shorter lines contain the same information.* We aren't *forbidden* to use a starred line again, but it's seldom useful to do so and we can always avoid doing it. When doing a proof, we focus on *unstarred* lines and what we can derive from them using the S- and I-rules.

By the definition of *formal proof* in the next section, the stars, line numbers, blocked-off conclusion, and justifications aren't strictly part of the proof. Rather, these are unofficial helps. Our proof, minus nonessentials, looks like this:

$$(T \supset (B \lor M))$$
$$T \qquad\qquad\qquad \text{premises}$$
$$(M \supset H)$$
$$\sim H$$

asm: $\sim B$ ← assumption
∴ $(B \lor M)$
∴ M   ← derived
∴ H   ← steps

∴ B

Let's try another proof:

1. If Eleanor saw the butler putting the white tablet into the drink and the white tablet was poison, then the butler poisoned the deceased.
2. Eleanor saw the butler putting the white tablet into the drink.
∴ If the white tablet was poison, then the butler poisoned the deceased.

We first assume that the conclusion is false. Then we show that, given our premises, it couldn't be false—since this would lead to a contradiction. So, given our premises, the conclusion must be true. The proof goes as follows:

* **1.** $((E \cdot W) \supset B)$ VALID
  2.   E
        $[\therefore (W \supset B)$
* **3.**   asm: $\sim(W \supset B)$
  4.   ∴ W [from 3]
  5.   ∴ $\sim B$ [from 3]
* **6.**   ∴ $\sim(E \cdot W)$ [from 1 and 5]
  7.   ∴ $\sim W$ [from 2 and 6, contradicting 4]
  8. ∴ $(W \supset B)$ [from 3, 4, and 7]

First we block off the conclusion (showing that we can't use it in deriving further lines) and in line 3 assume its opposite. Then we derive further lines using the S- and I-rules

---

* In the first case, the truth of "$(A \cdot B)$" would follow from the truth of "A" and "B." In the second case, the truth of "$(A \supset B)$" would follow from the truth of "B." A starred line contains no additional information, since everything that follows from the starred line also follows from other unstarred not-blocked-off lines.

until we get a contradiction. The false if-then in line 3 simplifies into lines 4 and 5 (and then gets starred). The if-then in 1 combines with 5 to give us 6 (and then 1 gets starred). The not-both in 6 combines with 2 to give us 7 (and then 6 gets starred). But 4 and 7 contradict. Since assumption 3 leads to a contradiction, we apply RAA. We block off lines 3 through 7 to show that we can't use them in deriving further lines. We derive conclusion 8 (our original conclusion). Thus we prove our argument valid.

If we try to prove an invalid argument, we won't succeed. Instead, we'll discover a refutation of the argument. We won't deal with such cases until Section 3.3.

3.2 FORMAL PROOFS

Now we'll learn the rules behind our method of doing proofs. Here are three preliminary definitions:

- Two wffs that are exactly alike except that one starts with an additional "~" are *contradictories*.
- "→" in the inference rules which follow indicates that, given *each* wff on the *left* as the wff of a not yet blocked-off line, one may infer a derived step consisting of "∴" followed by *any* of the wffs on the *right*.
- "↔" in the inference rules which follow indicates that, given *each* wff on *either side* as the wff of a not yet blocked-off line, one may infer derived steps consisting of "∴" followed by any of the wffs on the *other side*.

You can read "(P ⊃ Q), P → Q" as "from (P ⊃ Q) and P, one may derive Q." You can read "~ ~P ↔ P" as "from ~ ~P one may derive P, and from P one may derive ~ ~P."

The following inference rules hold regardless of what two pairs of contradictory wffs replace "P"/"~P" and "Q"/"~Q":

S-RULES: S-1 TO S-6	I-RULES: I-1 TO I-6
~ ~ P ↔ P	~(P · Q), P → ~Q
	~(P · Q), Q → ~P
(P · Q) ↔ P, Q	
~(P ∨ Q) ↔ ~P, ~ Q	(P ∨ Q), ~ P → Q
~(P ⊃ Q) ↔ P, ~ Q	(P ∨ Q), ~ Q → P
(P ≡ Q) ↔ (P ⊃ Q), (Q ⊃ P)	(P ⊃ Q), P → Q
~(P ≡ Q) ↔ (P ∨ Q), ~(P · Q)	(P ⊃ Q), ~ Q → ~P

S-1 is new and allows us to eliminate "~ ~" from (or add it to) the beginning of a wff:

$$\frac{\sim \sim P}{\therefore P} \qquad\qquad \frac{P}{\therefore \sim \sim P}$$

S-2 to S-4 are familiar from the last chapter. Now we can use them in either direction. So from a conjunction we may infer each conjunct, and from two wffs we may infer their conjunction:

$$\frac{(P \cdot Q)}{\begin{array}{l} \therefore P \\ \therefore Q \end{array}} \qquad\qquad \frac{\begin{array}{l} P \\ Q \end{array}}{\therefore (P \cdot Q)}$$

We'll usually use the S-rules in the simplifying direction. S-5 and S-6 are new. S-5 lets us break a *bi*conditional into two conditionals. S-6 lets us break up the negation of a biconditional. Regarding S-6, note that "(P ≡ Q)" says that "P" and "Q" have the *same* truth value. Then "~(P ≡ Q)" says that "P" and "Q" have *different* truth values—that one or the other is true, but not both. We'll seldom use S-1, S-5, and S-6. We covered all the I-rules in the last chapter.

Now we define *proof* more precisely:

- A *premise* is a line consisting of a wff by itself (with no preceding "∴" or "asm:").
- An *assumption* is a line consisting of "asm:" and then a wff.
- A *derived step* is a line consisting of "∴" and then a wff.
- A (*formal*) *proof* is a vertical sequence of zero or more premises followed by one or more assumptions and/or derived steps such that each derived step follows from previously not-blocked-off lines by one of the inference rules (S-1 to S-6, I-1 to I-6, or RAA) and each assumption is blocked off in accord with RAA.

Our proofs normally use the RAA rule. By RAA, an assumption that leads to a contradiction must be false. Here's a provisional form of RAA for one-assumption arguments:

> *RAA* Suppose that some pair of not-blocked-off lines have contradictory wffs. Then block off all the lines from the assumption on down. Infer a step consisting of "∴" followed by a contradictory of the wff of that assumption. (Provisional form of RAA)

The contradictory wffs could occur anywhere in the proof (as premises or assumptions or derived steps), as long as neither wff is blocked off. Study this example:

```
*  1.    (A · ~B)        VALID
*  2.    (C ⊃ B)
        [∴ ~C
   3.    ⎡asm: C
   4.    │∴ B [from 2 and 3]
   5.    │∴ A [from 1]
   6.    ⎣∴ ~B [from 1, contradicting 4]
   7.  ∴ ~C [from 3, 4, and 6]
```

First we assume the simpler contradictory of the conclusion. We might have assumed "~~C," but it's easier to assume "C." We derive a few things and find that lines 4 and 6 contradict. So we block off the wffs from the assumption on down and infer the opposite of the assumption; in other words, we infer the original conclusion. Here's another example:

   1.   A   VALID
* **2.**   (A ⊃ B)
       [∴ (A · B)
* **3.**   ⌈asm: ~(A · B)
   4. | ∴ B [from 1 and 2]
   5.   ⌊∴ ~A [from 3 and 4, contradicting 1]
   6. ∴ (A · B) [from 3, 1, and 5]

Here our derived step 5 contradicts a premise. In other examples, a derived step may contradict an assumption.

    Don't use the original conclusion to derive further steps or obtain a contradiction. Blocking off this conclusion reminds us not to use it in the proof.

    Assume a genuine contradictory of the conclusion—not like this:

    . . . .
  [∴ (~A ⊃ B)
    asm: (A ⊃ B) [wrong!]

"(~A ⊃ B)" and "(A ⊃ B)" *aren't* contradictories, since they aren't exactly alike except that one *starts* with an additional "~." Neither starts with "~"; each starts with "(." This mistake is more likely if you're loose with parentheses:

    . . . .
  [∴ ~A ⊃ B [not a wff!]
    asm: A ⊃ B [wrong!]

To form the contradictory of a wff, you may add a squiggle to the beginning. If the wff already begins with a squiggle (and not with a left-hand parenthesis), you may instead omit the squiggle. Here's the correct assumption:

    . . . .
  [∴ (~A ⊃ B)
    asm: ~(~A ⊃ B) [right!]

I suggest this strategy for proving a PC argument:

  1. Block off the conclusion. Add "asm:" followed by the simpler contradictory of the conclusion to form the next line.

2. Derive new steps using the S- and I-rules on *unstarred*, not-blocked-off wffs. Star the appropriate lines as you apply these rules. (Use the S-rules only in the simplifying direction.)

3. When some not-blocked-off pair of lines contradict, apply RAA. Your proof is done!

This strategy is enough to prove *most* valid PC arguments. Some arguments require further assumptions. I promise not to spring any of these on you until we get to Section 3.4.

3.2A Symbolic Argument Exercises

Each of these is a famous valid argument form with a special name. For each argument, first read it over and see whether it's valid intuitively. Then prove it valid.

0. Implication:

(A ⊃ B)

∴ (~A ∨ B)

ANSWER:

* 1.   (A ⊃ B)   VALID

        [∴ (~A ∨ B)

* 2.   ⎡ asm: ~(~A ∨ B)
  3.   │ ∴ A [from 2]
  4.   │ ∴ ~B [from 2]
  5.   ⎣∴ B [from 1 and 3, contradicting 4]
  6.   ∴ (~A ∨ B) [from 2, 4, and 5]

1. Contraposition:

(A ⊃ B)

∴ (~B ⊃ ~A)

3. DeMorgan's Laws:

~(A · B)

∴ (~A ∨ ~B)

5. Association:

(A · (B · C))

∴ ((A · B) · C)

7. Exportation:

((A · B) ⊃ C)

∴ (A ⊃ (B ⊃ C))

9. Hypothetical Syllogism:

(A ⊃ B)

(B ⊃ C)

∴ (A ⊃ C)

2. Addition:

A

∴ (A ∨ B)

4. DeMorgan's Laws:

~(A ∨ B)

∴ (~A · ~B)

6. Commutation:

(A · B)

∴ (B · A)

8. Exportation:

(A ⊃ (B ⊃ C))

∴ ((A · B) ⊃ C)

10. Dilemma:

(A ∨ B)

(A ⊃ C)

(B ⊃ D)

∴ (C ∨ D)

All the one-premise arguments here (except 2) remain valid if we reverse the premise and the conclusion. In addition, 5 and 6 remain valid if we substitute "∨" for "·." "Association" shows that it doesn't matter how we bracket a series of conjunctions (or disjunctions).

3.2B English Argument Exercises

Each of these arguments is valid. For each argument, first read it over and see whether it's valid intuitively. Then translate it into PC and prove it valid.

0. If Genesis gives the literal facts, then the earth existed three days before the sun was created.
"Day" is defined by reference to the sun.
If "day" is defined by reference to the sun, then the earth can't have existed three days before the sun was created.
∴ Genesis doesn't give the literal facts.
[Use L, E, and D. This argument against biblical fundamentalism is from Origen, a third-century Christian thinker.]

ANSWER: * 1.   (L ⊃ E)   VALID
             2.   D
        * 3.   (D ⊃ ~E)
              [∴ ~L
          4.   ⌈asm: L
          5.   │∴ ~E [from 2 and 3]
          6.   ⌊∴ E [from 1 and 4, contradicting 5]
          7.   ∴ ~L [from 4, 5, and 6]

1. If we have sensations of alleged material objects and yet no material objects exist, then God is a deceiver.
God isn't a deceiver.
∴ If we have sensations of alleged material objects, then material objects exist.
[Use S, M, and D. From René Descartes, who based our knowledge of the external world on theological premises.]

2. If knowing is a state of mind (like feeling a pain), then I could always tell by introspection whether I know.
If I could always tell by introspection whether I know, then I'd never mistakenly think that I know.
I sometimes mistakenly think that I know.
∴ Knowing isn't a state of mind.
[Use S, I, and M.]

3. If this logic pretest consists of simple arguments and untrained intuition is generally reliable in determining logical validity, then students taking this pretest before studying logic would get almost all the answers right.
This logic pretest consists of simple arguments.
Students taking this pretest before studying logic *don't* get almost all the answers right.
∴ Untrained intuition *isn't* generally reliable in determining logical validity.
[Use P, U, and S. I begin my lower-level logic course with a pretest. My students get about half the problems right.]

4. If God doesn't want to prevent evil, he isn't all good.
If God isn't able to prevent evil, he isn't all powerful.
Either God doesn't want to prevent evil, or he isn't able to.
∴ Either God isn't all powerful, or he isn't all good.
[Use W, G, A, and P. This is the classic problem of evil.]

5. We've examined a large and varied group of orderly things.
Most orderly things we've examined have intelligent designers.
The universe is orderly.
If most orderly things we've examined have intelligent designers and we've examined a large and varied group of orderly things, then probably most orderly things have intelligent designers.

If the universe is orderly and probably most orderly things have intelligent designers, then probably the universe has an intelligent designer.
∴ Probably the universe has an intelligent designer.
[Use W, M, U, P, and D. This is a form of the controversial argument from design (or *teleological argument*) for the existence of God. The last two premises are instances of these two principles of inductive logic: "If most A's we've examined are B's and we've examined a large and random group of A's, then probably most A's are B's" and "If *x* is A and probably most A's are B's, then probably *x* is a B."]

6. If time stretches back infinitely, then today wouldn't have been reached.
If today wouldn't have been reached, then today wouldn't exist.
Today exists.
If time doesn't stretch back infinitely, then there was a first moment of time.
∴ There was a first moment of time.
[Use I, R, T, and F.]

7. If she loves you and won't forgive you, then she doesn't love you.
∴ If she loves you, then she'll forgive you.
[Use L and F.]

8. If there are already laws preventing discrimination against women, then if the Equal Rights Amendment (ERA) would rob women of many current privileges then passage of the ERA would be against women's interests and women ought to work for its defeat.
The ERA would rob women of many current privileges (draft exemption, for example).
∴ If there are already laws preventing discrimination against women, then women ought to work for the defeat of the ERA.
[Use L, R, A, and W. This and the next give a mini-debate on the ERA.]

9. If women ought never to be discriminated against, then we should work for current laws against discrimination and also prevent future generations from imposing discriminatory laws against women.
The only way to prevent future generations from imposing discriminatory laws against women is to pass an Equal Rights Amendment (ERA).
If we should prevent future generations from imposing discriminatory laws against women and the only way to do this is to pass an ERA, then we ought to pass an ERA.
∴ If women ought never to be discriminated against, then we ought to pass an ERA.
[Use N, C, F, O, and A.]

10. [A *prime number* is a whole number greater than 1 that isn't evenly divisible by any such number other than itself. The first five prime numbers are 2, 3, 5, 7, and 11. Euclid in ancient Greece proved that there's no greatest prime number. His proof followed an RAA pattern. Assume that there's a greatest prime number, and call it x. Then x is the greatest prime. Let y be one-plus-the-product-of-all-the-prime-numbers-up-to-x. Then we can prove that x isn't the greatest prime.]
y is greater than x.
Either y is prime or y is composite.
If y is prime and y is greater than x, then x isn't the greatest prime.
Any prime divisor that y has is greater than x.
If y is composite, then y has a prime divisor.
If y has a prime divisor and any prime divisor that y has is greater than x, then y has a prime divisor greater than x.
If y has a prime divisor greater than x, then x isn't the greatest prime.
∴ x isn't the greatest prime.
[Use G, P, C, X, A, D, and H.]

3.3 REFUTATIONS

Now we'll see what happens when we try to do a formal proof on an invalid argument. Consider this invalid version of our butler argument (Section 3.1):

1. If the only people in the mansion at the time of the killing were the butler and the maid, then either the butler or the maid did it.
2. The only people in the mansion at the time of the killing were the butler and the maid.
3. If the maid did it, then the maid had a motive for the killing.
∴ The butler did it.

To make it invalid, I cut out premise 4 ("The maid *didn't* have a motive for the killing"). If we try to prove this argument, we soon reach a point at which we can apply the S- and I-rules no further but yet have no contradiction:

```
*  1.     (T ⊃ (B ∨ M))
   2.     T
*  3.     (M ⊃ H)
      [∴ B
   4.       asm: ~B
*  5.       ∴ (B ∨ M) [from 1 and 2]
   6.       ∴ M [from 4 and 5]
   7.       ∴ H [from 3 and 6]
```

We can't prove the argument valid, but we can show it to be invalid. To do this, we take the *simple wffs* (letters or their negations) from the not-blocked-off lines and put them in a box. This box gives a *refutation* of the argument—truth conditions making all the premises true but the conclusion false:

$$\boxed{\text{T, H, M, } \sim\text{B}}$$

The simple wffs in the box come from lines 2, 4, 6, and 7. The order of the wffs doesn't matter. I like to put the positive wffs first. We could also write the refutation in any of these three ways:

$$\boxed{T=1, H=1, M=1, B=0} \qquad \boxed{T^1, H^1, M^1, B^0}$$

> The only people in the mansion at the time of the killing were the butler and the maid (T).
> The maid had a motive for the killing (H).
> The maid did it (M).
> The butler didn't do it (~B).

Under these truth conditions, the premises are all true but the conclusion false:

$(T^1 \supset (B^0 \lor M^1)) = 1$ INVALID
$T^1 = 1$
$(M^1 \supset H^1) = 1$
$\therefore B^0 = 0$

This shows our argument to be invalid.

Lawyers often use similar reasoning to defend their clients. They sketch a possible situation that is consistent with the evidence and in which their client was innocent. They conclude that the evidence doesn't establish their client's guilt.

Let's take another invalid argument:

1. If the butler shot Jones, then he knew how to use a gun.
2. If the butler was a former marine, then he knew how to use a gun.
3. The butler was a former marine.
$\therefore$ The butler shot Jones.

If we try to prove this, we again won't be able to derive a contradiction:

1. $(S \supset K)$
* **2.** $(M \supset K)$
3.   M
　　[$\therefore$ S
4.   asm: $\sim$ S
5. $\therefore$ K [from 2 and 3]

At this point we are stuck. We might be tempted to use line 1 with line 4 or 5 to derive a further conclusion:

$(S \supset K)$ $(S \supset K)$
$\sim S$ K
─────────   ─────────
$\therefore \sim K$ $\therefore$ S

We should resist the temptation. Neither inference is valid! To infer with an if-then, we need the first part true or the second part false. If we make this wrong step, we'll end up "proving" the argument valid by misapplying the rules. It's important to know the S- and I-rules thoroughly. With one false step, the whole thing can go wrong!

Here once again we can take the *simple wffs* (letters or their negations) from the not-blocked-off lines and put them in a box. This box gives a *refutation* of the argument—truth conditions making all the premises true but the conclusion false:

$(S^0 \supset K^1) = 1$ **INVALID**
$(M^1 \supset K^1) = 1$
$M^1 = 1$ ┌─────────────┐
$\therefore S^0 = 0$ │ M, K, $\sim$S │
                         └─────────────┘

It could be that the butler was a former marine and knew how to use a gun, and yet didn't shoot Jones. This possibility makes the premises true and the conclusion false.

Let's summarize. Suppose we want to show that, given certain data, the butler must have done it. First we assume that he *didn't* do it. Then we try to show that, given our data, this is impossible—because it leads to a contradiction. If we find a contradiction, then it's impossible for our data to be true and yet for the butler not to have done it. Given our data, then, he *must* have done it; the argument against him is valid. Or we might find no contradiction. Then, we hope, we can find a consistent explanation of how the data might be true and yet he *didn't* do it. Then the argument against him is invalid.

I suggest this strategy for arriving at a proof or refutation of a PC argument:

1. Block off the conclusion. Add "asm:" followed by its simpler contradictory to form the next line.
2. Derive whatever new steps you can, using the S- and I-rules, just like before. Go to step 3 if you get a contradiction. Go to step 4 if you get no contradiction and yet can't derive anything else.
3. In this case you have a contradiction. Apply RAA. The argument is valid and you've proved it.
4. In this case you have no contradiction and yet can't derive anything else. Draw a box containing the simple wffs (letters or their negation) from the not-blocked-off lines. With luck, this box will give a *refutation* of the argument—truth conditions making the premises true and the conclusion false. This would show the argument to be invalid.

Applying these directions to a PC argument *normally* yields either a proof or a refutation. A few PC arguments require multiple assumptions. We won't see any of these until Section 3.4.

When you reach an alleged refutation and plug in the values, you should get all the premises true and the conclusion false:

$$\#\#\#\# = 1$$
$$***** = 1$$
$$\therefore \ \$\$\$\$ = 0$$

If you *don't* get this, you did something wrong. The line that comes out wrong (a premise that's 0 or ?, or a conclusion that's 1 or ?) is the source of the problem. Maybe you derived something incorrectly from this line. Or maybe you didn't derive something from it that you should have. So our proof strategy has a self-correcting feature. It can tell us that something went wrong—and where to look to fix the problem.

3.3A Symbolic Argument Exercises

Say whether valid (then give a proof) or invalid (then give a refutation). The first five are all invalid.

0. (A ⊃ B)   ANSWER: * 1. $(A^1 \supset B^1) = 1$ **INVALID**
   (C ⊃ D)   2. $(C^0 \supset D^1) = 1$
   ~(B · C)   * 3. $\sim(B^1 \cdot C^0) = 1$ ┌─────────────┐
 ∴ ~(A · D)   [∴ $\sim(A^1 \cdot D^1) = 0$]   │ A, B, D, ~C │
                                             └─────────────┘

[*] 4.   asm: (A · D)
5.   ∴ A [from 4]
6.   ∴ D [from 4]
7.   ∴ B [from 1 and 5]
8.   ∴ ~C [from 3 and 7]

1. (A ⊃ B)
∴ (B ⊃ A)

3. ((A · B) ⊃ C)
    ((C ∨ D) ⊃ ~E)
∴ ~(A · E)

5. ((A ⊃ B) ⊃ (C ⊃ D))
    (B ⊃ D)
    (A ⊃ C)
∴ (A ⊃ D)

7. (A ≡ B)
    (C ⊃ B)
    ~(D · C)
    D
∴ ~A

9. ∴ (A ∨ ~A)

2. ((A · B) ⊃ C)
∴ (A ⊃ C)

4. (A ⊃ (B · C))
    (~C ⊃ D)
∴ ((B · ~D) ⊃ A)

6. ((A ⊃ B) · (C ⊃ D))
    ~(B ≡ D)
    C
∴ ~A

8. (A ⊃ B)
    (~C · A)
    (B ⊃ (D ∨ C))
∴ (B · D)

10. ∴ ((A ∨ B) ⊃ A)

[For 9 and 10, a PC argument without premises is valid if and only if the conclusion is a tautology (a wff true in every possible case—see Section 2.6). To do a premiseless proof, assume the opposite of the conclusion and then try to derive a contradiction.]

3.3B English Argument Exercises

For each argument, first evaluate intuitively and then translate into PC and work it out. Say whether valid (then give a proof) or invalid (then give a refutation). Afterwards, try to reword the invalid arguments slightly or add further premises to produce plausible valid arguments.

0. If virtue can be taught, then either there are professional virtue-teachers or there are amateur virtue-teachers.
If there are professional virtue-teachers, then the Sophists can teach their students to be virtuous.
If there are amateur virtue-teachers, then the noblest Athenians can teach their children to be virtuous.
The Sophists can't teach their students to be virtuous (look at their students!) and the noblest Athenians can't teach their children to be virtuous (look at Pericles's son!).
∴ Virtue can't be taught.
[Use V, P, A, S, and N. From Plato's *Meno*. Socrates thought that we couldn't teach another person to live rightly. Virtue has to come from within.]

ANSWER: [*] 1.   (V ⊃ (P ∨ A))   VALID
           [*] 2.   (P ⊃ S)
           [*] 3.   (A ⊃ N)
           [*] 4.   (~S · ~N)
              [∴ ~V

```
      5.    ┌ asm: V
 * 6.    │   ∴ (P ∨ A) [from 1 and 5]
      7.    │   ∴ ~S [from 4]
      8.    │   ∴ ~N [from 4]
      9.    │   ∴ ~P [from 2 and 7]
     10.    │   ∴ ~A [from 3 and 8]
     11.    └   ∴ A [from 6 and 9, contradicting 10]
     12.        ∴ ~V [from 5, 10, and 11]
```

1. There are two hands. (Look!)
   If there are two hands, then there are external objects.
   If there are external objects, then there's an external world.
∴ There's an external world.
   [Use T, O, and W. From G. E. Moore.]

2. If we aren't willing to endanger the entire planet to save ourselves from an invasion, then we are rational.
   If we intend to respond to a Soviet attack with a massive nuclear strike, then we are willing to endanger the entire planet to save ourselves from an invasion.
   We do intend to respond to a Soviet attack with a massive nuclear strike.
∴ We aren't rational.
   [Use W, R, and I.]

3. If predestination is true, then God causes us to sin.
   If God causes us to sin and yet damns sinners to eternal punishment, then God isn't good.
∴ If God is good, then either predestination isn't true or else God doesn't damn sinners to eternal punishment.
   [Use P, C, D, and G. This attacks the views of the colonial thinker Jonathan Edwards, who held P, D, and G.]

4. It would be equally wrong for a sadist (through a drug injection) to have blinded you before or after your birth.
   If it would be equally wrong for a sadist (through a drug injection) to have blinded you before or after your birth, then it's false that one's moral right to equal consideration begins at birth.
   If infanticide is wrong and abortion isn't wrong, then one's moral right to equal consideration begins at birth.
   Infanticide is wrong.
∴ Abortion is wrong.
   [Use E, R, I, and A.]

5. If you hold a moral belief and don't act on it, then you're inconsistent.
   If you're inconsistent, then you're doing wrong.
∴ If you hold a moral belief and act on it, then you aren't doing wrong.
   [Use M, A, I, and W. Is the conclusion true?]

6. If Socrates escapes from jail, then he's willing to obey the state only when it pleases him.
   If he's willing to obey the state only when it pleases him, then he doesn't really believe what he says and he's inconsistent.
∴ If Socrates really believes what he says, then he won't escape from jail.
   [Use E, W, R, and I. From Plato's *Crito*. Socrates had been jailed and sentenced to death for teaching philosophy. He discussed with his friends whether he ought to escape from jail instead of suffering the death penalty.]

7. Either Socrates's death will be perpetual sleep, or if the gods are good then Socrates's death will be an entry into a better life.
If Socrates's death will be perpetual sleep, then Socrates shouldn't fear death.
If Socrates's death will be an entry into a better life, then Socrates shouldn't fear death.
∴ Socrates shouldn't fear death.
[Use P, G, B, and F. From Plato's *Crito*. Except for a dropped premise, this reflects Socrates's argument that he had nothing to fear from death. What premise was dropped?]

8. If Socrates escapes and does what is wrong, then he causes his friends to suffer and is rightfully regarded as an outcast.
If Socrates causes his friends to suffer, then he does wrong.
∴ If Socrates escapes, then he does wrong.
[Use E, W, S, and R. This uses ideas from Plato's *Crito*.]

9. The function of government is to protect life, liberty, and the pursuit of happiness.
The British colonial government doesn't protect these.
The only way to change it is by revolution.
If the function of government is to protect life, liberty, and the pursuit of happiness and the British colonial government doesn't protect these, then the British colonial government ought to be changed.
If the British colonial government ought to be changed and the only way to change it is by revolution, then we ought to have a revolution.
∴ We ought to have a revolution.
[Use F, B, O, C, and R. This summarizes the reasoning behind the Declaration of Independence. The first premise was claimed to be a self-evident truth, the second and third were backed up by historical data, and the fourth and fifth were implicit conceptual bridge premises.]

10. If determinism is true then we have no free will.
If Heisenberg's interpretation of quantum physics is correct, then there are events not causally necessitated by prior events.
If there are events not causally necessitated by prior events, then determinism is false.
∴ If Heisenberg's interpretation of quantum physics is correct, then we have free will.
[Use D, F, H, and E.]

11. The apostles' teaching either is from God or is a human invention.
If it's from God and we kill the apostles, then we'll be killing God's messengers.
If it's a human invention, then it'll collapse of its own accord.
If it'll collapse of its own accord and we kill the apostles, then our killings will be unnecessary.
∴ If we kill the apostles, either our killings will be unnecessary or we'll be killing God's messengers.
[Use G, H, K, M, C, and U. From Rabbi Gamaliel, as mentioned in the Acts of the Apostles.]

12. If you know that you don't exist, then you don't exist.
If you know that you don't exist, then you know some things.
If you know some things, then you exist.
∴ You exist.
[Use K, E, and S.]

13. If you're smart, then you get good test scores.
If you praise the teacher and aren't smart, then the teacher likes you.
If the teacher likes you or you get good test scores, then you pass.
∴ You pass.
[Use S, T, P, L, and A.]

14. We have an idea of a perfect being.
  If we have an idea of a perfect being, then this idea is from the world or from a perfect being.
  If this idea is from a perfect being, then there is a God.
∴ There is a God.
  [Use I, W, P, and G. Except for a dropped premise, this is from René Descartes.]

15. Life is worth living if and only if Michigan beats Ohio State.
∴ If Michigan beats Ohio State and loses the Rose Bowl, then life is worth living.
  [Use W, B, and L.]

16. If the universe contains unavoidable wrong actions, then we ought to regret the universe as a whole.
  If determinism is true and regretting cruelty is wrong, then the universe contains unavoidable wrong actions.
∴ If determinism is true, then either we ought to regret the universe as a whole (the pessimism option) or else cruelty isn't wrong and regretting cruelty isn't wrong (the "nothing matters" option).
  [Use D, C, U, O, and R. This sketches the reasoning in William James's "The Dilemma of Determinism." James thought that when we couldn't prove one side or the other to be correct (as on the issue of determinism) it was more rational to pick our beliefs in accord with practical considerations. He argued that these weighed against determinism.]

17. If belief in God has scientific backing, then it's rational.
  No conceivable scientific experiment could decide the issue of whether there is a God.
  If belief in God has scientific backing, then some conceivable scientific experiment could decide the issue of whether there is a God.
∴ Belief in God isn't rational.
  [Use B, R, and D.]

18. If the world isn't ultimately absurd, then conscious life will go on forever and the world process will culminate in an eternal personal goal.
  If there is no God, then conscious life won't go on forever.
∴ If the world isn't ultimately absurd, then there is a God.
  [Use A, F, C, and G. This is from the Jesuit scientist, Pierre Teilhard de Chardin. He didn't think we could prove whether the world was ultimately absurd.]

19. Every event with finite probability eventually takes place.
  If the world powers don't get rid of their nuclear weapons, then there's a finite probability that humanity will eventually destroy the world.
  If every event with finite probability eventually takes place and there's a finite probability that humanity will eventually destroy the world, then humanity will eventually destroy the world.
∴ Either the world powers will get rid of their nuclear weapons, or humanity will eventually destroy the world.
  [Use E, R, F, and H.]

20. If materialism is true, then idealism is false.
  If idealism is true, then materialism is false.
  If mental events exist, then materialism is false.
  If the materialist *thinks* that his or her theory is true, then mental events exist.
∴ If the materialist *thinks* that his or her theory is true, then idealism is true.
  [Use M, I, E, and T.]

21. God is all powerful.

If God is all powerful, he could have created the world in any logically possible way and the world has no necessity.

If the world has no necessity, then we can't know the way the world is by abstract speculation apart from experience.

∴ We can't know the way the world is by abstract speculation apart from experience.

[Use A, C, N, and K. This is from William of Ockham, the late medieval thinker. His claim that knowing the world requires sense experience had a theological basis.]

22. If a belief is proved, then it's worthy of acceptance.

If a belief isn't disproved but is of practical value to our lives, then it's worthy of acceptance.

If a belief is proved then it isn't disproved.

∴ If a belief is proved or is of practical value to our lives, then it's worthy of acceptance.

[Use P, W, D, and V. The second premise is from the pragmatist philosophy of William James.]

23. If the ethical claim "X is good" is deducible from some descriptive claim about X (for example, "X is socially approved"), then "good" has the same meaning as "F" (where "F" stands for some descriptive phrase like "socially approved").

If "good" has the same meaning as "F," then "This is F but is it good?" is an open question if and only if "This is F but is it F?" is also an open question.

"This is F but is it good?" is an open question.

"This is F but is it F?" isn't an open question.

∴ The ethical claim "X is good" isn't deducible from some descriptive claim about X.

[Use D, M, A, and B. From G. E. Moore.]

24. If motion is real, then one can move from A to B in a finite time.

If one can cross only one spatial point at a time, then one can't cross an infinity of spatial points in a finite time.

One can cross only one spatial point at a time.

The distance from A to B can be divided into an infinity of spatial points.

If one can move from A to B in a finite time, then either the distance from A to B can't be divided into an infinity of spatial points or else one can cross an infinity of spatial points in a finite time.

∴ Motion isn't real.

[Use M, F, O, I, and D. From the ancient Greek Zeno of Elea, who denied the reality of motion.]

25. [X is the *square root* of y, provided that $x^2 = y$. So 3 is the square root of 9, since $3^2 = 9$. Is the square root of 2 equal to some fraction x/y, where x and y are positive whole numbers? There's an RAA argument that tries to prove that the answer is no. Assume that the square root of 2 equals x/y (where x and y are positive whole numbers). Assume further that the fraction x/y is simplified as far as it can be (like 1/3, but unlike 2/6 or 3/9). Then we can argue that the square root of 2 *doesn't* equal x/y.]

x/y is simplified as far as it can be.

If the square root of 2 equals x/y, then $2 = x^2/y^2$.

If $2 = x^2/y^2$, then $2y^2 = x^2$.

If $2y^2 = x^2$, then x is even.

If x is even and $2y^2 = x^2$, then y is even.

If x is even and y is even, then x/y isn't simplified as far as it can be.

∴ The square root of 2 *doesn't* equal x/y.

[Use S, Q, A, B, X, and Y. This argument tries to show that the square root of 2 is "irrational" in the sense of not being a *ratio* of positive whole numbers.]

3.4 MULTIPLE-ASSUMPTION BUTLERS

Sometimes one assumption isn't enough to prove that the butler did it. Consider this argument:

1. If the butler was at the party, then he fixed the drinks and poisoned the deceased.
2. If the butler wasn't at the party, then the detective would have seen him leaving the mansion and would have reported this to the jury.
3. The detective didn't report this to the jury.
∴ The butler poisoned the deceased.

1.	(P ⊃ (D · B))
2.	(~P ⊃ (S · R))
3.	~R
[∴ B	
4.	asm: ~B

We're stuck. We can't apply the S- or I-rules or RAA. And we don't have enough simple wffs to construct a refutation. So what can we do?

We need to make another assumption. Was the butler at the party? It would be nice to know that he was. Then we could use line 1 to draw further conclusions. To prove that he *was* at the party, we assume that he *wasn't*. (We assume the opposite of what we want to prove.) If the butler wasn't at the party, then by line 2 the detective saw him leaving and *reported* this. But line 3 says that he *didn't report* it. So assuming that the butler *wasn't* at the party leads to a contradiction:

1.	(P ⊃ (D · B))
2.	(~P ⊃ (S · R))
3.	~R
[∴ B	
4.	asm: ~B
5.	asm: ~P
6.	∴ (S · R) [from 2 and 5]
7.	∴ S [from 6]
8.	∴ R [from 6, contradicting 3]

This contradiction doesn't show that our argument is valid. It shows that it's impossible to have the premises true while *both* assumptions (lines 4 and 5) are also true. If we go on assuming that the butler *didn't* do it (line 4), we have to conclude that the butler *was* at the party. We conclude this using RAA:

1.   (P ⊃ (D · B))
2.   (~P ⊃ (S · R))
3.   ~R
     [∴ B
4.   asm: ~B
5.   ⎡asm: ~P
6.   ⎢∴ (S · R) [from 2 and 5]
7.   ⎢∴ S [from 6]
8.   ⎣∴ R [from 6, contradicting 3]
9.   ∴ P [from 5, 3, and 8]

As we apply RAA, we block off all the lines from the last assumption on down. Lines 5 to 8 follow *presuming* that the butler *wasn't* at the party. Now we've concluded that he *was* at the party. So we can't use lines 5 through 8 to derive further steps. We can use only lines that aren't presently blocked off.

For multiple-assumption proofs like this one, we have to expand rule RAA:

> **. RAA** Suppose that some pair of not-blocked-off lines have contradictory wffs. Then block off all the lines from the *last not-blocked-off* assumption on down. Infer a step consisting of "∴" followed by a contradictory of the wff of that assumption.

In applying RAA, we hold on to the earlier assumptions (for example, that the butler didn't do it) as long as we can. When we find a contradiction, we always get rid of the last assumption that we made but haven't yet gotten rid of.

Given that the butler *was* at the party, it's easy to derive a second contradiction. By line 1, the butler fixed the drinks and *poisoned* the deceased. Line 4 says he *didn't poison* the deceased. So we get a second contradiction and apply RAA again:

1.   (P ⊃ (D · B))   VALID
2.   (~P ⊃ (S · R))
3.   ~R
     [∴ B
4.   ⎡asm: ~B
5.   ⎢   ⎡asm: ~P
6.   ⎢   ⎢∴ (S · R) (from 2 and 5)
7.   ⎢   ⎢∴ S [from 6]
8.   ⎢   ⎣∴ R [from 6, contradicting 3]
9.   ⎢∴ P [from 5, 3, and 8]
10.   ⎢∴ (D · B) [from 1 and 9]
11.   ⎢∴ D [from 10]
12.   ⎣∴ B [from 10, contradicting 4]
13. ∴ B [from 4 and 12]

This finishes our proof that the argument is valid.

Study this example carefully. When you understand it well, you're ready to learn the general strategy for doing multiple-assumption proofs.

3.5 MULTIPLE-ASSUMPTION PROOFS

The most difficult part of doing multiple-assumption proofs is knowing *when* to make another assumption and *what* to assume.

This rule explains *when* to make another assumption:

> Make a further assumption when (and only
> when) you can't further apply the S- or I-rules or
> RAA and yet there are still *unstarred*
> not-blocked-off complex wffs.

Use the regular S- and I-rules and RAA as far as you can. *Make further assumptions only as a last resort.* You'll rarely need more than two assumptions to prove an argument valid.

This rule explains *what* to assume:

> Go to one of these *unstarred* complex wffs. Ask
> yourself, "What would be nice to have—the left
> part or its negation—to infer something about
> the truth or falsity of the right part?" Assume
> the opposite of what would be nice to have.

You figure out what to assume by looking at what would be nice to have in order to infer something from an unstarred complex wff. This unstarred complex wff will have one of these three forms: "~(A · B)," "(A $\lor$ B)," or "(A $\supset$ B)." So there are three cases:

Suppose the unstarred complex wff has the form "~(A · B)." It would be nice to have "A" (to use to derive "~B"). So assume "~A":

~(A · B)
asm: ~A   [nice to have "A"]

Suppose the unstarred complex wff has the form "(A $\lor$ B)." It would be nice to have "~A" (to use to derive "B"). So assume "A":

(A $\lor$ B)
  asm: A   [nice to have "~A"]

Suppose the unstarred complex wff has the form "(A $\supset$ B)." It would be nice to have "A" (to use to derive "B"). So assume "~A":

(A $\supset$ B)
asm: ~A   [nice to have "A"]

Always assume the *opposite* of what you want to prove.

Don't assume the denial of an entire line. Suppose the unstarred complex wff is "(A ⊃ B)." Don't assume "~(A ⊃ B)":

> (A ⊃ B)
> asm: ~(A ⊃ B)   [useless!]

You have an instant contradiction between the two wffs! So you can block off the assumption and apply RAA:

> (A ⊃ B)
> [asm: ~(A ⊃ B)   [useless!]
> ∴ (A ⊃ B)   [useless!]

The net effect is to derive "(A ⊃ B)"—which you already had! Always assume the *opposite* of what you'd like to have to continue the proof.

At this point we have to expand our starring system. So far, we've interpreted a star in front of a line as meaning "We can henceforth ignore this line." And we've used these rules:

> Star any line that you simplify using an S-rule.
>
> When you use an I-rule to derive a line from previous lines, star the *longer* previous line used.

We'll still star lines in these two cases. But sometimes we'll use multiple stars. Multiple stars mean *"We can ignore this line for the present."* Multiple stars are sometimes erased later, showing that we can no longer ignore the line. Here are the rules for how many stars to use and when to erase stars:

> Use a number of stars to equal the number of currently not-blocked-off assumptions.
>
> When an assumption leads to a contradiction and is blocked off, erase star strings having more stars than the number of currently not-blocked-off assumptions.

Suppose we have *two* currently not-blocked-off assumptions. Then a line we star will get *two* stars ("**"):

		. . . .	
Two stars →	** 20.	(A ⊃ ~D)	
(2 live	→ 21.	asm: C	
assumptions)	→ 22.	asm: D	
	23.	∴ ~A [from 20 and 22]	

We can ignore line 20 as long as it's starred. The information from line 20 now exists in line 23. (The truth of 23 would guarantee the truth of 20.) So line 20 is redundant and thus not needed. Suppose that later we find a contradiction and thus block off line 23:

```
                    . . . . . . . . .
Erase stars →   20.     (A ⊃ ~D)
   (1 live      → 21.    asm: C
assumption)     22.     ⌈asm: D
                23.     | ∴ ~A [from 20 and 22]
                24.     | ∴ B [from ??]
                25.     ⌊∴ ~B [from ??]
                26.      ∴ ~D [from 22, 24, and 25]
```

Then there's only *one* currently not-blocked-off assumption. We erase star strings having more than *one* star. So we erase "**" on line 20. This means that we can no longer ignore this line. Since line 23 is now blocked off, the information in line 20 is no longer redundant.

When we make an additional assumption, we can always for the time being ignore the complex line we used to figure out what to assume. So we add another rule about starring:

> When you make an additional assumption, star the complex line you used to figure out what to assume.

Use a number of stars to equal the number of currently not-blocked-off assumptions (including the assumption you just added).

Let's repeat the proof of the last section, but using the new rules. Recall that in doing the proof we soon got stuck:

```
 1.     (P ⊃ (D · B))
 2.     (~P ⊃ (S · R))
 3.     ~R
      [∴ B
 4.     asm: ~B
```

We can't further apply the S- or I-rules or RAA. And yet there are unstarred complex not-blocked-off wffs (those in lines 1 and 2). So we need to make another assumption. We take one of these unstarred complex wffs—for example, the one in line 1. We ask ourselves this question:

> What would be nice to have, "P" or "~P," in order to infer something about "(D · B)"?

It would be nice to have "P." (We couldn't validly infer anything about "(D · B)" if we had "~P.") We assume the opposite of what would be nice to have. So we assume "~P" (line 5). Then we derive our contradiction:

** **1.** (P ⊃ (D · B))
** **2.** (~P ⊃ (S · R))
  3. ~R
   [∴ B
  4. asm: ~B
  5.   asm: ~P [nice to have "P" for 1]
** **6.**   ∴ (S · R) [from 2 and 5]
  7.   ∴ S [from 6]
  8.   ∴ R [from 6, contradicting 3]

Lines starred after we make our second assumption get *two* stars, since now there are *two* not-blocked-off assumptions. Then we apply RAA to derive "P":

Erase **1.** (P ⊃ (D · B))
double **2.** (~P ⊃ (S · R))
stars   **3.** ~R
   [∴ B
  4. asm: ~B
  5. ⎡asm: ~P [nice to have "P" for 1]
  6. |∴ (S · R) [from 2 and 5]
  7. |∴ S [from 6]
  8. ⎣∴ R [from 6, contradicting 3]
  9. ∴ P [from 5, 3, and 8]

Now we have just *one* not-blocked-off assumption. So we erase strings of more than *one* star. We can't ignore lines 1 and 2 any longer. In fact, we can use line 1 right away. (Recall that it was *nice to have* "P" for line 1.) Line 1 gets *one* star, since there's now only *one* not-blocked-off assumption. We soon get a second contradiction and complete the proof:

* **1.** (P ⊃ (D · B))   VALID
  2. (~P ⊃ (S · R))
  3. ~R
   [∴ B
  4. ⎡asm: ~B
  5. |⎡asm: ~P [nice to have "P" for 1]
  6. ||∴ (S · R) [from 2 and 5]
  7. ||∴ S [from 6]
  8. |⎣∴ R [from 6, contradicting 3]
  9. |∴ P [from 5, 3, and 8]
* **10.** |∴ (D · B) [from 1 and 9]
 11. |∴ D [from 10]
 12. ⎣∴ B [from 10, contradicting 4]
13. ∴ B [from 4 and 12]

Flow Chart 1. To prove a valid PC argument, first block off the conclusion. Add "asm:" followed by the conclusion's simpler contradictory to form the next line. Then go to START.

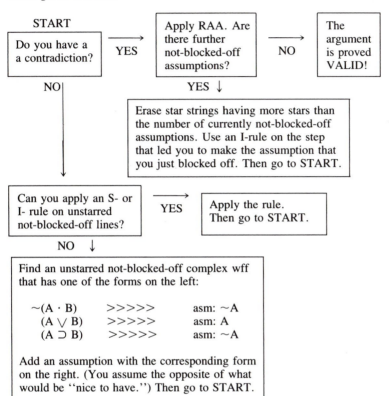

The strategy in Flow Chart 1 can prove *any* valid PC argument. We don't have to follow this strategy slavishly. Sometimes it's easier to follow other strategies, as in our next example. But the directions are there if we need them.

Consider this argument:

1. If the butler put the white tablet into the drink and the white tablet was poison, then the butler killed the deceased and the butler is guilty.
2. The butler put the white tablet into the drink.
3. The white tablet was poison.
∴ The butler is guilty.

Our proof starts this way:

1.   ((P · W) ⊃ (K · G))
2.   P
3.   W
   [∴ G
4.   asm: ~G

We are stuck. We can't apply the S- or I-rules or RAA. We might try to construct a refutation from the simple wffs we have. But this won't work; the first premise comes out false. What can we do? On our normal strategy, we'd make another assumption:

> It would be nice to have "(P · W)" [to derive "(K · G)" from the first premise]. So let's assume "~(P · W)."

We make this assumption and quickly get a contradiction:

** 1.   ((P · W) ⊃ (K · G))
  2.   P
  3.   W
   [∴ G
  4.   asm: ~G
** 5.   asm: ~(P · W) [nice to have "(P · W)" for 1]
  6.   ∴ ~W [from 2 and 5, contradicting 3]

Then we apply RAA and erase each "******":

1.   ((P · W) ⊃ (K · G))
2.   P
3.   W
   [∴ G
4.   asm: ~G
5.   ⌈asm: ~(P · W) [nice to have "(P · W)" for 1]
6.   ⌊∴ ~W [from 2 and 5, contradicting 3]
7.   ∴ (P · W) [from 5, 6, and 3]

We continue until we find our second contradiction and thus complete the proof:

* 1.   ((P · W) ⊃ (K · G))   VALID
  2.   P
  3.   W
   [∴ G

  4.   asm: ~G
  5.   ⌈asm: ~(P · W) [nice to have "(P · W)" for 1]
  6.   ⌊∴ ~W [from 2 and 5, contradicting 3]
* **7.**   ∴ (P · W) [from 5, 6, and 3]
* **8.**   ∴ (K · G) [from 1 and 7]
  9.   ∴ P [from 7]
10.   ∴ W [from 7]
11.   ∴ K [from 8]
12.   ∴ G [from 8, contradicting 4]
13. ∴ G [from 4 and 12]

We could shorten this if we followed a different strategy. We could omit steps 9 through 11, since we don't need these to get the contradiction. We could also avoid the second assumption. Let's return to just before we made this assumption:

  1.   ((P · W) ⊃ (K · G))
  2.   P
  3.   W
     [∴ G
  4.   asm: ~G

At this point we could say to ourselves:

It would be nice to have "(P · W)" [to derive "(K · G)" from the first premise]. We already have "P" and "W." So we can use our S-rules backwards to derive "(P · W)":

    P
    W
    ————
  ∴ (P · W)

Then our proof would go this way:

* **1.**   ((P · W) ⊃ (K · G))   VALID
  2.   P
  3.   W
    [∴ G
  4.   ⌈asm: ~G
  5.   ∴ (P · W) [from 2 and 3, nice to have for 1]
  6.   ∴ (K · G) [from 1 and 5]
  7.   ⌊∴ G [from 6, contradicting 4]
  8. ∴ G [from 4 and 7]

We also save a step by deriving just "G" from line 6. We don't need "K" to get a contradiction.

So we can sometimes save steps by using S-rules backwards or by deriving only what we need to get a contradiction. Sometimes we can also save steps by being crafty about which line to use to get our assumption. It's more useful to make an assumption that gives us more information. These short-cuts are optional. If you like, you can just follow the previously outlined proof strategy.

3.5A Symbolic Argument Exercises

Each of these is valid. Give a formal proof for each one. (The last two problems are the hardest.)

0. $(\sim(A \lor B) \supset (C \supset D))$
   $(\sim A \cdot \sim D)$
∴ $(\sim B \supset \sim C)$

ANSWER: * 1. $(\sim(A \lor B) \supset (C \supset D))$ VALID
        * 2. $(\sim A \cdot \sim D)$
              [∴ $(\sim B \supset \sim C)$
        * 3.   ⌐ asm: $\sim(\sim B \supset \sim C)$
         4.   │ ∴ $\sim B$ [from 3]
         5.   │ ∴ C [from 3]
         6.   │ ∴ $\sim A$ [from 2]
         7.   │ ∴ $\sim D$ [from 2]
         8.   │ ⌐ asm: $(A \lor B)$ [nice to have "$\sim(A \lor B)$" for 1]
         9.   │ └ ∴ B [from 6 and 8, contradicting 4]
        10.   │ ∴ $\sim(A \lor B)$ [from 8, 4, and 9]
      * 11.   │ ∴ $(C \supset D)$ [from 1 and 10]
        12.   └ ∴ D [from 5 and 11, contradicting 7]
        13.   ∴ $(\sim B \supset \sim C)$ [from 3, 7, and 12]

Infer whatever you can (for example, lines 4 through 7) *before* making another assumption. Make additional assumptions only as a last resort. We give line 1 two stars when we do line 8—and erase these two stars when we do line 10. Line 1 gets one star back again when we use it to derive line 11. How could we prove this argument without making a second assumption?

1. $(B \lor A)$
   $(B \supset A)$
∴ $\sim(A \supset \sim A)$

2. $(A \supset B)$
   $(A \lor (A \cdot C))$
∴ $(A \cdot B)$

3. $(((A \cdot B) \supset C) \supset (D \supset E))$
   D
∴ $(C \supset E)$

4. $((A \supset B) \supset C)$
   $(C \supset (D \cdot E))$
∴ $(B \supset D)$

5. $((A \cdot B) \lor (A \cdot C))$
∴ $(A \cdot (B \lor C))$

6. $(\sim A \equiv B)$
∴ $\sim(A \equiv B)$

7. ~(A · (C ∨ D))
    (~A ⊃ (B · ~C))
∴ ~C
9. ((A ∨ ~B) ⊃ (C ·D))
    ((C ∨ E) ⊃ (D ⊃ B))
∴ B

8. (A ∨ (D · E))
    (A ⊃ (B ⊃ C))
∴ (D ∨ C)
10. (A ⊃ (B · C))
∴ ((A ⊃ B) · (A ⊃ C))

3.5B English Argument Exercises

Each of these arguments is valid. For each argument, first read it over and see whether it is valid intuitively. Then translate it into PC and prove it valid.

0. If Nixon knew about the massive Watergate coverup, then he lied to the American people on national television and he should resign.
If Nixon didn't know about the massive Watergate coverup, then he was incompetent in not knowing what his administration was doing and he should resign.

∴ Nixon should resign.
    [Use K, L, R, and I.]

ANSWER: * 1.   (K ⊃ (L · R))   VALID
          2.   (~K ⊃ (I · R))
               [∴ R
          3.   ⌐ asm: ~R
          4.   ⌐ asm: ~K [nice to have "K" for 1]
          5.   ∴ (I · R) [from 2 and 4]
          6.   ∴ I [from 5]
          7.   L∴ R [from 5, contradicting 3]
          8.   ∴ K [from 4, 3, and 7]
      * 9.   ∴ (L · R) [from 1 and 8]
       10.   ∴ L [from 9]
       11.   L∴ R [from 9, contradicting 3]
       12. ∴ R [from 3 and 11]

We give lines 1, 2, and 5 two stars when we use them in lines 4 through 7. We erase these two stars when the second assumption dies, in line 8.

1. If Cindy's new computer has a hard disk and we give the computer the SHIP command, then it'll read "Prepare for shipping" and *not* "No such device."
We give the computer the SHIP command.
It reads "No such device."
∴ Cindy's new computer doesn't have a hard disk.
    [Use H, S, P, and N. Use S-rules backwards for this one.]

2. If Michigan either won or tied, then Michigan is going to the Rose Bowl and Gensler is happy.

∴ If Gensler isn't happy, then Michigan didn't tie.

[Use W, T, R, and H.]

3. You'll get an A in the course if and only if you either get a hundred on the final exam or else bribe the teacher.

You won't get a hundred on the final exam.

∴ You'll get an A in the course if and only if you bribe the teacher.

[Use A, H, and B.]

4. If you get an A or a B on the final, then you'll get an A in the course and make the dean's list.

∴ If you get a B on the final, then you'll make the dean's list.

[Use A, B, C, and D.]

5. There are moral obligations.

God's existence wouldn't explain moral obligation.

If there are moral obligations and moral obligations are explainable, then either there's an explanation besides God's existence or else God's existence explains moral obligations.

∴ Either moral obligations aren't explainable, or else there's an explanation besides God's existence.

[Use M, G, E, and B.]

6. You'll be elected if and only if you get a lot of money in campaign contributions.

You'll get a lot of money in campaign contributions if and only if you compromise your principles.

∴ You'll be elected if and only if you compromise your principles.

[Use E, M, and C.]

7. If determinism is true and Dr. Freudlov correctly predicts (using deterministic laws) exactly what I'll do, then if she tells me her prediction then I'll do something else.

If Dr. Freudlov tells me what she thinks I'll do and yet I'll do something else, then Dr. Freudlov doesn't correctly predict (using deterministic laws) exactly what I'll do.

∴ If determinism is true, then either Dr. Freudlov doesn't correctly predict (using deterministic laws) exactly what I'll do or else she won't tell me her prediction.

[Use D, P, T, and E.]

8. If I'm coming down with a cold and I exercise, then I'll get worse and feel awful.

If I don't exercise, then my body will suffer from exercise deprivation and I'll feel awful.

∴ If I'm coming down with a cold, then I'll feel awful.

[Use C, E, W, A, and D.]

9. Moral judgments express either truth claims or feelings.

If moral judgments express truth claims, then "ought" expresses either a concept from sense experience or else an objective concept that isn't from sense experience.

"Ought" doesn't express a concept from sense experience.

"Ought" doesn't express an objective concept that isn't from sense experience.

∴ Moral judgments don't express truth claims but do express feelings.

[Use T, F, S, and O.]

10. [Here the parents tell their son that the precondition for financing his graduate school is that he leave Suzy.]

If you make this demand on your son and he leaves Suzy, then he'll regret being forced to leave her and he'll always resent you.

If you make this demand on your son and he doesn't leave Suzy, then he'll regret not going to graduate school and he'll always resent you.

∴ If you make this demand on your son, then he'll always resent you.

[Use D, L, F, A, and G. A friend of mine thus talked the parents out of their demand. This one is difficult.]

3.6 MULTIPLE-ASSUMPTION REFUTATIONS

Let's take another invalid butler argument:

1. If the butler was at the party, then the butler fixed the drinks and the butler poisoned the deceased.
2. If the butler wasn't at the party, then the butler was at a friend's house.
∴ The butler poisoned the deceased.

```
**  1.    (P⁰ ⊃ (D · B⁰)) = 1              INVALID
```
$$** \quad 1. \quad (P^0 \supset (D \cdot B^0)) = 1 \qquad \textbf{INVALID}$$
$$** \quad 2. \quad (\sim P^0 \supset F^1) = 1$$
$$\quad\quad [\therefore \; B^0 = 0 \qquad \boxed{F, \sim P, \sim B}$$
$$\quad\quad 3. \quad \text{asm: } \sim B$$
$$\quad\quad 4. \quad\quad \text{asm: } \sim P \text{ [nice to have ``P'' for 1]}$$
$$\quad\quad 5. \quad\quad \therefore F \text{ [from 2 and 4]}$$

> The butler was at a friend's house (F).
> The butler wasn't at the party (~P).
> The butler didn't poison the deceased (~B).

This goes much like our example in Section 3.4. But now we reach no contradiction. Instead we produce a refutation—a possible situation making the premises true and the conclusion false. This refutation shows the argument to be invalid.

Suppose we are trying to prove a PC argument. We should continue with the attempted proof until one of these happens:

- Every assumption leads to a contradiction. Then we have a *proof* that the argument is *valid*.
- Every complex not-blocked-off line is starred and yet there's no contradiction. Then we can use the simple not-blocked-off wffs to give us a *refutation* showing that the argument is *invalid*.

The second case is modified from before. With an invalid argument we now keep working until all complex not-blocked-off lines are starred. We sometimes end by making a series of assumptions that seem to lead nowhere (but actually give us our refutation).

The strategy in Flow Chart 2 can prove *or refute* any valid *or invalid* PC arguments. The strategy always gives a definite result either way. We don't have to follow the

Flow Chart 2. To prove or disprove a PC argument, first block off the conclusion. Add "asm:" followed by the conclusion's simpler contradictory to form the next line. Then go to START.

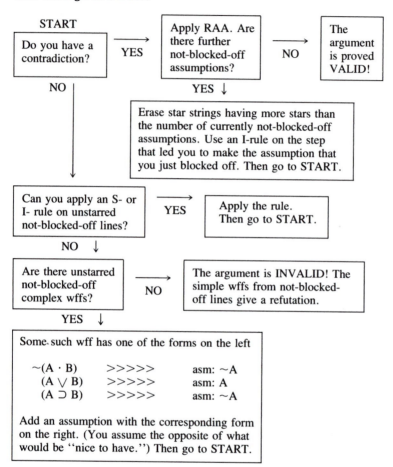

directions slavishly. Sometimes it's easier to use our own strategy. But the directions are there if we need them.

Here's another invalid argument—and the result of applying our strategy in a mechanical way to produce a refutation:

1. If the maid prepared the drink, then the butler didn't prepare it.

2. The maid didn't prepare the drink.

3. If the butler prepared the drink, then the butler poisoned the drink and the butler is guilty.

∴ The butler is guilty.

** **1.** $(M^0 \supset \sim B^0) = 1$ **INVALID**
  2. $\sim M^0 = 1$
*** **3.** $(B^0 \supset (P \cdot G^0)) = 1$ $\boxed{\sim M, \sim B, \sim G}$
        $[\therefore \ G^0 = 0$
  4.   asm: $\sim G$
  5.   asm: $\sim M$ [nice to have "M" for 1]
  6.   asm: $\sim B$ [nice to have "B" for 3]

> The maid didn't prepare the drink ($\sim M$).
> The butler didn't prepare the drink ($\sim B$).
> The butler isn't guilty ($\sim G$).

We give line 1 *two* stars when we make our *second* assumption, and line 3 *three* stars when we make our *third* assumption. We find no contradiction. Instead we produce a refutation—showing the argument to be invalid. (But who *did* prepare the drink?)

Line 5 is redundant (line 2 already has "$\sim M$"). We can always skip redundant assumptions or derived steps. Even so, we can star the lines that we'd star if we included these steps.

Our strategy if applied correctly always leads to a proof or refutation. What proof or refutation it leads to depends on how we work out the problem. Proofs and refutations might differ but still be correct.

3.6A Symbolic Argument Exercises

Say whether each is valid (then give a proof) or invalid (then give a refutation). The first seven are all invalid.

0. $(A \lor \sim(B \supset C))$
    $(D \supset (A \supset B))$
$\therefore (C \supset \sim(D \lor A))$

ANSWER: ** **1.** $(A^1 \lor \sim(B \supset C^1)) = 1$ **INVALID**
    *** **2.** $(D^0 \supset (A^1 \supset B)) = 1$
            $[\therefore (C^1 \supset \sim (D^0 \lor A^1)) = 0$ $\boxed{A, C, \sim D}$
    * **3.**   asm: $\sim(C \supset \sim(D \lor A))$
      4. $\therefore C$ [from 3]
    *** **5.** $\therefore (D \lor A)$ [from 3]
      6.   asm: A [nice to have "$\sim A$" for 1]
      7.   asm: $\sim D$ [nice to have "D" for 2]
      8. $\therefore A$ [from 5 and 7 (redundant)]

1. $\sim(A \cdot B)$ **2.** $(\sim A \lor \sim B)$
$\therefore (\sim A \cdot \sim B)$ $\therefore \sim(A \lor B)$
3. $(A \supset (B \cdot C))$ **4.** $(A \supset B)$
    $((D \supset E) \supset A)$ $(C \supset (\sim D \cdot S))$
$\therefore (E \lor C)$ $\therefore (D \lor P)$

5. $(\sim A \supset B)$
∴ $\sim(A \supset B)$
7. $((A \cdot B) \supset \sim(C \cdot D))$
$\sim C$
$(E \supset B)$
∴ $\sim E$
9. $(A \supset (B \supset C))$
$(B \lor \sim(C \supset D))$
∴ $(D \supset \sim(A \lor B))$
11. $(A \supset B)$
∴ $((A \cdot B) \equiv A)$
13. $((A \cdot \sim B) \supset C)$
$((B \cdot \sim A) \supset D)$
$((C \lor D) \supset E)$
∴ E
15. $\sim(\sim A \cdot \sim B)$
$(C \lor D)$
$\sim(D \cdot A)$
∴ $\sim(C \equiv B)$

6. $\sim(A \cdot B)$
∴ $\sim(A \equiv B)$
8. $(A \supset B)$
$(\sim C \supset \sim B)$
$\sim(C \cdot \sim A)$
∴ $(A \equiv C)$
10. $(\sim(A \cdot B) \supset \sim(C \cdot D))$
C
∴ $(\sim A \supset \sim D)$
12. $((A \lor B) \supset \sim(C \supset D))$
∴ $(D \supset \sim B)$
14. $(A \supset B)$
$\sim(B \cdot C)$
$(A \lor C)$
∴ $(B \equiv A)$
16. $(A \supset (B \supset C))$
$((C \cdot D) \supset E)$
$((F \cdot D) \supset \sim E)$
∴ $(A \supset \sim(B \cdot F))$

3.6B English Argument Exercises

For each argument, first evaluate intuitively and then translate into PC and work it out. Say whether each is valid (then give a proof) or invalid (then give a refutation).

0. If I hike the Appalachian Trail and go during late spring, then I'll get maximum daylight and maximum mosquitoes.
   If I'll get maximum mosquitoes, then I won't be comfortable.
   If I go right after school, then I'll go during late spring.
∴ If I hike the Appalachian Trail and don't go right after school, then I'll be comfortable.
   [Use A, L, D, M, C, and S.]

ANSWER:	**	1.	$((A^1 \cdot L^0) \supset (D \cdot M^0)) = 1$	**INVALID**
	***	2.	$(M^0 \supset \sim C^0) = 1$	
	**	3.	$(S^0 \supset L^0) = 1$	A, ~L, ~S
			[∴ $((A^1 \cdot \sim S^0) \supset C^0) = 0$]	~M, ~C
	*	4.	asm: $\sim((A \cdot \sim S) \supset C)$	
	*	5.	∴ $(A \cdot \sim S)$ [from 4]	
		6.	∴ $\sim C$ [from 4]	
		7.	∴ A [from 5]	
		8.	∴ $\sim S$ [from 5]	
	**	9.	asm: $\sim(A \cdot L)$ [nice to have "$(A \cdot L)$" for 1]	
		10.	∴ $\sim L$ [from 7 and 9]	
		11.	∴ $\sim S$ [from 3 and 10 (redundant)]	
		12.	asm: $\sim M$ [nice to have "M" for 2]	

[In line 12 we say "It would be nice to have maximum mosquitoes"!] The refutation is a possible situation in which I hike the Appalachian Trail, don't go during late spring, don't go right after school, don't get maximum mosquitoes, and still am not comfortable. Maybe I get rain or blisters!

1. If the world contains moral goodness, then the world contains free beings and these free beings sometimes do wrong.
   If these free beings sometimes do wrong, then the world is imperfect and the creator is imperfect.
   ∴ If the world doesn't contain moral goodness, then the creator is imperfect.
   [Use M, F, S, W, and C.]
2. If you take chemistry or physics, then you'll complete your science requirement and be able to graduate in January.
   ∴ If you take chemistry, then you'll be able to graduate in January.
   [Use C, P, S, and G.]
3. If you take chemistry or physics, then you'll complete your science requirement and be able to graduate in January.
   ∴ If you don't take chemistry, then you won't be able to graduate in January.
   [Use C, P, S, and G.]
4. If Socrates didn't approve of the laws of Athens, then he'd either have left Athens or else have tried to change the laws.
   If Socrates didn't leave Athens and didn't try to change the laws, then he agreed to obey the laws.
   Socrates didn't leave Athens.
   ∴ If Socrates didn't try to change the laws, then he approved of the laws and agreed to obey them.
   [Use A, L, C, and O. From Plato's *Crito*.]
5. If it rains or snows, it's miserable weather for football.
   If it's miserable weather for football and I stay the whole game, then I'm a Michigan-football fanatic.
   ∴ If I'm a Michigan-football fanatic, then I'll stay the whole game.
   [Use R, S, M, W, and F.]
6. If you say you favor Michigan, then Gensler will think you're insincere and you'll fail his logic course.
   If you don't say you favor Michigan, then Gensler will think you favor Ohio State and you'll fail his logic course.
   ∴ You'll fail Gensler's logic course.
   [Use S, I, F, and O.]
7. If feelings can be based on false information and false information can be criticized, then feelings can be criticized.
   If feelings can be criticized and ethical judgments are based on feelings, then ethical judgments can be criticized and ethics is partly rational.
   ∴ If feelings can be based on false information, then ethics is partly rational.
   [Use B, C, F, E, J, and R.]
8. [Let LP = *logical positivism* = the view that every claim that makes sense (in the sense of being true or false) is either true by definition or experimentally testable.]

If LP is true and makes sense, then it itself either is true by definition or is experimentally testable.

LP isn't true by definition.

LP isn't experimentally testable.

If LP doesn't make sense, then it isn't true.

∴ LP isn't true.

[Use T, M, D, and E. Largely because of such arguments, few philosophers these days hold logical positivism.]

9. If I give a test, then students do either well or poorly.

If students do well, then I think I made the test too easy and I'm frustrated.

If students do poorly, then I think they didn't learn any logic and I'm frustrated.

∴ If I give a test, then I'm frustrated.

[Use T, W, P, E, F, and L. This is from a class who tried to talk me out of giving a test. I responded by saying that, while their argument might be sound, if *I don't* give a test then they don't study, and if they don't study then I'm frustrated. So if I *don't* give a test then I'm also frustrated. What follows from "If I give a test then I'm frustrated and if I don't give a test then I'm frustrated"?]

10. If the final loading capacitor in the radio transmitter is arcing, then the standing wave ratio (SWR) is too high and the efficiency is lowered.

If you hear a cracking sound, then the final loading capacitor in the radio transmitter is arcing.

∴ Either the SWR isn't too high, or you hear a cracking sound.

[Use A, H, L, and C.]

11. We'll find a cause for your action, if and only if your action has a cause and we look hard enough.

If all events have causes, then your action has a cause.

All events have causes.

∴ We'll find a cause for your action, if and only if we look hard enough.

[Use F, H, L, and A.]

12. If smoke in the office is a problem, then either there's bad ventilation or else some people in the office smoke excessively.

If there's bad ventilation, then there would be problems even without cigarettes and banning smoking wouldn't solve the problem of bad air.

∴ If smoking in the office is a problem, then banning smoking wouldn't solve the problem of bad air.

[Use P, V, E, W, and S.]

13. Herman sees that the piece of chalk is white.

The piece of chalk is the smallest thing on the desk.

Herman doesn't see that the smallest thing on the desk is white. (He can't see the whole desk and hence can't tell that the piece of chalk is the smallest thing on the desk.)

If Herman sees a material thing, then if he sees that the piece of chalk is white and the piece of chalk is the smallest thing on the desk then he sees that the smallest thing on the desk is white.

If Herman doesn't see a material thing, then he sees a sense datum.

∴ Herman doesn't see a material thing, but he does see a sense datum.

[Use H, P, X, M, and S. This attacks the "direct realism" view that we (directly) perceive material things and not just sense data. Be careful with the translation.]

14. If we can directly know that God exists, then we can know God by experience.

If we can indirectly know that God exists, then we can know God by logical inference from experience.

If we can know that God exists, then we can directly know that God exists or we can indirectly know that God exists.

If we can't know God empirically, then we can't know God by experience and we can't know God by logical inference from experience.

If we can know God empirically, then "God exists" is a scientific hypothesis and is empirically falsifiable.

"God exists" isn't empirically falsifiable.

∴ We can't know that God exists.

[Use D, E, I, L, K, M, S, and F. From Bertrand Russell. (Would a similar argument prove that I can't know that there are other thinking beings besides myself?) This one is hard to prove unless you make the additional assumption that would give you the most information.]

15. If I perceive, then my perception is either delusive (I perceive no material object) or veridical (I perceive a material object).

If my perception is delusive, then I directly perceive something and yet I don't directly perceive a material object.

If I directly perceive something and yet I don't directly perceive a material object, then I directly perceive a sensation (sense datum).

If my perception is veridical, then I directly perceive something.

If my perception is veridical and I directly perceive a material object, then my experience in veridical perception would always differ qualitatively from my experience in delusive perception.

My experience in veridical perception doesn't always differ qualitatively from my experience in delusive perception.

∴ If I perceive, then I directly perceive a sensation and I don't directly perceive a material object.

[Use P, D, V, T, M, S, and Q. This complicated argument is a version of the "argument from illusion."]

3.7 DIRECT AND CONDITIONAL PROOFS

So far we've used only indirect proofs. An *indirect proof* works by deriving a contradiction from the assumption that the conclusion is false. Our rules also allow *direct proofs*, in which the conclusion is derived from the premises without making any assumptions. Compare the following proofs:

	Indirect Proof		Direct Proof
1.	A	**1.**	A
2.	(A ⊃ B)	**2.**	(A ⊃ B)
3.	(B ⊃ C)	**3.**	(B ⊃ C)
	[∴ C		[∴ C

	Indirect Proof		Direct Proof

4.	⌈asm: ~C
5.	│ ∴ B [from 1 and 2]
6.	│ ∴ C [from 3 and 5, contra-
	└ dicting 4]
7.	∴ C [from 4 and 6]

| 4. | ∴ B [from 1 and 2] |
| 5. | ∴ C [from 3 and 4] |

The direct proof here is simpler and two steps shorter. It involves blocking off the conclusion and then (in two easy steps) deriving the conclusion from the premises.

Conditional proofs are also common. These use rule CP (which holds regardless of what wffs substitute for "A" and "B"):

> *CP* If you assume a wff A and then derive a wff B, then
> you can block off all the lines from this assumption on
> down and infer a derived step "∴ (A ⊃ B)."

Compare these proofs:

	Indirect Proof		Conditional Proof
1.	(A ⊃ B)	1.	(A ⊃ B)
2.	(B ⊃ C)	2.	(B ⊃ C)
	[∴ (A ⊃ C)		[∴ (A ⊃ C)
3.	⌐asm: ~ (A ⊃ C)	3.	⌐asm: A
4.	│ ∴ A [from 3]	4.	│ ∴ B [from 1 and 3]
5.	│ ∴ ~C [from 3]	5.	└∴ C [from 2 and 4]
6.	│ ∴ B [from 1 and 4]	6.	∴ (A ⊃ C) [from 3 and 5 by CP]
7.	│ ∴ C [from 2 and 6, contra-		
	└ dicting 5]		
8.	∴ (A ⊃ C) [from 3, 5, and 7]		

In the shorter conditional proof, we assume "A" and then derive "C." Rule CP then allows us to block off steps 3 through 5 and derive "∴ (A ⊃ C)."

Why use *indirect* proofs? There are three main reasons:

1. The indirect method is *more powerful*. The first exercise set in this chapter (Section 3.2A) had *ten* valid arguments. Using just the S- and I-rules, we could prove only *three* of these using direct proofs and only *four* using conditional proofs. but we can easily prove all ten using indirect proofs. If we switched to direct and conditional proofs only, we'd need to add many more inference rules to prove these arguments.

2. Our indirect method incorporates a practical *automatic proof strategy* that always works. The direct and conditional methods often involve much guesswork and frustration.

3. The indirect method is also a *disproof method*. It automatically refutes invalid arguments. The direct and conditional methods can't do this.

Despite the advantages of indirect proofs, it's useful to know about the other two methods. Direct or conditional proofs are sometimes slightly easier and save two steps.

4

Basic Quantificational Logic

4.1 QUANTIFICATIONAL FORMULAS

Our quantificational calculus (QC) is a language for testing those arguments whose validity depends on one or more of these ideas: *all, some, equals, not, and, or, if-then,* and *if and only if.* Besides the PC symbols, QC contains "∃," " = ," and small letters.

Translating from English to QC is difficult. We'll learn it in stages. First we'll learn how to translate most sentences that begin with "all," "some," "not all," or "no." We'll move to harder translations in Section 4.5 and in the next chapter.

Capital and small letters in QC each have various uses. We'll use capital letters to represent statements, general terms, and relations:

- A capital letter alone (not followed by small letters) represents a *statement.* "S" might mean "It's snowing."
- A capital letter followed by a single small letter represents a *general term* (a term that does or could describe more than one person or thing). "Ir" might mean "Romeo is Italian."
- A capital letter followed by two or more small letters represents a *relation.* "Lrj" might mean "Romeo loves Juliet" and "Grbj" might mean "Romeo gave the book to Juliet."

For now, we'll focus on the second use (general terms).

We'll use small letters as constants and variables:

- Small letters from "a" to "v" are *constants.* Constants stand for singular terms (terms that stand for a single person or thing) like "Romeo" or "the current president."

- Small letters from "w" to "z" are *variables*. Variables represent any arbitrary member of a class of things.

You're probably familiar with the constant/variable distinction from math. Here "2" and "4" are constants that represent particular numbers, while "x" and "y" are variables that represent any unspecified number.

QC wffs are typographical strings constructable using the PC rules plus two rules from this chapter and two more from the next. Our first rule tells us how to form wffs using constants and variables:

1. Any capital letter followed by a single constant or variable is a wff.

This rule gives us wffs like these two:

Ir = Romeo is Italian.
Ix = x is Italian.

The first wff uses "r" as a *constant* to represent a particular person named Romeo. The second wff uses "x" as a *variable* to represent some unspecified person; this formula makes no definite claim, since we haven't said whom we are talking about.

We can use the PC rules to generate further wffs like these:

~Ir	= Romeo isn't Italian.
~Ix	= x isn't Italian.
(Ix ⊃ Lx)	= If x is Italian then x is a lover.
(Lx · Rx)	= x is a lover and x is rich.

Quantifiers are sequences of the form "(x)" or "(∃x)"—where any variable may replace "x":

SYMBOL	MEANING	NAME
(x)	for all x	universal quantifier
(∃x)	for some x	existential quantifier

Quantifiers tell us in *how many* cases the following formula is true. Quantifiers can use any variable, but for now we'll just use "x." This rule tells us how to form wffs using quantifiers:

2. Suppose we have a wff containing a given variable but not containing any quantifier using that variable. The result of prefixing this wff with a quantifier using that variable is a wff.

Applying this rule to "Ix" gives us "(x)Ix" and "(∃x)Ix":

(x)Ix
= For all x, x is Italian.
= All are Italians.
= Everyone is Italian.

(∃x)Ix
= For some x, x is Italian.
= There exists an x such that x is Italian.
= Some are Italian.
= One or more are Italian.

Negating these yields further wffs:

~(x)Ix
= It is not the case that, for all x, x is Italian.
= Not all are Italians.

~(∃x)Ix
= It is not the case that, for some x, x is Italian.
= No one is Italian.

Here are some longer wffs:

(x)(Ix ⊃ Lx)
= For all x, if x is Italian then x is a lover.
= All Italians are lovers.

(∃x)(Lx · Rx)
= For some x, x is a lover and x is rich.
= Some lovers are rich.

~(x)(Ix ⊃ Lx)
= It is not the case that, for all x, if x is Italian then x is a lover.
= Not all Italians are lovers.

~(∃x)(Lx · Rx)
= It is not the case that, for some x, x is a lover and x is rich.
= No lovers are rich.

These two *aren't* wffs: "(x)Ir" and "(x)(∃x)Ix." The wff after "(x)" has to contain the variable "x" but can't contain another quantifier using "x."
   Parentheses in QC can be tricky. Here's a rule for how many pairs of parentheses to use in a wff:

Use one pair of parentheses for each "·," "⊃," "∨," and "≡."
Use one pair of parentheses for each quantifier.
Use no other parentheses.

A wff must use a proper number of parentheses. Here are some examples:

WFF	NON-WFF
(Ix · Lx)	(~Ix)
(Ix ⊃ Lx)	~(Ix)
(Ix ∨ Lx)	(x)(Ix)
(Ix ≡ Lx)	((x)Ix)
~Ix	(~(x)Ix)
(x)Ix	~((x)Ix)
(∃x)Ix	~(x)(Ix)
(x)(Ix ⊃ Lx)	(x)Ix ⊃ Lx
(x)((Ix · Rx) ⊃ Lx)	((x)(Ix ⊃ Lx))

Also, don't put "~" *within* a quantifier. Each quantifier is an indivisible unit. This *isn't* a wff:

(~∃x)Ix

Instead, we write "~(∃x)Ix."
    Some wffs cannot express true or false statements. Consider these four QC wffs:

WFF	STATEMENT?	POSSIBLE MEANING
Ix	No	x is Italian.
Ir	Yes	Romeo is Italian.
(x)Ix	Yes	Everyone is Italian.
(∃x)Ix	Yes	Someone is Italian.

"Ix" isn't true or false, since we haven't specified whom we are talking about. The variable "x" in "Ix" is said to be "free" since it doesn't refer to any quantifier. QC wffs that contain free variables can't express true or false statements.

4.2 BEGINNING TRANSLATIONS

We'll now focus on English sentences that translate into wffs starting with a quantifier, or with "~" and then a quantifier. I'll give two translation rules that usually work. Our first rule tells what quantifier to use and where to put it:

IF THE ENGLISH BEGINS WITH	THEN THE WFF BEGINS WITH
all (every)	(x)
some	(∃x)
not all	~(x)
no	~(∃x)

Here are some simple examples:

> *All* are Italian.
> = For all x, x is Italian.
> = (x)Ix

> *Someone* is Italian.
> = For some x, x is Italian.
> = (∃x)Ix

> *Not all* are Italian.
> = It is not the case that, for all x, x is Italian.
> = ~(x)Ix

> *No* one is Italian.
> = It is not the case that, for some x, x is Italian.
> = ~(∃x)Ix

Here are some further examples:

> *All* are non-Italians.
> = (x)~Ix

> *Some* aren't rich.
> = (∃x)~Rx

> *Not everyone* is rich or Italian.
> = ~(x)(Rx ∨ Ix)

> *No* one is rich and non-Italian.
> = ~(∃x)(Rx · ~Ix)

When the English begins with "all," "some," "not all," or "no," the quantifier must go *outside* all parentheses. Don't translate the last two as "~((x)Rx ∨ Ix)" and "~((∃x)Rx · ~Ix)."

Use whatever propositional connective ("·," "∨," "⊃," or "≡") the English sentence specifies. So use "∨" if the English contains "or":

> All are German *or* Italian.
> = (x)(Gx ∨ Ix)

Often the English sentence doesn't specify the connective:

> All Italians are lovers.
> = (x)(Ix ??? Lx)

In such a case, follow these two rules:

With the "all . . . is . . . ." form, use "⊃" as the middle connective:
all . . . is . . . (x)(. . . ⊃ . . .)

Otherwise, if the English doesn't specify the connective, use "·":	
some . . . is . . . (∃x)(. . . · . . .)	no . . . is . . . (∃x)(. . . · . . .)

Here are some simple examples:

    All Italians *are* lovers.
= For all x, *if* x is Italian *then* x is a lover.
= (x)(Ix ⊃ Lx)

    Some Italians are lovers.
= For some x, x is Italian *and* x is a lover.
= (∃x)(Ix · Lx)

    No Italians are lovers.
= It is not the case that, for some x, x is Italian *and* x is a lover.
= ~(∃x)(Ix · Lx)

Here are two further examples:

    All non-Italians *are* lovers.
= For all x, *if* x isn't Italian *then* x is a lover.
= (x)(~Ix ⊃ Lx)

    Some rich Italians are lovers.
= For some x, x is rich *and* Italian *and* a lover.
= (∃x)((Rx · Ix) · Lx)

In the last example, "rich Italian" is "rich *and* Italian" ["(Rx · Ix)"]. ("Rich" here might mean "rich for an Italian.") Note carefully this example:

    All rich Italians *are* lovers.
= For all x, *if* x is rich *and* Italian, *then* x is a lover.
= (x)((Rx · Ix) ⊃ Lx)

"⊃" is the *middle* connective ("*If* rich Italian *then* lover"). Between "rich" and "Italian" we use the rule about *otherwise* using "·."
    Avoid these common mistranslations of "all . . . is . . . .":

all A is B
wrong: (x)(Ax · Bx) [Use "⊃"!]
wrong: ((x)Ax ⊃ Bx) [Put "(x)" outside parentheses!]
correct: (x)(Ax ⊃ Bx)

The first mistranslation means "All things are *both* A *and* B" (for example, "Everyone is *both* Italian *and* a lover"). The second isn't a statement (since the last "x" doesn't refer to any quantifier).

Avoid these common mistranslations of "some . . . is . . .":

some A is B
wrong: (∃x)(Ax ⊃ Bx) [Use "·"!]
wrong: ((∃x)Ax · Bx) [Put "(∃x)" outside parentheses!]
correct: (∃x)(Ax · Bx)

The first mistranslation means "Something is such that *if* it's A *then* it's B." The second isn't a statement.

Suppose you have a sentence that doesn't have "is" or "are" as the main verb. First paraphrase it using "is" or "are." Then translate it into QC. Here's an example:

　　All dogs hate cats.
= All dogs are cat-haters.
= For all x, if x is a dog then x is a cat-hater.
= (x)(Dx ⊃ Cx)

Often we'll use a restricted universe of discourse. Consider the statement "No one is evil"—meaning "No *person* is evil." We might translate it this way:

　　~(∃x)(Px · Ex)
= It is not the case that, for some x, x is a person and x is evil.

Or we might just presume that we are talking about persons and use this simpler translation:

　　~(∃x)Ex
= It is not the case that, for some (person) x, x is evil.

This restricts the universe of discourse to *persons*. Here we take "(∃x)" to mean "for *some person*." The universe of discourse is the set of entities that words like "all," "some," and "no" cover in a given discussion. In translating arguments about the same kind of entity (persons, arguments, duties, logicians, and so on), we'll often simplify our translations by restricting the universe of discourse. Often we'll implicitly assume the universe of discourse of persons.

4.2A Exercises

Using these equivalences, translate each English sentence into QC:

Ex = x is evil
Lx = x is a logician
Cx = x is crazy

Remember that our two rules are rough guides that sometimes don't work. Our rules will give a wrong translation for a couple of these sentences. Read your QC wff carefully to make sure it means the same as the English sentence.

0. Not all logicians are evil.   ANSWER: $\sim$(x)(Lx $\supset$ Ex)
["It is not the case that, for all x, if x is a logician then x is evil."]

1. Someone isn't evil.
2. All are evil.
3. No one is evil.
4. If x is a logician, then x is evil.
5. All logicians are evil.
6. Some logicians are evil.
7. No logician is evil.
8. Some logicians are evil and crazy.
9. No logician who is crazy is evil.
10. Not all are evil.
11. Every logician is crazy or evil.
12. Some who are crazy aren't evil logicians.
13. All crazy logicians are evil.
14. Not all nonlogicians are evil.
15. All who aren't logicians are evil.
16. Some logicians who aren't crazy are evil.
17. No one who is crazy or evil is a logician.
18. Not everyone is crazy or evil.
19. All evil people and crazy people are logicians.
20. All are evil logicians.

4.3 PROOFS

QC proofs work much like PC proofs. But we need four new inference rules for quantifiers and two more (that we'll take in the next chapter) for identity.

Q1 and Q2 are *reverse squiggle rules*. These hold regardless of what variable replaces "x" and what pair of contradictory wffs replaces "Fx"/"~Fx":

Q1   ~(∃x)Fx   ↔   (x)~Fx

Q2   ~(x)Fx   ↔   (∃x)~Fx

By Q1, "It is not the case that someone is funny" = "Everyone is nonfunny." Each means "No one is funny." By Q2, "Not everyone is funny" = "Someone isn't funny." We'll usually use these from left to right, to give us a wff beginning with a quantifier:

~(∃x)Fx
∴ (x)~Fx

~(x)Fx
∴ (∃x)~Fx

So we usually go from "there isn't someone who is" to "everyone isn't," and from "not everyone is" to "someone isn't."

Q3 is the *universal dropping rule*. It holds regardless of what variable, constant, and wffs, respectively, replace "x," "a," "Fx," and "Fa"—provided that the replacements for "Fx" and "Fa" are identical except that wherever "x"'s replacement occurs in the former, "a"'s replacement occurs in the latter:

Q3   (x)Fx   →   Fa

If *everyone* is funny, then Al is funny, Bob is funny, and so forth. From "(x)Fx" we can derive "Fa," "Fb," and so forth:

(x)Fx	
∴ Fa	(use *any* constant for "a")

Q3 is often called *universal instantiation*.

Replace *all* occurrences of the variable by the *same* constant. Compare these two derivations:

(x)(Fx ⊃ Cx)
∴ (Fa ⊃ Ca)   [right!]

(x)(Fx ⊃ Cx)
∴ (Fa ⊃ Cb)   [wrong!]

If all funny people are comedians, it follows that if Al is funny, then Al is a comedian. But it doesn't follow that if Al is funny then *Bob* is a comedian.

The universal quantifier must begin the wff. If a squiggle begins the wff, we can't use Q3:

~(x)Fx
∴ ~Fa   [wrong!]

Here we should use the reverse squiggle rule to go from "~(x)Fx" to "(∃x)~Fx."
Then we can use the existential dropping rule.

Q4 is the *existential dropping rule*. It holds regardless of what variable, constant, and wffs, respectively, replace "x," "a," "Fx," and "Fa"—provided that the replacements for "Fx" and "Fa" are identical except that wherever "x"'s replacement occurs in the former, "a"'s replacement occurs in the latter:

> *Q4*   (∃x)Fx   →   Fa   Here "a"'s replacement must be a *new* constant (one not occurring in earlier lines). There also must be a not-blocked-off assumption.

> | (∃x)Fx | |
> |---|---|
> | ∴ Fa | (use a *new* constant for "a") |

This rule is strange. It seems to lead to inferences like this:

Someone robbed the bank.	(∃x)Rx
∴ Al robbed the bank.	∴ Ra

If *someone* robbed the bank, it doesn't follow that it was Al. Maybe Betty or Charlie did it! But Q4 works because it tells us to invent an *arbitrary name* for the derived step. We say, "Someone robbed the bank—let's call this person (or one such person) 'Al'—so Al robbed the bank." Q4 works fine if we keep the name arbitrary. The constant we substitute for "x" must be *new* (it cannot have occurred in earlier lines of the proof) and some assumption must be in effect. Q4 is often called *existential instantiation*.

With Q4, the existential quantifier must begin the wff and we must replace "x" by the same *new* constant throughout. Consider this pair:

	Fa	Fa	
	(∃x)Gx	(∃x)Gx	
[*wrong!*]	∴ Ga	∴ Gb	[*right!*]

In the "wrong" case, we can't go from "(∃x)Gx" to "Ga." Here "a" occurs in previous lines and hence is *old*. A constant once used becomes "old." We could instead derive "Gb" (as in the "right" case). Here are more examples:

	(∃x)Fx	(∃x)Fx	
	(∃x)Gx	(∃x)Gx	
[*right!*]	∴ Fa	∴ Fa	[*right!*]
[*wrong!*]	∴ Ga	∴ Gb	[*right!*]

Use a new and different constant each time you drop a "(∃x)."

In doing QC proofs, use the new rules to get rid of the quantifiers. Then proceed as in a PC proof. Here's an English version of a QC proof:

1.   All material things are spatial.
2.   All things are material.
      [∴ All things are spatial.
3.   ⌈asm: Not all things are spatial.
4.   │ ∴ Something isn't spatial. [from 3]
5.   │ ∴ a isn't spatial. [from 4—call thing "a"]
6.   │ ∴ If a is material then a is spatial. [from 1]
7.   │ ∴ a is material. [from 2]
8.   ⌊∴ a is spatial. [from 6 and 7, contradicting 5]
9. ∴ All things are spatial. [from 3, 5, and 8]

The steps here should seem natural. This symbolic version is a QC formal proof:

1.   (x)(Mx ⊃ Sx)   VALID
2.   (x)Mx
        [∴ (x)Sx
* 3.   ⌈asm: ~(x)Sx
* 4.   │ ∴ (∃x)~Sx [from 3]
5.   │ ∴ ~Sa [from 4]
* 6.   │ ∴ (Ma ⊃ Sa) [from 1]
7.   │ ∴ Ma [from 2]
8.   ⌊∴ Sa [from 6 and 7, contradicting 5]
9.   ∴ (x)Sx [from 3, 5, and 8]

After making the assumption (line 3), we reverse the squiggle to move the quantifier to the outside (line 4). We drop the existential quantifier using a new constant (line 5). We drop the universal quantifiers using this same constant (lines 6 and 7). Then we apply the PC rules to get a contradiction (lines 5 and 8). RAA gives us the original conclusion (line 9).

I suggest this strategy for doing a QC proof or refutation:

1. Block off the conclusion. Add "asm:" followed by its simpler contradictory to form the next line.

2. *Reverse squiggles.* Apply Q1 or Q2 to each not-blocked-off line starting with "~(∃x)" or "~(x)." This will give you lines with initial quantifiers. Star (and ignore) the lines used.

3. *Drop each initial existential quantifier.* Apply Q4 to each not-blocked-off line starting with "(∃x)." Use a *new* constant for each one. Star (and ignore) the lines used.

4. *Drop each initial universal quantifier.* Apply Q3 to each not-blocked-off line starting with "(x)." Drop each "(x)" *once for each old constant.* [Use a single new constant if you have no old constant.] Don't star the formula, since sometimes you have to drop the "(x)" again when new constants appear later in the proof.

5. You should now be rid of the quantifiers. Finish the proof using the PC strategy. Derive what you can using the S- and I-rules, and make further assumptions if needed. Apply

RAA when you get a contradiction. You should get a proof of validity or a refutation showing invalidity.

Suppose we have an argument containing only the beginning QC wffs that we've used so far. Correctly applying these directions will then yield *either* a proof of validity *or* a refutation showing invalidity. (We won't see invalid arguments until Section 4.4.)
Here's another example of a proof:

All things are material and spatial.
∴ All things are material.

    1.   (x)(Mx · Sx)   VALID
           [∴ (x)Mx
* **2.**   ⌈asm: ~(x)Mx
* **3.** |∴ (∃x)~Mx [from 2]
   4. |∴ ~Ma [from 3]
* **5.** |∴ (Ma · Sa) [from 1]
   6.   ⌊∴ Ma [from 5, contradicting 4]
   7. ∴ (x)Mx [from 2, 4, and 6]

We assume the opposite of the conclusion in 2. We reverse squiggles in 3, drop the existential quantifier using a new constant in 4, and drop the universal quantifier using this constant in 5. We get a contradiction and then complete the proof.

Reversing lines 4 and 5 would be wrong. Suppose we first go from "(x)(Mx · Sx)" to "(Ma · Sa)." Then "a" becomes an *old* constant. So we can't then go from "(∃x)~Mx" to "~Ma." Drop existentials *before* universals.

4.3A Symbolic Argument Exercises

Each of these is valid. Give a formal proof for each.

0. (x)(Fx ⊃ Gx)
   ~(x)Gx
∴ (∃x)~(Fx ∨ Gx)
   ANSWER: 1.   (x)(Fx ⊃ Gx)   VALID
          *   2.   ~(x)Gx
                  [∴ (∃x)~(Fx ∨ Gx)
          *   3.   ⌈asm: ~(∃x)~(Fx ∨ Gx)
              4. |∴ (x)(Fx ∨ Gx) [from 3]
          *   5. |∴ (∃x)~Gx [from 2]
              6. |∴ ~Ga [from 5]
          *   7. |∴ (Fa ⊃ Ga) [from 1]
          *   8. |∴ (Fa ∨ Ga) [from 4]
              9. |∴ ~Fa [from 6 and 7]
             10.   ⌊∴ Fa [from 6 and 8, contradicting 9]
             11.   ∴ (∃x)~(Fx ∨ Gx) [from 3, 9, and 10]

The jump from 3 to 4 might be confusing. We could reverse the squiggle in two ways:

$$\frac{\sim(\exists x)\sim(Fx \lor Gx)}{\therefore (x)\sim \sim(Fx \lor Gx)} \qquad \frac{\sim(\exists x)\sim(Fx \lor Gx)}{\therefore (x)(Fx \lor Gx)}$$

On the left, we simply replace "$\sim(\exists x)$" with "$(x)\sim$" and end up with "$\sim \sim$" in the middle. Later we'd have to apply the double negation rule S-1. On the right, we collapse "$\sim \sim$" right away. This shortens the proof.

(*Don't* try to shorten things further by assuming the opposite of the conclusion *and* reversing the squiggle *all in the same step*. This is a difficult maneuver and usually leads to disaster!)

1. $(\exists x)(Fx \cdot Gx)$
∴ $(\exists x)Fx$
3. $(x)(Fx \supset Gx)$
∴ $\sim(\exists x)(Fx \cdot \sim Gx)$
5. $(x)((Fx \lor Gx) \supset Hx)$
∴ $(x)(Fx \supset Hx)$
7. $\sim(\exists x)(Fx \cdot Gx)$
  $(\exists x)Fx$
∴ $\sim(x)Gx$
9. $(x)(Fx \supset Gx)$
  $(x)(Gx \supset Hx)$
∴ $(x)(Fx \supset Hx)$

2. $(x)\sim(Fx \lor Gx)$
∴ $(x)\sim Fx$
4. $(x)\sim Fx$
∴ $\sim(\exists x)(Fx \cdot Gx)$
6. $\sim(\exists x)(Fx \cdot \sim Gx)$
∴ $(x)(Fx \supset Gx)$
8. $(x)(Fx \supset Gx)$
  $(\exists x)\sim Gx$
∴ $(\exists x)\sim Fx$
10. $(x)(Fx \supset Gx)$
  $(x)(\sim Fx \supset Gx)$
∴ $(x)Gx$

4.3B English Argument Exercises

Each of these arguments is valid. For each, first read it over and see whether it's valid intuitively. Then translate it into QC and prove it valid.

0. No pure water is burnable.
Some Cuyahoga River water is burnable.
∴ Some Cuyahoga River water isn't pure water.
[Use Px, Bx, and Cx.]

ANSWER: * 1. $\sim(\exists x)(Px \cdot Bx)$ VALID
       * 2. $(\exists x)(Cx \cdot Bx)$
          [∴ $(\exists x)(Cx \cdot \sim Px)$]
     * 3.   ⌈ asm: $\sim(\exists x)(Cx \cdot \sim Px)$
       4.   │ ∴ $(x)\sim(Cx \cdot \sim Px)$ [from 3]
       5.   │ ∴ $(x)\sim(Px \cdot Bx)$ [from 1]
     * 6.   │ ∴ $(Ca \cdot Ba)$ [from 2]
     * 7.   │ ∴ $\sim(Ca \cdot \sim Pa)$ [from 4]
     * 8.   │ ∴ $\sim(Pa \cdot Ba)$ [from 5]
       9.   │ ∴ Ca [from 6]
      10.   │ ∴ Ba [from 6]
      11.   │ ∴ Pa [from 7 and 9]
      12.   ⌊ ∴ $\sim Ba$ [from 8 and 11, contradicting 10]
      13. ∴ $(\exists x)(Cx \cdot \sim Px)$ [from 3, 10, and 12]

1. Everyone who deliberates about alternatives believes in free will.
   Everyone deliberates about alternatives.
   ∴ Everyone believes in free will.
   [Use Dx and Bx. From William James.]

2. All (in the electoral college) who do their jobs are useless.
   All (in the electoral college) who don't do their jobs are dangerous.
   ∴ All (in the electoral college) are useless or dangerous.
   [Use Jx for "x does his or her job," Ux for "x is useless," and Dx for "x is dangerous." Use the universe of discourse of electoral college members. Howard Pospesel used this example.]

3. Everyone makes mistakes.
   ∴ Every logic teacher makes mistakes.
   [Use Mx and Lx.]

4. All that's known is experienced through the senses.
   Nothing that is experienced through the senses is known.
   ∴ Nothing is known.
   [Use Kx and Ex. Empiricism (the first premise) plus skepticism about the senses (the second premise) yields general skepticism.]

5. No feeling of pain is publicly observable.
   All chemical processes are publicly observable.
   ∴ No feeling of pain is a chemical process.
   [Use Fx, Ox, and Cx. This attacks a form of materialism which identifies mental events with material events. This argument could also be tested using syllogistic logic. See Chapter 2 of my more basic book, *Logic: Analyzing and Appraising Arguments*.]

6. All students in this course who do their homework know how to do this problem.
   All students in this course do their homework. (??)
   ∴ All students in this course know how to do this problem.
   [Use Sx, Hx, and Kx.]

7. Everything has a cause.
   No first cause has a cause.
   ∴ There's no first cause.
   [Use "Hx" for "x has a cause" and "Fx" for "x is a first cause." A first cause (e.g. God?) is a being that caused other things to exist but yet nothing caused it to exist.]

8. Everyone who isn't with me is against me.
   ∴ Everyone who isn't against me is with me.
   [Use Wx and Ax. The premise and conclusion are claims from the Gospels that are sometimes thought to be incompatible.]

9. Some lies in unusual circumstances aren't wrong.
   ∴ Not all lies are wrong.
   [Use Lx, Ux, and Wx]

10. Nothing based on sense experience is certain.
    Some logical inferences are certain.
    All certain things that aren't based on sense experience are truths of reason.
    ∴ Some truths of reason are certain.
    [Use Bx, Cx, Lx, and Rx.]

4.4 REFUTATIONS

Applying our proof strategy on an *invalid* argument leads to a refutation:

  1.   All material things are spatial.
  2.   *Something* is material.
   [∴ *All* things are spatial.
  3.   asm: Not all things are spatial.
  4.   ∴ Something isn't spatial. [from 3]
  5.   ∴ a isn't spatial. [from 4—call this thing "a"]
  6.   ∴ b is material. [from 2—call thing thing "b"]
  7.   ∴ If a is material then a is spatial. [from 1]
  8.   ∴ If b is material then b is spatial. [from 1]
  9.   ∴ a isn't material. [from 5 and 7]
 10.   ∴ b is spatial. [from 6 and 8]

Imagine a little world with two things, a and b:

> a isn't material.
> a isn't spatial.
> b is material.
> b is spatial.

Here *each* material thing (namely, b) is spatial, and *something* (namely, b) is material. So the premises of our argument are true. But the conclusion that *all* things are spatial is false, since a isn't spatial. So in this world the premises are true but the conclusion false. So our argument is invalid.
      Our symbolic version goes this way:

  1.   (x)(Mx ⊃ Sx)   **INVALID**
 * **2.**   (∃x)Mx   **a, b**
       [∴ (x)Sx
 * **3.**   asm: ~(x)Sx   ┌─────────────┐
 * **4.**   ∴ (∃x)~Sx [from 3]   │ ~Ma, ~Sa   │
  5.   ∴ ~Sa [from 4]   │ Mb, Sb   │
  6.   ∴ Mb [from 2]   └─────────────┘
 * **7.**   ∴ (Ma ⊃ Sa) [from 1]
 * **8.**   ∴ (Mb ⊃ Sb) [from 1]
  9.   ∴ ~Ma [from 5 and 7]
 10.   ∴ Sb [from 6 and 8]

After making the assumption (line 3), we reverse a squiggle to move a quantifier to the outside (line 4). Then we drop the two existential quantifiers, using a new and different constant each time (lines 5 and 6). We drop the universal quantifier twice, first using "a" and then using "b" (lines 7 and 8). We apply the PC rules (lines 9 and 10). We reach no contradiction. So we gather the simple pieces to provide a refutation.

Use these rules to evaluate universal and existential wffs:

A *universal* wff is true if and only if *all* cases are true.
An *existential* wff is true if and only if *at least one* case is true.

In this imagined world, our *universal* premise "(x)(Mx ⊃ Sx)" is true, since *all* cases are true:

$$(Ma \supset Sa) = (0 \supset 0) = 1$$
$$(Mb \supset Sb) = (1 \supset 1) = 1$$

And our *existential* premise "(∃x)Mx" is true, since *at least one* case is true:

$$Mb = 1$$

But our *universal* conclusion "(x)Sx" is false, since *at least one* case is false:

$$Sa = 0$$

So we get a possible world in which the premises are true and the conclusion false. This shows our argument to be invalid.

Here's another example:

> Nothing material is spiritual.
> Something is material.
> ∴ Nothing is spiritual.

```
*  1.      ~(∃x)(Mx · Sx)              INVALID
*  2.      (∃x)Mx                        a, b
           [∴ ~(∃x)Sx
*  3.      asm: (∃x)Sx                 ┌─────────────┐
   4.      ∴ (x)~(Mx · Sx) [from 1]  · │ Ma, ~Sa     │
   5.      ∴ Ma [from 2]               │ ~Mb, Sb     │
   6.      ∴ Sb [from 3]               └─────────────┘
*  7.      ∴ ~(Ma · Sa) [from 4]
*  8.      ∴ ~(Mb · Sb) [from 4]
   9.      ∴ ~Sa [from 5 and 7]
  10.      ∴ ~Mb [from 6 and 8]
```

In this world, some things are material and some are spiritual. But nothing is both at once. In evaluating the ''~(∃x)(Mx · Sx)'' premise, evaluate first the part starting with the quantifier. ''(∃x)(Mx · Sx)'' is false since no case is true:

$$(Ma · Sa) = (1 · 0) = 0$$
$$(Mb · Sb) = (0 · 1) = 0$$

So the denial ''~(∃x)(Mx · Sx)'' is true. The ''(∃x)Mx'' premise is true since at least one case is true:

$$Ma = 1$$

In evaluating the ''~(∃x)Sx'' conclusion, we again first evaluate the part starting with the quantifier. ''(∃x)Sx'' is true since at least one case is true:

$$Sb = 1$$

So the denial ''~(∃x)Sx'' is false. So we have a possible world where the premises are true but the conclusion false. This shows our argument to be invalid.

These two rules are crucial in doing proofs and refutations:

> Use a new and different constant each time you drop an existential quantifier.

> Drop each universal quantifier once for each old constant.

Consider our previous argument. We'd violate the first rule if we derived ''Sa'' in step 6, since ''a'' at this point is an old constant. Then we'd incorrectly ''prove'' the argument to be valid. We'd violate the second if we didn't derive ''~(Mb · Sb)'' in line 8. Then we wouldn't get a truth value for ''Mb'' in our attempted refutation, and so the truth value of the first premise would be unknown. This would show that we had to do something further with the first premise.

The possible worlds we use for refutations must contain at least one entity. Seldom do we need more than two entities.

4.4A Symbolic Argument Exercises

Say whether each is valid (then give a proof) or invalid (then give a refutation). The first five are all invalid.

0. (∃x)Fx
   (∃x)Gx
∴ (∃x)(Fx · Gx)

ANSWER: * 1.   (∃x)Fx   **INVALID**
       * 2.   (∃x)Gx   **a, b**
          [∴ (∃x)(Fx · Gx)
       * 3.   asm: ~(∃x)(Fx · Gx)   Fa, ~Ga
         4.   ∴ (x)~(Fx · Gx) [from 3]   ~Fb, Gb
         5.   ∴ Fa [from 1]
         6.   ∴ Gb [from 2]
       * 7.   ∴ ~(Fa · Ga) [from 4]
       * 8.   ∴ ~(Fa · Gb) [from 4]
         9.   ∴ ~Ga [from 5 and 7]
      10.   ∴ ~Fb [from 6 and 8]

This is like arguing that, since someone is male and someone is female, it follows that someone is both male and female.

1. (∃x)Fx
∴ (x)Fx
3. (∃x)Fx
    ~(x)Gx
∴ ~(x)(Fx ⊃ Gx)
5. ~(x)Fx
    (∃x)~Gx
∴ ~(x)(Fx ∨ Gx)
7. (x)(Fx ∨ Gx)
    (∃x)~Gx
∴ (x)Fx

9. (x)((Fx · Gx) ⊃ Hx)
    (∃x)Fx
    (∃x)Gx
∴ (∃x)Hx

2. (∃x)Fx
∴ (∃x)~Fx
4. (∃x)(Fx ∨ Gx)
    (∃x)~Fx
∴ (∃x)Gx
6. (x)(Fx ⊃ Gx)
    (x)~Gx
∴ (x)~Fx
8. (x)(Fx ⊃ Gx)
    (x)(Gx ⊃ Hx)
    ~(∃x)Hx
∴ ~(∃x)Fx
10. (∃x)(Fx ⊃ Gx)
    (x)(~Gx ⊃ Hx)
    (∃x)(Fx ⊃ Hx)
∴ (∃x)Hx

4.4B English Argument Exercises

For each argument, first evaluate intuitively and then translate into QC and work it out. Say whether each is valid (then give a proof) or invalid (then give a refutation).

0. No pure water is burnable.
    Some Cuyahoga River water isn't burnable.
∴ Some Cuyahoga River water is pure water.
    [Use Px, Bx, and Cx.]

ANSWER: * 1.   ~(∃x)(Px · Bx)   **INVALID**
         * 2.   (∃x)(Cx · ~Bx)   **a**
              [∴ (∃x)(Cx · Px)
         * 3.   asm: ~(∃x)(Cx · Px)   Ca, ~Ba
           4.   ∴ (x)~(Cx · Px) [from 3]   ~Pa
           5.   ∴ (x)~(Px · Bx) [from 1]
         * 6.   ∴ (Ca · ~ Ba) [from 2]
         * 7.   ∴ ~(Ca · Pa) [from 4]
           8.   ∴ ~(Pa · Ba) [from 5]
           9.   ∴ Ca [from 6]
          10.   ∴ ~Ba [from 6]
          11.   ∴ ~Pa [from 7 and 9]

Here entity a might be the whole river, imagined to be Cuyahoga River water, not burnable, and not pure water. Are you tempted to misapply the I-rules and derive "Pa" from 8 and 10? If so, you might need to review the S- and I-rules.

1. No material thing is infinite.
Not all things are material.
∴ Something is infinite.
[Use Mx and Ix.]
2. Every wide receiver who plays well is fast.
∴ Every wide receiver who is fast plays well.
[Use Rx, Wx, and Fx.]
3. Some people smoke.
Not all people have clean lungs.
∴ Some who smoke don't have clean lungs.
[Use Sx and Cx.]
4. All rigorous duties are based on the categorical imperative.
All meritorious duties are based on the categorical imperative.
All duties are rigorous or meritorious.
∴ All duties are based on the categorical imperative.
[Use Rx, Bx, Mx, and the universe of discourse of duties. A duty is "meritorious" or "rigorous" depending on whether or not it has exceptions. The "categorical imperative" is Kant's basic moral principle. Kant's writings suggest this argument.]
5. All who aren't crazy agree with me.
∴ No one who is crazy agrees with me.
[Use Cx and Ax.]
6. Everything can be conceived.
Everything that can be conceived is an idea.
∴ Everything is an idea.
[Use Cx and Ix. From the idealist George Berkeley. Berkeley attacked materialism by trying to show that matter doesn't exist apart from our sensations. He claimed that a chair was just a collection of experiences. Bertrand Russell thought that the second premise was confused.]

7. All sound arguments are valid.
∴ All invalid arguments are unsound.
[Use Sx and Vx and the universe of discourse of arguments. "Invalid" means "not valid," and "unsound" means "not sound."]

8. Some Marxists plot violent revolution.
Some members of the faculty are Marxists.
∴ Some members of the faculty plot violent revolution.
[Use Mx, Rx, and Fx.]

9. All valid arguments with "ought" in the conclusion also have "ought" in the premises.
All arguments seeking to reduce the moral to the nonmoral have "ought" in the conclusion but don't have "ought" in the premises.
∴ No argument seeking to reduce the moral to the nonmoral is valid.
[Use Vx for "x is valid," Cx for "x has 'ought' in the conclusion," Px for "x has 'ought' in the premises," Rx for "x seeks to reduce the moral to the nonmoral," and the universe of discourse of arguments. Ethical naturalists claim we can deduce basic moral principles from descriptive facts alone.]

10. All trespassers are eaten.
∴ Some trespassers are eaten.
[Use Tx and Ex. The premise is from a sign on the Appalachian Trail in northern Virginia. Many traditional logic books take "all A is B" to entail "some A is B."]

11. Some necessary being exists.
All necessary beings are perfect beings.
∴ Some perfect being exists.
[Use Nx and Px. Kant claimed that the cosmological argument for the existence of God *at most* proves just the first premise. It doesn't prove the existence of God (a perfect being) unless we add the second premise. But the second premise, by the next argument, presupposes the central claim of the ontological argument for the existence of God—that a perfect being necessarily exists. So, Kant claimed, the cosmological argument presupposes the ontological argument.]

12. All necessary beings are perfect beings.
∴ Some perfect beings are necessary beings.
[Use Nx and Px. Kant followed the traditional logic of his day in thinking that "all A is B" entails "some B is A." See problem 10.]

13. All who can think clearly would do well in logic.
Some who can't think clearly ought to study logic.
All who would do well in logic ought to study logic.
∴ All ought to study logic.
[Use Cx, Wx, and Ox.]

14. No one who isn't a logical positivist holds the verifiability criterion of meaning.
∴ All who hold the verifiability criterion of meaning are logical positivists.
[Use Px and Vx. The verifiability criterion of meaning says that a statement is meaningful (in the sense of being true or false) if and only if it's true by definition or empirically verifiable. The first premise of the next argument depends on this view.]

15. All things true are either true by definition or empirically verifiable.
No basic moral principles are true by definition.
No basic moral principles are empirically verifiable.
∴ No basic moral principles are true.
[Use Tx, Dx, Ex, and Mx. From A. J. Ayer.]

4.5 INTERMEDIATE TRANSLATIONS

At the beginning of this chapter, we mentioned *statement letters* and *constants*:

- A capital letter alone (not followed by small letters) represents a *statement*. "S" might mean "It's snowing."
- Small letters from "a" to "v" are *constants*. Constants stand for singular terms (terms that stand for a single person or thing) like "Romeo" or "the current president."

Here's an example using both:

$(S \supset Cr)$ = If it's snowing, then Romeo is cold.

We'll now start using such things in our translations and proofs.

We'll also start using multiple and noninitial quantifiers. This requires more explanation. Consider this expanded version of our first translation rule (Section 4.2):

WHEREVER THE ENGLISH HAS	THE QC WFF HAS
all (every)	(x)
some	$(\exists x)$
not all	$\sim(x)$
no	$\sim(\exists x)$

Here's an example:

> *If all* are Italians, *then all* are lovers.
> $= ((x)Ix \supset (x)Lx)$

"If" normally translates as "(." Since the sentence begins with "if all," our wff begins with "(" and then "(x)." The placement of the quantifier in the wff generally mirrors the placement of the quantifier word in the English sentence:

All . . .	=	(x) . . .
If all . . .	=	$((x)$ . . .
Both all . . .	=	$((x)$ . . .
Either all . . .	=	$((x)$ . . .
Not all . . .	=	$\sim(x)$ . . .
If not all . . .	=	$(\sim(x)$ . . .

Our wff also uses *two* quantifiers, mirroring the *two* "all"'s in the English sentence. A wff normally uses the same number of quantifiers as the English sentence has quantifier words ("all," "every," "any," "some," "no," and so on).

Here are some further examples:

If not everyone is Italian, *then some* aren't lovers.
= (∼(x)Ix ⊃ (∃x)∼Lx)

If no one is Italian, *then no* lover is Italian.
= (∼(∃x)Ix ⊃ ∼(∃x)(Lx · Ix))

If all Italians are lovers, *then if all* are Italians *then all* are lovers.
= ((x)(Ix ⊃ Lx) ⊃ ((x)Ix ⊃ (x)Lx))

Someone is an Italian and *someone* is a lover.
= ((∃x)Ix · (∃x)Lx)

In the last case, the "someone" who is Italian may or may not be the same "someone" who is a lover.

"Any" differs from "all" in subtle ways. To translate "all" we just write "(x)" to mirror where the "all" occurs in the English sentence. But to translate "any" we'll follow either of these two rules:

To translate "any" in an English sentence:
Put a "(x)" at the *beginning* of the QC wff, outside of all parentheses.
Or rephrase the sentence so that it doesn't use "any"; then translate into QC. These equivalences hold: not any = no if any = if some any (alone) = all

Here's an example:

Not all are rich.
= ∼(x)Rx

Not any are rich.
= For all x, x isn't rich.
= (x)∼Rx [first rule]
= *No one* is rich.
= ∼(∃x)Rx [second rule]

Here our two rules for "any" give different, but equivalent, translations. You can use whichever rule you find easier. Here are two further examples:

If all are lovers, there will be peace.
= ((x)Lx ⊃ P)

If any are lovers, there will be peace.
= For all x, if x is a lover then there will be peace.
= (x)(Lx ⊃ P) [first rule]
= *If some* are lovers, there will be peace.
= ((∃x)Lx ⊃ P) [second rule]

Not all Italians are lovers.
= ~(x)(Ix ⊃ Lx)

Not any Italians are lovers.
= For all x, x is not an Italian lover.
= (x)~(Ix · Lx) [first rule]
= *No* Italians are lovers.
= ~(∃x)(Ix · Lx) [second rule]

Sometimes "all" and "any" are interchangeable:

All Italians are lovers.
= *Any* Italian is a lover.
= (x)(Ix ⊃ Lx)

4.5A Exercises

Using these equivalences, translate each English sentence into a QC wff. Remember that our rules are rough guides that sometimes don't work. Read your QC wff carefully to make sure it means the same as the English sentence.

Ex = x is evil   g = Gensler
Lx = x is a logician   R = It will rain
Cx = x is crazy

0. If everyone is evil then Gensler is evil.   ANSWER: ((x)Ex ⊃ Eg)

1. Gensler is evil.
2. If Gensler is a logician, then some logicians are evil.
3. Gensler is either crazy or evil.
4. Not everyone is evil.
5. Not anyone is evil.
6. If all logicians are evil, then some logicians are evil.
7. If no one is a logician, then everyone is evil.
8. If someone is evil, it will rain.
9. If everyone is evil, it will rain.
10. If anyone is evil, it will rain.
11. If some are logicians then some are evil.
12. If Gensler is a logician, then he's evil.

13. If someone is a logician, then this person is evil.
14. Everyone is an evil logician.
15. Not any logician is evil.

4.6 PROOFS AND REFUTATIONS

We need to change our strategy to do proofs involving intermediate wffs. Recall the proof strategy we gave in Section 4.3:

- Assume the opposite of what you want to prove.
- * Reverse squiggles.
- * Drop initial existential quantifiers, using a *new* constant for each one you drop.
- Drop each initial universal quantifier once for each old constant.
- * Use the S- and I-rules.
- * If you have to, make another assumption.
- You eventually get a proof or a refutation.

Here "*" shows that we star (and then ignore) the wffs used.

We'll now make one change in our strategy. From now on we'll use the S- and I-rules whenever we can. Often we'll have to use these rules *before* dropping quantifiers.

Our rule to drop only *initial* quantifiers remains in effect. If a left-hand parenthesis comes before the quantifier, then the quantifier isn't *initial* and we can't drop it. Suppose that this wff is a line of a proof that we are doing:

$$((x)Fx \supset (x)Gx)$$

The first thing in the wff is "(" coming before the first quantifier. So neither quantifier here is *initial*. Thus these inferences are wrong:

$((x)Fx \supset (x)Gx)$
∴ $(Fa \supset (x)Gx)$ [Wrong!]

$((x)Fx \supset (x)Gx)$
∴ $(Fa \supset Ga)$ [Wrong!]

Our wff is an if-then, where the first and second parts of the if-then happen to contain universal quantifiers. There are two ways to infer with an if-then. If we have the first part true, then we can infer that the second part is true. So if we have a line that says "(x)Fx," we can do this:

* $((x)Fx \supset (x)Gx)$
   $(x)Fx$
∴ $(x)Gx$

If we have the second part false, then we can infer that the first part is also false. So if we have a line that says "~(x)Gx," we can do this:

  *   ((x)Fx ⊃ (x)Gx)
     ~(x)Gx
∴ ~(x)Fx

Lacking either of these options, we might in the end have to make an additional assumption:

  ** ((x)Fx ⊃ (x)Gx)
     asm: ~(x)Fx [nice to have "(x)Fx"]

    Suppose that we reach no contradiction but instead find a refutation. Then we may have to determine the truth value of "((x)Fx ⊃ (x)Gx)." We'd first find the truth value of the parts starting with quantifiers. Here we'd first evaluate "(x)Fx" and "(x)Gx." Then we'd plug these truth values into the wff to find the truth value of the whole. So if (x)Fx = 1 and (x)Gx = 0 then ((x)Fx ⊃ (x)Gx) = (1 ⊃ 0) = 0.
    Here's a proof using our expanded strategy:

    If some are enslaved, then all have their freedom threatened.
∴ If this person is enslaved, then I have my freedom threatened.

 * **1.**   ((∃x)Sx ⊃ (x)Tx)   VALID
        [∴ (St ⊃ Ti)
 * **2.**   ⌈asm: ~(St ⊃ Ti)
  3. | ∴ St [from 2]
  4. | ∴ ~Ti [from 2]
  5. | ⌈asm: ~(∃x)Sx [nice to have "(∃x)Sx" for 1]
  6. | | ∴ (x)~Sx [from 5]
  7. | ⌊∴ ~St [from 6, contradicting 3]
 * **8.** | ∴ (∃x)Sx [from 5, 3, and 7]
  9. | ∴ (x)Tx [from 1 and 8]
 10. | ∴ Sb [from 8 (not needed)]
 11.   ⌊∴ Ti [from 9, contradicting 4]
 12. ∴ (St ⊃ Ti) [from 2, 4, and 11]

After making the assumption, we apply an S-rule to get lines 3 and 4. Then we are stuck, since we can't drop the noninitial quantifiers in line 1. So we make a second assumption in line 5, get a contradiction, and derive line 8. We soon get a second contradiction to complete the proof.
    We just noted that we don't need line 10. We get line 10 by dropping the existential quantifier in "(∃x)Sx" and using a new constant to get "Sb." However, note that we already have "St" in line 3. Whenever we already have an instance of an existential wff, we can skip the step about dropping "(∃x)" and writing in a new constant.
    Here's an example of how the expanded strategy leads to a refutation of an invalid argument (note the change from "some" to "all" in the first premise):

If *all* are enslaved, then all have their freedom threatened.
∴ If this person is enslaved, then I have my freedom threatened.

** 1.   ((x)Sx ⊃ (x)Tx)   **INVALID**
       [∴ (St ⊃ Ti)   **t, i, a**
* 2.   asm: ~(St ⊃ Ti)
   3.   ∴ St [from 2]   ┌─────────────────┐
   4.   ∴ ~Ti [from 2]   │ St, ~Ti, ~Sa   │
* 5.   asm: ~(x)Sx [nice to have "(x)Sx" for 1]   └─────────────────┘
* 6.   ∴ (∃x)~Sx [from 5]
   7.   ∴ ~Sa [from 6]

In evaluating the premise, we should first evaluate the parts starting with quantifiers. "(x)Sx" and "(x)Tx" are both false. So the first premise is "(0 ⊃ 0)," or "1." The conclusion is "(1 ⊃ 0)," or "0." In this possible world, the premise of the argument is true while the conclusion is false. So the argument is invalid.

4.6A Symbolic Argument Exercises

Say whether each is valid (then give a proof) or invalid (then give a refutation).

0. (x)(Fx ∨ Gx)
∴ ((x)Fx ∨ (x)Gx)   ANSWER: 1.   (x)(Fx ∨ Gx)
                                          [∴ ((x)Fx ∨ (x)Gx)
   INVALID   * 2.   asm: ~((x)Fx ∨ (x)Gx)
      a, b   * 3.   ∴ ~(x)Fx [from 2]
                                   * 4.   ∴ ~(x)Gx [from 2]
┌──────────────┐   * 5.   ∴ (∃x)~Fx [from 3]
│ Ga, ~Fa   │   * 6.   ∴ (∃x)~Gx [from 4]
│ Fb, ~Gb   │   7.   ∴ ~Fa [from 5]
└──────────────┘   8.   ∴ ~Gb [from 6]
                                   * 9.   ∴ (Fa ∨ Ga) [from 1]
                                   * 10.   ∴ (Fb ∨ Gb) [from 1]
                                      11.   ∴ Ga [from 7 and 9]
                                      12.   ∴ Fb [from 8 and 10]

This is like arguing that, since everyone is male or female, therefore either everyone is male or everyone is female.

1. (x)(Fx ⊃ P)
∴ ((∃x)Fx ⊃ P)
3. (x)((Fx ∨ Gx) ⊃ Hx)
Fm
∴ Hm
5. ((∃x)Fx ⊃ (x)Gx)
~Gp
∴ ~Fp
7. ((∃x)Fx ∨ (∃x)Gx)
∴ (∃x)(Fx ∨ Gx)
9. ((x)Fx ⊃ P)
∴ (x)(Fx ⊃ P)
11. (x)(Fx ⊃ P)
Fr
∴ P
13. ((∃x)Fx ⊃ (∃x)Gx)
∴ (x)(Fx ⊃ Gx)
15. (x)((Fx · Gx) ⊃ P)
~P
Fo
∴ ~(x)(Fx ⊃ Gx)
17. (P ⊃ (x)Fx)
∴ (x)(P ⊃ Fx)
19. ((∃x)Fx ⊃ P)
∴ (x)(Fx ⊃ P)

2. (∃x)Fx
∴ Fc
4. (x)(Fx ∨ Gx)
~Fe
∴ (∃x)Gx
6. ~(∃x)(Fx · Gx)
~Fd
∴ Gd
8. (∃x)(Fx ∨ Gx)
∴ ((x)~Gx ⊃ (∃x)Fx)
10. (x)(Fx ⊃ P)
∴ ((x)Fx ⊃ P)
12. (x)(Fx · Gx)
∴ ((x)Fx · (x)Gx)
14. (∃x)(Fx ∨ Gx)
∴ ((∃x)Fx ∨ (∃x)Gx)
16. Fj
(∃x)Gx
(x)((Fx · Gx) ⊃ Hx)
∴ (∃x)Hx
18. ((x)Fx ∨ (x)Gx)
∴ (x)(Fx ∨ Gx)
20. (∃x)(Fx · (Gx ∨ Hx))
∴ (~Hk ⊃ (∃x)(Fx · Gx))

4.6B English Argument Exercises

For each argument, first evaluate intuitively and then translate into QC and work it out. Say whether each is valid (then give a proof) or invalid (then give a refutation).

0. Gensler is evil.
∴ If Gensler is a logician, then some logicians are evil.
[Use g, Ex, and Lx.]

   ANSWER: 1.   Eg   VALID
            [∴ (Lg ⊃ (∃x)(Lx · Ex))
  * 2.   ⌈asm: ~(Lg ⊃ (∃x)(Lx · Ex))
    3.   │ ∴ Lg [from 2]
  * 4.   │ ∴ ~(∃x)(Lx · Ex) [from 2]
    5.   │ ∴ (x)~(Lx · Ex) [from 4]
  * 6.   │ ∴ ~(Lg · Eg) [from 5]
    7.   ⌊ ∴ ~Eg [from 3 and 6, contradicting 1]
    8. ∴ (Lg ⊃ (∃x)(Lx · Ex)) [from 2, 1, and 7]

1. Everything has a cause.
If the world has a cause, then there is a God.

∴ There is a God.

[Use Cx for "x has a cause," w for "the world," and G for "There is a God." In other arguments, we might have to break down "There is a God" into "(∃x)Gx" ("For some x, x is a God") to prove validity. A beginning philosophy student suggested this argument. The next example shows that the first premise could easily lead to the opposite conclusion.]

2. Everything has a cause.

If there is a God, then something doesn't have a cause (namely, God).

∴ There is no God.

[Use Cx and G. (You don't have to translate "(namely, God)" in the second premise.) Some have suggested a more complex phrasing of the first premise to avoid this problem: "Every contingent being or set of contingent beings has a cause."]

3. If everyone litters, the world will be dirty.

∴ If you litter, then the world will be dirty.

[Use Lx, D, and u.]

4. Anything enjoyable is either immoral or fattening.

Nothing is immoral.

∴ Anything that isn't fattening isn't enjoyable.

[Use Ex, Ix, and Fx.]

5. Anything that can be explained either can be explained as caused by scientific laws or can be explained as resulting from a free choice of a rational being.

The totality of basic scientific laws can't be explained as caused by scientific laws (since this would be circular).

∴ Either the totality of basic scientific laws can't be explained or else it can be explained as resulting from a free choice of a rational being.

[Use Ex for "x can be explained," Sx for "x can be explained as caused by scientific laws," Fx for "x can be explained as resulting from a free choice of a rational being," and t for "the totality of scientific laws." From R. G. Swinburne, a philosopher of religion.]

6. If there are no necessary beings, then there are no contingent beings.

∴ All contingent beings are necessary beings.

[Use Nx and Cx. St. Thomas Aquinas accepted the premise but not the conclusion.]

7. Everyone who has sense will vote against my opponent.

∴ If everyone has sense, then everyone will vote against my opponent.

[Use Sx and Vx.]

8. Any senator or House member is a member of Congress.

No Communist is a member of Congress.

Jones is a Communist.

∴ Jones isn't a senator.

[Use Sx, Hx, Mx, Cx, and j.]

9. Anything not disproved that is of practical value to one's life to believe in ought to be believed.

Free will isn't disproved.

∴ If free will is of practical value to one's life to believe in, then it ought to be believed.

[Use Dx, Vx, Ox, f for "free will," and the universe of discourse of beliefs. From William James.]

10. If anyone can learn logic, then you can learn logic.

All who can learn physics can learn logic.

Einstein can learn physics.

∴ You can learn logic.
[Use Lx, u, Px, and e.]

11. If someone knows the future, then no one has free will.
∴ No one who knows the future has free will.
[Use Kx and Fx.]

12. If everyone teaches philosophy, then everyone will starve.
∴ Everyone who teaches philosophy will starve.
[Use Tx and Sx.]

13. If the world had no temporal beginning, then some series of moments before the present moment is a completed infinite series.
There's no completed infinite series.
∴ The world had a temporal beginning.
[Use Tx for "x had a temporal beginning," w for "the world," Mx for "x is a series of moments before the present moment," and Ix for "x is a completed infinite series." Is this argument just an instance of *modus tollens*: "(P ⊃ Q), ~Q ∴ ~P"? This one and the next are from Immanuel Kant. Kant thought our intuitive metaphysical principles lead to conflicting conclusions and thus can't be trusted.]

14. Everything that had a temporal beginning was caused to exist by something previously in existence.
If the world had a temporal beginning, then there was no time before the world began.
If the world was caused to exist by something previously in existence, then there was time before the world began.
∴ The world didn't have a temporal beginning.
[Use Tx for "x had a temporal beginning," Cx for "x was caused to exist by something previously in existence," w for "the world," and B for "There was time before the world began."]

15. If someone talks, all go to jail.
If Suzy is picked up by the police, she'll talk.
∴ If Suzy is picked up by the police, then I'll go to jail.
[Use Tx, Jx, Px, s, and i.]

16. No proposition with factual content is necessary (= "self-contradictory to deny").
∴ Either no mathematical proposition has factual content, or no mathematical proposition is necessary.
[Use Fx for "x has factual content," Nx for "x is necessary," Mx for "x is mathematical," and the universe of propositions. From the logical positivist A. J. Ayer.]

17. Either everyone rises from the dead, or else no one rises from the dead.
If Christ didn't rise from the dead, we are liars and our faith is in vain.
∴ If our faith isn't in vain, then everyone rises from the dead.
[Use Rx, c, L, and V. This reconstructs an argument from St. Paul in I Corinthians.]

18. Every material thing has a cause.
Not everything has a cause.
∴ Not everything is material.
[Use Mx and Cx.]

19. If everyone lies, the results will be disastrous.
∴ If anyone lies, the results will be disastrous.
[Use Lx and D.]

20. Any basic social rule that people would agree to if they were free, rational, and fully informed but ignorant of their exact place in society (whether rich or poor, white or black, male or female) is a principle of justice.

The equal-liberty principle and the difference principle are basic social rules that people would agree to if they were free, rational, and fully informed but ignorant of their exact place in society.

∴ The equal-liberty principle and the difference principle are principles of justice.

[Use Ax, Jx, e, and d. From John Rawls. The equal-liberty principle says that each person is entitled to the greatest liberty compatible with an equal liberty for all others. The difference principle says that wealth is to be distributed equally, except for incentives that ultimately benefit everyone and that are equally open to all.]

21. Everyone makes moral judgments.

Moral judgments logically presuppose judgments about God.

If moral judgments logically presuppose judgments about God, then everyone who makes moral judgments believes (at least implicitly) that there is a God.

∴ Everyone believes (at least implicitly) that there is a God.

[Use Mx for "x makes moral judgments," L for "Moral judgments logically presuppose judgments about God," and Bx for "x believes (at least implicitly) that there is a God." From the Jesuit theologian Karl Rahner.]

22. If everyone deliberates about what to do, then everyone believes (at least implicitly) in free will.

∴ Everyone who deliberates about what to do believes (at least implicitly) in free will.

[Use Dx and Bx.]

23. Nothing that isn't caused can be integrated into the unity of our experience.

Everything can be integrated into the unity of our experience.

∴ Everything is caused.

[Use Cx and Ix (for "x can be integrated into the unity of our experience"). Kant thought we should qualify the second premise to "Everything *that we could experientially know* can be integrated into the unity of our experience." Then the conclusion would be, "Everything *that we could experientially know* is caused." Kant thought the unqualified "Everything is caused" leads to contradictions (see problems 1 and 2).]

24. Any consistent person who thinks that abortion is permissible will consent to the idea of himself or herself having been aborted.

No person with normal desires will consent to the idea of himself or herself having been aborted.

You have normal desires.

∴ If you're consistent, then you won't think that abortion is permissible.

[Use Cx for "x is consistent," Px for "x thinks that abortion is permissible," Kx for "x consents to the idea of himself or herself having been aborted," Nx for "x has normal desires," and u for "you." See my article in *Philosophical Studies*, January 1986.]

25. If there is no God, then all experiencers are finite.

Whatever is true is experienced.

If all experiencers are finite and whatever is true is experienced, then there's some experiencer who experiences that all experiencers are finite.

No finite being experiences that all experiencers are finite.

∴ There is a God.

[Use G for "There is a God," Ex for "x is an experiencer," Fx for "x is finite," T for "Whatever is true is experienced," and Ax for "x experiences that all experiencers are finite." From the idealist Josiah Royce. The second premise denies that truth exists independently of experience. This one is difficult.]

5

Advanced Quantificational Logic

5.1 IDENTITY TRANSLATIONS

In this chapter, we crank QC up to full power. Later in the chapter, we'll add relational statements. Now we'll add identity statements. Our rule 3 for forming QC wffs introduces "$=$" ("equals"):

> **3.** The result of writing a variable or constant, then "$=$," and then a variable or constant is a wff.

Here are two examples:

$r=l$ = Romeo is the lover of Juliet.
$x=y$ = x equals y.

To negate an identity wff, we write "$\sim$" in front of it:

$\sim r=l$ = Romeo *isn't* the lover of Juliet.

(Some people prefer "$r \neq l$.") We don't use parentheses with "$=$." So these *aren't* legitimate wffs:

$(r=l)$ $\sim(r=l)$ $(\sim r=l)$ $(\sim(r=l))$

There's no need for parentheses here—no ambiguity that we need parentheses to clear up.

The simplest use of "=" is to translate an "is" that goes between singular terms. Singular terms differ from general terms:

- A *singular term* is a term that stands for a single person or thing. Examples include names ("Romeo") and phrases of the form "the so and so" or "this so and so" ("the lover of Romeo" or "this woman"). Singular terms translate into small letters.
- A *general term* is a term that does or could describe more than one person or thing. Examples include adjectives ("green") and phrases of the form "a so and so" ("an Italian"). General terms translate into capital letters.

Compare these two cases:

Romeo is *the lover of Juliet.*
= r = l

Romeo is *an Italian.*
= Ir

The "is" in the first case is flanked by singular terms and so translates as "=." Here "is" signifies identity—that *Romeo* and *the lover of Juliet* are the same person. The "is" in the second case is followed by a general term. Here "is" signifies predication—that Romeo can be correctly characterized as Italian.

We can also translate "besides," "other than," and "alone" using identity:

Someone *besides* Romeo is rich.
= Someone *other than* Romeo is rich.
= Someone *who isn't* Romeo is rich.
= For some x, x, $\neq$ Romeo and x is rich.
= $(\exists x)(\sim x = r \cdot Rx)$

Romeo *alone* is rich.
= Romeo is rich and no one *besides* Romeo is rich.
= Romeo is rich, and there is no x such that x$\neq$Romeo and x is rich.
= $(Rr \cdot \sim(\exists x)(\sim x = r \cdot Rx))$

We can also translate some numerical notions using identity. Here's an example:

At least two people are rich.
= For some x and some y: x$\neq$y, x is rich, and y is rich.
= $(\exists x)(\exists y)(\sim x = y \cdot (Rx \cdot Ry))$

The pair of quantifiers "$(\exists x)(\exists y)$" ("for some x and for some y") doesn't say whether or not x and y are identical. So we need "$\sim x = y$" ("x$\neq$y") to say that they *aren't* identical.

Note the variables "x" and "y." From now on, we'll often need multiple variables to keep the references straight. It doesn't matter what variables we use. These two wffs are equivalent:

$(\exists x)Rx$ = For some x, x is rich.
$(\exists y)Ry$ = For some y, y is rich.

Each means "*At least one* person is rich."
Here's how we translate "*Exactly one* person is rich" and "*Exactly two* persons are rich":

Exactly one person is rich.
= For some x: x is rich and no one besides x is rich.
= For some x: x is rich and there is no y such that $y \neq x$ and y is rich.
= $(\exists x)(Rx \cdot \sim(\exists y)(\sim y = x \cdot Ry))$

Exactly two persons are rich.
= For some x and for some y: x is rich and y is rich and $x \neq y$ and no one besides x and y is rich.
= For some x and for some y: x is rich and y is rich and $x \neq y$ and there is no z such that $z \neq x$ and $z \neq y$ and z is rich.
= $(\exists x)(\exists y)(((Rx \cdot Ry) \cdot \sim x = y) \cdot \sim(\exists z)((\sim z = x \cdot \sim z = y) \cdot Rz))$

It's possible (but awkward) to express notions like "There are exactly n F's" (where n is any whole number) in our QC.
We can express *addition* in QC. Consider this English paraphrase of "1 + 1 = 2" and the corresponding wff:

1 + 1 = 2
If exactly one thing is F and exactly one thing is G, and nothing is F and G, then exactly two things are F or G.
$((((\exists x)(Fx \cdot \sim(\exists y)(\sim y = x \cdot Fy)) \cdot (\exists x)(Gx \cdot \sim(\exists y)(\sim y = x \cdot Gy))) \cdot \sim(\exists x)(Fx \cdot Gx)) \supset (\exists x)(\exists y)((((Fx \lor Gx) \cdot (Fy \lor Gy)) \cdot \sim x = y) \cdot \sim(\exists z)((\sim z = x \cdot \sim z = y) \cdot (Fz \lor Gz))))$

We could prove QC's version of "1 + 1 = 2" by assuming its denial and deriving a contradiction. We *won't* do this. But it's interesting that it *could* be done. In principle, we could prove "2 + 2 = 4" and "5 + 7 = 12"—and even the additions on your income tax form! Some mean logic teachers assign such things for homework.

5.1A Exercises

Translate each sentence into QC. For the first ten examples, use these equivalences:

Ex = x is evil   a = Aristotle
Lx = x is a logician   p = Plato
                        t = the greatest logician

0. Everyone besides Aristotle and Plato is evil.
ANSWER: $(x)((\sim x = a \cdot \sim x = p) \supset Ex)$

1. Aristotle is the greatest logician.
2. Aristotle is a logician.
3. Aristotle isn't Plato.
4. Someone besides Aristotle is a logician.
5. Aristotle alone is a logician.
6. All logicians other than Aristotle are evil.
7. No one besides Aristotle is evil.
8. There are at least two logicians.
9. There is exactly one logician.
10. There is exactly one evil logician.
11. If the thief is intelligent, then you aren't the thief.
12. Someone besides the boss is arrogant.
13. Kurt isn't a philosopher.
14. Judy is the teacher, but she isn't Italian.
15. Carol is my only sister.

5.2 IDENTITY ARGUMENTS

We need two new rules for identity. Q5 is the *self-identity* rule. It holds regardless of what constant replaces "a":

$$Q5 \rightarrow a = a$$

Q5 is an axiom. An *axiom* is a basic assertion of a system that isn't proved but is used to prove other things. Unlike a rule of inference, an axiom doesn't tell us that we can go *from* certain claims *to* certain other claims. Rather, an axiom tells us that we may assert certain claims in their own right. Q5 tells us that we may assert a self-identity as a "derived step" at any point in a proof, no matter what the previous lines are.

Q6 is the *equals may substitute for equals* rule. It holds regardless of what constants replace "a" and "b" and what wffs replace "Fa" and "Fb"—provided that the two wffs are alike except that the two constants are interchanged in one or more occurrences:

$$Q6 \quad Fa, a = b \quad \rightarrow \quad Fb$$

Here's an example:

Fg	Gensler is a fanatical backpacker.
g = a	Gensler is the author of this book.
∴ Fa	The author of this book is a fanatical backpacker.

The "g = a" premise allows us to take the "Fg" premise and substitute "a" for "g."
Here's a simple proof using our new inference rules:

I weigh 170 pounds.

My mind doesn't weigh 170 pounds.

∴ I am not identical to my mind.

1.   Wi   VALID
2.   ~Wm
         [∴ ~i = m
3.   ⌈ asm: i = m
4.   ⌊ ∴ Wm [from 1 and 3, contradicting 2]
5. ∴ ~i = m [from 3, 2, and 4]

Line 4 follows using Q6. If x and y are identical, then whatever can be truly said of one can be truly said of the other. Likewise, if something can be truly said of x but *not* of y, then x and y *aren't* identical.

Here's a simple invalid argument and its refutation:

The bankrobber wears size-twelve hiking boots.

You wear size-twelve hiking boots.

∴ You're the bankrobber.

1.   Wb   **INVALID**   **b, u**
2.   Wu
  . [∴ u = b   ┌─────────────────────┐
3.   asm: ~u = b   │ Wb, Wu, ~u = b   │
                       └─────────────────────┘

Since we can't infer anything, we set up a universe to refute the argument. This universe contains two distinct persons, the bankrobber and you. Each wears size-twelve hiking boots. Here the premises are true but the conclusion is false. So the argument is invalid.

Our next example involves *pluralism* and *monism*:

Pluralism

= There is more than one being.

= For some x and for some y: x ≠ y.

= (∃x)(∃y)~x = y

Monism

= There is exactly one being.

= For some x: every y is identical to x.

= (∃x)(y)y = x

Here's a proof that pluralism entails the falsity of monism:

> There is more than one being.
> ∴ It's false that there is exactly one being.

* **1.** $(\exists x)(\exists y){\sim}x = y$ VALID
           [∴ ${\sim}(\exists x)(y)y = x$
* **2.**   ⌐asm: $(\exists x)(y)y = x$
* **3.** | ∴ $(\exists y){\sim}a = y$ [from 1]
  4. | ∴ ${\sim}a = b$ [from 3]
  5. | ∴ $(y)y = c$ [from 2]
  6. | ∴ $a = c$ [from 5]
  7. | ∴ $b = c$ [from 5]
  8.   L∴ $a = b$ [from 6 and 7, contradicting 4]
  9. ∴ ${\sim}(\exists x)(y)y = x$ [from 2, 4, and 8]

Lines 1 and 2 have back-to-back quantifiers. We can drop only a quantifier that is *initial* and hence *outermost*. So we have to drop the quantifiers one at a time, starting from the outside. After dropping quantifiers, we use our "equals may substitute for equals" rule to get line 8. Here the "$b = c$" premise allows us to take "$a = c$" and substitute "b" for the "c," thus getting "$a = b$."

In this proof, we didn't bother to derive "$c = c$" from "$(y)y = c$" in line 5. From now on, dropping universal quantifiers using *every* old constant would often be too tedious. We'll switch to the policy of deriving from universal wffs *only* steps likely to be useful for our proof or refutation.

Our "equals may substitute for equals" rule, Q6, seems to hold without restriction in arguments about mathematics or about material phenomena. But the rule doesn't always work in arguments about mental phenomena. Consider this argument (where "Bx" is for "Jones believes that x denounced Cataline"):

> Bc   Jones believes that *Cicero* denounced Cataline.
> $c = t$ *Cicero* is the same person as *Tully*.
> ∴ Bt   Jones believes that *Tully* denounced Cataline.

Q6 says this is *valid*. But it's not; it's *invalid*! The premises might be true while the conclusion is false, especially if Jones knows just a little ancient Roman history. Jones, while believing that Cicero denounced Cataline, might not know that Cicero is the same person as Tully. So Jones might not believe that Tully denounced Cataline.

To avoid the problem, we'll disallow translating into QC any predicates or relations that violate the "equals may substitute for equals" rule, Q6. We hence forbid letting "Bx" stand for "Jones believes that x denounced Cataline." Statements about beliefs and other mental phenomena often violate Q6. We have to be careful translating such statements into QC. Some systems of belief logic introduce special symbols for talking about beliefs and explicitly restrict the use of Q6.

So the mental seems to follow somewhat different logical patterns from the material. Does this show that the materialist scheme of reducing the mental to the material won't work? Philosophers dispute this question.

5.2A Symbolic Argument Exercises

Say whether each is valid (then give a proof) or invalid (then give a refutation):

0. a = b   ANSWER: 1.   a = b   VALID
∴ b = a   [∴ b = a
                          2.   ⌈asm: ~b = a
                          3.   │∴ ~b = b [from 1 and 2]
                          4.   ⌊∴ b = b [from Q5, contradicting 3]
                          5.   ∴ b = a [from 2, 3, and 4]

1. Fa   **2.** Fa
∴ (~Fb ⊃ ~b = a)   ∴ (~b = a ⊃ ~Fb)
3. a = b   **4.** ~a = b
   b = c   c = b
∴ a = c   ∴ ~a = c
5. a = b   **6.** ~a = b
   (x)(Fx ⊃ Gx)   ~c = b
   ~Ga   ∴ a = c
∴ ~Fb
7. a = b   **8.** Fa
∴ (Fa ≡ Fb)   ∴ (x)(x = a ⊃ Fx)
9. ∴ (∃x)(y)y = x   **10.** ∴ (∃x)(∃y)~x = y

Problem 9 is about *monism* (the claim that there is exactly one being), while 10 is about *pluralism* (the claim that there is more than one being). Is either provable in QC?

5.2B English Argument Exercises

For each argument, first evaluate intuitively and then translate into QC and work it out. Say whether each is valid (then give a proof) or invalid (then give a refutation). In this exercise, you have to figure out most of your own letters and decide whether to use small letters or capitals.

0. Some are logicians.
   Some aren't logicians.
∴ There is more than one being. *(pluralism)*
         ANSWER: * 1.   (∃x)Lx   VALID
                   * 2.   (∃x)~Lx
                           [∴ (∃x)(∃y)~x = y
                   * 3.   ⌈asm: ~(∃x)(∃y)~x = y
                     4.   │∴ (x)~(∃y)~x = y [from 3]
                     5.   │∴ La [from 1]
                     6.   │∴ ~Lb [from 2]
                   * 7.   │∴ ~(∃y)~a = y [from 4]
                     8.   │∴ (y)a = y [from 7]
                     9.   │∴ a = b [from 8]
                     10.   ⌊∴ Lb [from 5 and 9, contradicting 6]
                     11.   ∴ (∃x)(∃y)~x = y [from 3, 6, and 10]

Here we apply the reverse-squiggle and quantifier-dropping rules to only one quantifier at a time. Only the quantifier that is outermost at a given time can be dropped at that time.

1. Keith is my only nephew.
My only nephew knows more about BASIC than I do.
Keith is a ten-year-old.
∴ Some ten-year-olds know more about BASIC than I do.
[Use k, m, Bx, and Tx.]
2. Judy isn't the world's best cook.
The world's best cook lives in Chicago.
∴ Judy doesn't live in Chicago.
3. This chemical process is publicly observable.
This pain isn't publicly observable.
∴ This pain isn't identical to this chemical process.
[This attacks the "identity theory" of the mind, that mental processes are identical to chemical processes.]
4. The person who left a lighter is the murderer.
The person who left a lighter is a smoker.
No smokers are backpackers.
∴ The murderer isn't a backpacker.
5. The man of Suzy's dreams is either rich or handsome.
You aren't rich.
∴ If you're handsome, then you're the man of Suzy's dreams.
6. Patricia lives in North Dakota.
Blondie lives in North Dakota.
∴ At least two people live in North Dakota.
7. Alice stole money.
The nastiest person at the party stole money.
Alice isn't the nastiest person at the party.
∴ At least two people stole money.
8. If Speedy Jones looks back to the quarterback just before the hike, then Speedy Jones is the primary receiver.
The primary receiver is the receiver you should try to cover.
∴ If Speedy Jones looks back to the quarterback just before the hike, then Speedy Jones is the receiver you should try to cover.
9. Either you knew where the money was, or the thief knew where it was.
You didn't know where the money was.
∴ You aren't the thief.
10. Exactly one person lives in North Dakota.
Paul lives in North Dakota.
Paul is a farmer.
∴ Everyone who lives in North Dakota is a farmer.

5.3 RELATIONAL TRANSLATIONS

Our last rule for forming QC wffs adds relational statements:

4. Any capital letter followed by two or more small letters (constants or variables or some combination of these) is a wff.

Here are two examples:

> Lrj   [Romeo loves Juliet.]
> Gxyz   [x gave y to z.]

Capital letters followed by two or more small letters stand for relations.

Translating relational sentences into QC is difficult. There are few rules to help you. You have to see a lot of examples and catch the patterns.

Here are some relational sentences without quantifiers:

> Romeo loves Juliet.
> = Lrj

> Juliet loves Romeo.
> = Ljr

> Juliet loves herself.
> = Juliet loves Juliet.
> = Ljj

> Juliet loves Romeo and Antonio.
> = Juliet loves Romeo and Juliet loves Antonio.
> = (Ljr · Lja)

And here are three easy examples with single quantifiers:

> Everyone loves Juliet.
> = For all x, x loves Juliet.
> = (x)Lxj

> Someone loves Juliet.
> = For some x, x loves Juliet.
> = (∃x)Lxj

> No one loves Juliet.
> = For no x, x loves Juliet.
> = ~(∃x)Lxj

If English has a quantifier right after the ''loves,'' we have to rephrase things so that the quantifier goes first:

> ''Juliet loves everyone (someone, no one)''
>
> means
>
> ''For all (some, no) x, Juliet loves x.''

Here are some examples:

> Juliet loves everyone.
> = For all x, Juliet loves x.
> = (x)Ljx

> Juliet loves someone.
> = For some x, Juliet loves x.
> = (∃x)Ljx

> Juliet loves no one.
> = For no x, Juliet loves x.
> = ~(∃x)Ljx

Here are similar but more complex examples:

> Juliet loves every Italian.
> = For all x, if x is Italian then Juliet loves x.
> = (x)(Ix ⊃ Ljx)

> Juliet loves some Italian.
> = For some x, x is Italian and Juliet loves x.
> = (∃x)(Ix · Ljx)

> Juliet loves no Italians.
> = For no x, x is Italian and Juliet loves x.
> = ~(∃x)(Ix · Ljx)

These common patterns can be confusing at first.
Here are two easy sentences with two quantifiers:

> Everyone loves everyone.
> = For all x and for all y, x loves y.
> = (x)(y)Lxy

> Someone loves someone.
> = For some x and for some y, x loves y.
> = (∃x)(∃y)Lxy

In the second case, the "someone" who loves may or may not be the same "someone" who is loved. "Someone loves someone" may be true because someone loves himself or herself ["(∃x)Lxx"] *or* because someone loves someone else ["(∃x)(∃y)(~x = y · Lxy)"].

The next two are more difficult. They differ only in the order of the quantifiers. In each case, the order of the quantifiers in QC mirrors the order of the quantifier words in the English sentence:

Everyone loves someone or other.
= For every x there is some y such that x loves y.
= $(x)(\exists y)Lxy$

There is someone that everyone loves.
= There is some y such that, for all x, x loves y.
= $(\exists y)(x)Lxy$

In the first case, we might all love different people. Perhaps each person loves his or her parents. In the second case we all love the *same* person. Perhaps everyone loves God. These pairs emphasize the difference:

Everyone loves someone or other.	There is someone that everyone loves.
Everyone lives in some place or other.	There is some place where everyone lives.
Everyone makes some error or other.	There is some error that everyone makes.

The sentence on the right in each case entails the sentence on the left, but not the other way around.

With back-to-back quantifiers, the order doesn't matter if the quantifiers are both universals or both existentials:

$(x)(y) = (y)(x)$ $(\exists x)(\exists y) = (\exists y)(\exists x)$

But the order usually matters if one quantifier is universal and the other is existential:

$(x)(\exists y) \neq (\exists y)(x)$

Also, it doesn't matter what variables we use, so long as the reference pattern is the same. These three are equivalent:

$(y)(\exists x)Lxy$ $(x)(\exists y)Lyx$ $(w)(\exists z)Lzw$

Each has a universal, then an existential, then "L," then the variable used in the existential, and finally the variable used in the universal.

Many two-term relations have special properties, such as reflexivity or symmetry. Here are some examples:

Everything is identical to itself.
= "Is identical to" is *reflexive*. [Identity is a relation but uses a special symbol.]
= $(x)x = x$

Nothing is taller than itself.
= "Taller than" is *irreflexive*.
= (x)~Txx

In all cases, if x is a relative of y, then y is a relative of x.
= "Being a relative of" is *symmetrical*.
= (x)(y)(Rxy ⊃ Ryx)

In all cases, if x is a parent of y then y isn't a parent of x.
= "Being a parent of" is *asymmetrical*.
= (x)(y)(Pxy ⊃ ~Pyx)

In all cases, if x is taller than y and y is taller than z, then x is taller than z.
= "Being taller than" is *transitive*.
= (x)(y)(z)((Txy · Tyz) ⊃ Txz)

In all cases, if x is a foot taller than y and y is a foot taller than z, then x isn't a foot taller than z.
= "Being a foot taller than" is *intransitive*.
= (x)(y)(z)((Fxy · Fyz) ⊃ ~Fxz)

Love fits none of these six categories. Love is neither reflexive nor irreflexive: Sometimes people love themselves and sometimes they don't. Love is neither symmetrical nor asymmetrical: If x loves y, then sometimes y loves x in return and sometimes y doesn't. Love is neither transitive nor intransitive: If x loves y and y loves z, then sometimes x loves z and sometimes x doesn't love z.
Here are some further examples:

Every Italian loves someone.
= For every x, if x is Italian then there is some y such that x loves y.
= (x)(Ix ⊃ (∃y)Lxy)

Everyone loves some Italian.
= For every x there is some y such that y is Italian and x loves y.
= (x)(∃y)(Iy · Lxy)

It's always the case that if a first person understands a second then the first loves the second.
= For every x and y: if x understands y then x loves y.
= (x)(y)(Uxy ⊃ Lxy)

Everyone loves a lover.
= For every x, if x loves someone then everyone loves x.
= (x)((∃y)Lxy ⊃ (z)Lzx)

    Juliet loves everyone besides herself.
= For every x, if x≠Juliet then Juliet loves x.
= (x)(~x=j ⊃ Ljx)

    Romeo loves all and only those who don't love themselves.
= For every x, Romeo loves x if and only if x doesn't love x.
= (x)(Lrx ≡ ~Lxx)

Study these and the other examples carefully. Focus on how to paraphrase the English sentences.

    I don't have any tidy rules for translating relational sentences into QC wffs. But if you get confused, I suggest you try these steps:

- If it would be helpful, rephrase the English sentence to make its meaning clearer.
- Put a different variable after each English quantifier.
- One by one, replace each English quantifier by a QC quantifier, rephrasing the sentence according to the quantificational idiom.
- Complete the translation.

Here are three examples:

    Every Italian loves some Italian.
= Every x who is Italian loves some y who is Italian.
= (x)(if x is Italian then x loves some y who is Italian)
= (x)(if x is Italian then (∃y)(y is Italian and x loves y))
= (x)(Ix ⊃ (∃y)(Iy · Lxy))

    There is an unloved lover.
= There is someone whom no one loves but who loves someone.
= There is some x whom no y loves but who loves some z.
= (∃x)(no y loves x but x loves some z)
= (∃x)(~(∃y)y loves x but x loves some z)
= (∃x)(~(∃y)y loves x but (∃z)x loves z)
= (∃x)(~(∃y)Lyx · (∃z)Lxz)

    Some Italian besides Romeo loves Juliet.
= Some x who is Italian and who isn't Romeo loves Juliet.
= (∃x)(x is Italian and x isn't Romeo and x loves Juliet)
= (∃x)((Ix · ~x=r) · Lxj)

The important thing is to paraphrase the English sentence according to the quantificational idiom.

    We now have four sorts of QC wffs that don't contain quantifiers or propositional connectives or negation:

undivided	entity-property	identity	relational
R	Ir	r = 1	Lrj
It's raining.	Romeo is Italian.	Romeo is the lover of Juliet.	Romeo loves Juliet.

5.3A Exercises

Using these equivalences, translate each of these into QC:

Cxy = x caused y   a = Aristotle
Gxy = x is greater than y   g = God
Ex = x is evil   w = the world

0. Aristotle caused nothing evil.   ANSWER: ~(∃x)(Ex · Cax)

This means "It's not the case that there is some x such that x is evil and Aristotle caused x."

1. God caused the world.
2. The world caused God.
3. Aristotle is greater than some evil things.
4. Aristotle is greater than no one.
5. It is always true that if a first thing is evil and a second thing isn't evil then the first is greater than the second.
6. If God caused the world then God is greater than the world.
7. It is not always true that if a first thing caused a second then the first is greater than the second.
8. Nothing is greater than itself.
9. Aristotle is greater than anything else.
10. x is an effect (in other words, something caused x).
11. God caused himself.
12. God had no cause.
13. If God had no cause, then the world had no cause.
14. There is something than which nothing is greater.
15. Every entity is greater than some entity or other.
16. Nothing caused itself.
17. It is always true that if a first thing is greater than a second then the second isn't greater than the first.
18. There is something that caused everything.
19. Everything is caused by something or other.
20. Something evil caused all evil things.
21. Everything caused all evil things.
22. God caused everything besides himself.
23. Nothing caused anything that caused itself.
24. It is always true that if a first thing caused a second and the second caused a third then the first caused the third.
25. There is a first cause (that is, there is a cause that itself has no cause).

5.4 RELATIONAL ARGUMENTS

Relational proofs require no further inference rules. Here's a proof of a sad but valid relational argument:

> Romeo loves Juliet.
> Juliet doesn't love Romeo.
> ∴ It's not always the case that if a first person loves a second then the second loves the first.

> **1.**   Lrj   VALID
> **2.**   ~Ljr
>   [∴ ~(x)(y)(Lxy ⊃ Lyx)
> **3.**   ┌asm: (x)(y)(Lxy ⊃ Lyx)
> **4.**   │ ∴ (y)(Lry ⊃ Lyr) [from 3]
> * **5.**   │ ∴ (Lrj ⊃ Ljr) [from 4]
> **6.**   └∴ Ljr [from 1 and 5, contradicting 2]
> **7.** ∴ ~(x)(y)(Lxy ⊃ Lyx) [from 3, 2, and 6]

Line 3 begins with back-to-back quantifiers. We can only drop a quantifier which is *initial* and hence *outermost*. So we have to drop the quantifiers one at a time, working from the outside. Our older proof strategy would tell us to drop *each* universal quantifier using *both* of the constants "r" and "j." But now we derive from the universal quantifiers only wffs likely to give us useful information for our proof or refutation.
    Here's another valid relational argument:

> There is someone that everyone loves.
> ∴ Everyone loves someone or other.

> (∃y)(x)Lxy
> ∴ (x)(∃y)Lxy

This argument is intuitively valid. If there is some one person (God, for example) that everyone loves, then everyone loves at least one person. Here's the proof:

> * **1.**   (∃y)(x)Lxy   VALID
>   [∴ (x)(∃y)Lxy
> * **2.**   │ asm: ~(x)(∃y)Lxy
> * **3.**   │ ∴ (∃x)~(∃y)Lxy [from 2]
> * **4.**   │ ∴ ~(∃y)Lay [from 3]
> **5.**   │ ∴ (y)~Lay [from 4]
> **6.**   │ ∴ (x)Lxb [from 1]
> **7.**   │ ∴ ~ Lab [from 5]
> **8.**   └∴ Lab [from 6, contradicting 7]
> **9.** ∴ (x)(∃y)Lxy [from 2, 7, and 8]

We handle the quantifiers in the normal way. We use the reverse-squiggle rules to get quantifiers into the initial position, drop initial quantifiers only, and drop existentials (using new constants) before universals.

Relational proofs raise interesting problems. Let's call a QC argument without identity or relations a *monadic* argument. Such arguments have two important features:

1. Our proof strategy of Section 4.6 leads in a mechanical way to the proof of any valid *monadic* argument and the refutation of any invalid *monadic* argument.

2. The refutation of an invalid *monadic* argument never requires a universe with an infinite number of entities. It's been proved that the refutation of an invalid *monadic* argument with n distinct predicates never requires a universe with more than 2^n distinct entities.

Neither feature holds for *relational* arguments. Contrary to feature 1, there's no possible mechanical strategy that always leads to a proof or refutation of a relational argument. This result is called Church's theorem, after Alonzo Church. It means that relational proofs and refutations sometimes require ingenuity and not just mechanical methods. It also means that our strategy can lead into an endless loop. And, contrary to feature 2, the refutation of invalid relational arguments sometimes requires a universe with an infinite number of entities.

Instructions lead into an *endless loop* if following them involves doing the same kind of thing over and over, endlessly. I've written computer programs with endless loops by mistake. These two glossary items produce an endless loop:

Endless loop. See entry for "loop, endless."

Loop, endless. See entry for "endless loop."

Our mechanical proof strategy of Section 4.6 can lead into such a loop. If you see this coming, quit the mechanical strategy! You then have to figure things out on your own.

Here's an identity argument that leads into such a loop:

∴ Not everything is identical to something or other.

    [∴ $\sim(y)(\exists x)y = x$ INVALID

  1.   asm: $(y)(\exists x)y = x$
* **2.**   ∴ $(\exists x)a = x$ [from 1]
  3.   ∴ $a = b$ [from 2]
* **4.**   ∴ $(\exists x)b = x$ [from 1]
  5.   ∴ $b = c$ [from 4]
* **6.**   ∴ $(\exists x)c = x$ [from 1]

....

We drop the universal quantifier in 1, using a new constant "a" (since there are no old ones) to get 2. Then we drop the existential quantifier in 2, using new constant "b" to get 3. We drop the universal quantifier in 1 again, using "b," to get 4. (Our mechanical strategy involves dropping universal quantifiers using *every* constant that appears in the proof.) Then we drop the existential quantifier in 4, using new constant "c" to get 5. And so on endlessly. This version is more graphic:

Assume: Everything is identical to something or other.

∴ a is identical to something or other.
  Call this thing "b."
∴ a is identical to b.
∴ b is identical to something or other.
  Call this thing "c."
∴ b is identical to c.
∴ c is identical to something or other.
....

To refute the argument, we need only a world with a single entity "a" that is identical to itself:

a

a = a

In this world, everything is identical to something or other—and hence the conclusion is false. We have to think up this world for ourselves. The strategy doesn't provide it automatically. Rather, the strategy leads into an endless loop.

Wffs beginning with a universal/existential quantifier combination, like "(y)(∃x)x = y" in this last argument, often lead into endless loops. Here's another example:

Everyone loves someone or other.
∴ There is someone that everyone loves.

1.   (x)(∃y)Lxy
    [∴ (∃y)(x)Lxy
2.   asm: ∼(∃y)(x)Lxy
    ....

Here, line 1 leads into an endless loop:

Line 1 Everyone loves someone or other.

∴ a loves someone or other.
  Call this person "b."
∴ a loves b.
∴ b loves someone or other.
  Call this person "c."
∴ b loves c.
∴ c loves someone or other.
....

(The assumption would lead us into a second endless loop!) Again, our strategy doesn't automatically give us a proof or refutation. We must improvise. With some ingenuity, we can construct this possible world which makes the premise true and the conclusion false:

a, b

| Laa, ~Lab | Egoistic |
| Lbb, ~Lba | World |

Here, each person loves only himself or herself. This makes "Everyone loves someone or other" true. But "Everyone loves a" and "Everyone loves b" are both false. So "There is someone that everyone loves" is false.*

When you try to refute an argument, try a universe with a small number of entities; often two entities is enough. Add statements to make the premises true and the conclusion false. Consider this example:

Everyone loves himself or herself.　　　(x)Lxx
∴ There is someone that everyone loves.　∴ (∃x)(y)Lyx

Let's try a universe with two entities, a and b. To make the premise true, we add "a loves a" and "b loves b." To make the conclusion false, we need to make sure that not everyone loves a (so we add "b doesn't love a") and that not everyone loves b (so we add "a doesn't love b"). Our refutation then looks like this:

a, b

| Laa, Lbb |
| ~Lba, ~Lab |

In this little world, everyone loves himself or herself, and yet there is no one that everyone loves. So our argument is invalid.

The refutation of a relational argument sometimes requires a universe with an *infinite* number of entities. Here's an example:

It is always the case that if x is greater than y and y is greater than z then x is greater than z.

It is always the case that if x is greater than y then y isn't greater than x.

b is greater than a.
∴ There is something than which nothing is greater.

---

* This refutation will also work:

a, b

| Lab, ~Laa | Altruistic |
| Lba, ~Lbb | World |

Here each person loves only the other. This makes "Everyone loves someone or other" true. But it also makes "There is someone that everyone loves" false.

$(x)(y)(z)((Gxy \cdot Gyz) \supset Gxz)$
$(x)(y)(Gxy \supset {\sim}Gyx)$
Gba
$\therefore (\exists x){\sim}(\exists y)Gyx$

In every world with a *finite* number of beings, if the premises are true then the conclusion is also true. Every finite universe has some being unsurpassed in greatness. But there could be an *infinity* of beings of ever increasing greatness. Then no being would be unsurpassed in greatness. So the argument is invalid.

We can't draw a box with an infinite number of entities to refute the argument. But we can refute it in another way. We can give an argument of the same form with true premises and a false conclusion. This shows the form to be invalid. Let's take the natural numbers (0, 1, 2, . . .) as the universe of discourse. Let "a" refer to 0 and "b" refer to 1 and "Gxy" mean "x > y." On this reinterpretation, the premises are all true. But the conclusion, which says "There is a number than which no number is greater," is false. This shows that the form is invalid.

So relational arguments raise problems involving infinity (endless loops and infinite worlds) that the other kinds of arguments we've studied don't raise.

5.4A Exercises

Say whether each is valid (then give a proof) or invalid (then give a refutation). Problems 12 and 14 are difficult.

0. $(\exists x)(\exists y)Lxy$
$\therefore (\exists y)(\exists x)Lxy$

     ANSWER: * 1. $(\exists x)(\exists y)Lxy$ VALID
                       $[\therefore (\exists y)(\exists x)Lxy$
           * 2.   asm: ${\sim}(\exists y)(\exists x)Lxy$
             3. $\therefore (y){\sim}(\exists x)Lxy$ [from 2]
           * 4. $\therefore (\exists y)Lay$ [from 1]
             5. $\therefore Lab$ [from 4]
           * 6. $\therefore {\sim}(\exists x)Lxb$ [from 3]
             7. $\therefore (x){\sim}Lxb$ [from 6]
             8. $\therefore {\sim}Lab$ [from 7, contradicting 5]
             9. $\therefore (\exists y)(\exists x)Lxy$ [from 2, 5, and 8]

1. $(x)(y)Lxy$	**2.** $(x)(y)(Lxy \supset x = y)$
$\therefore (\exists x)Lax$	$\therefore (x)Lxx$
3. $(x)Lxa$	**4.** $(\exists x)(y)Lxy$
$\therefore (x)Lax$	$\therefore (\exists x)Lxa$
5. $(x)Lxx$	**6.** $(x)(y)Lxy$
$\therefore (\exists x)(y)Lxy$	$\therefore (x)(y)((Fx \cdot Gy) \supset Lxy)$
7. $(x)Gaxb$	**8.** $(x)Lxx$
$\therefore (\exists x)(\exists y)Gxcy$	$\therefore (x)(y)(Lxy \supset x = y)$
9. Lab	**10.** $(x)(y)(Uxy \supset Lxy)$
Lbc	$(x)(\exists y)Uxy$
$\therefore (\exists x)(Lax \cdot Lxc)$	$\therefore (x)(\exists y)Lxy$

11. (∃x)Lxa
    ~Laa
∴ (∃x)(~x = a · Lxa)

12. (x)Lxa
    (x)(Lax ⊃ x = b)
∴ (x)Lxb

13. (x)(∃y)Lxy
∴ Laa

14. (x)(y)(Lxy ⊃ (Fx · ~Fy))
∴ (x)(y)(Lxy ⊃ ~Lyx)

15. (x)(y)(z)((Lxy · Lyz) ⊃ Lxz)
    (x)(y)(Lxy ⊃ Lyx)
∴ (x)Lxx

5.4B Exercises

For each argument, first evaluate intuitively and then translate into QC and work it out. Say whether each is valid (then give a proof) or invalid (then give a refutation).

0. Romeo loves either Lisa or Colleen.
Romeo doesn't love anyone who isn't Italian.
Colleen isn't Italian.
∴ Romeo loves Lisa.
[Use Lxy, r, l, and c.]
    ANSWER: * 1.   (Lrl ∨ Lrc)   VALID
              2.   (x)(~Ix ⊃ ~Lrx)
              3.   ~Ic
           [∴ Lrl
              4.   ⌈ ∴ asm: ~Lrl
        * 5. | ∴ (~Ic ⊃ ~Lrc) [from 2]
              6. | ∴ ~Lrc [from 3 and 5]
              7.   ⌊ ∴ Lrl [from 1 and 6, contradicting 4]
              8. ∴ Lrl [from 4 and 7]

1. Juliet loves everyone.
∴ Someone loves you.
[Use Lxy, j, and u]
2. Nothing caused itself.
∴ There is nothing that caused everything.
[Use Cxy.]
3. Everyone loves someone or other.
∴ Someone loves himself or herself.
[Use Lxy.]
4. Nothing caused itself.
This chemical brain process caused this pain.
∴ This chemical brain process isn't identical to this pain.
[Use Cxy, b, and p.]
5. Everyone loves my baby.
My baby doesn't love anyone besides me.
∴ My baby is me.
[Use Lxy, b, and m. Richard Jeffrey used this example.]

6. Romeo loves all females.
   No females love Romeo.
   Juliet is female.
   ∴ Romeo loves someone who doesn't love him.
   [Use Lxy, r, Fx, and j.]

7. Alice is older than Betty.
   Betty is older than Cheryl.
   ∴ Alice is older than Cheryl.
   [Use Oxy, a, b, and c. What further premise do we need?]

8. Everyone loves Juliet.
   ∴ There is someone that everyone loves.
   [Use Lxy and j.]

9. There is something that everything depends on.
   ∴ Everything depends on something or other.
   [Use Dxy.]

10. Everything depends on something or other.
    ∴ There is something that everything depends on.
    [Use Dxy. This is like arguing "Everyone lives in some house, so there must be some (one) house that everyone lives in." Some great minds have committed this fallacy. Aristotle argued "Every agent acts for an end, so there must be some (one) end for which every agent acts." St. Thomas Aquinas argued "If everything at some time fails to exist, then it must be that at some (one) time everything fails to exist." And John Locke argued "Everything is caused by something, so there must be some (one) thing that caused everything."]

11. It is always the case that if a first thing caused a second then the first exists before the second exists.
    Nothing exists before it exists.
    ∴ Nothing caused itself.
    [Use Cxy and Bxy (for "x exists before y exists").]

12. Alice is older than Betty.
    ∴ Betty isn't older than Alice.
    [Use Oxy, a, and b. What further premise do we need?]

13. There is someone that everyone loves.
    ∴ Someone loves himself or herself.
    [Use Lxy.]

14. Andy shaves all and only those who don't shave themselves.
    ∴ Andy isn't Andy.
    [Use Sxy and a. The "barber paradox" shows that the premise is self-contradictory. This relates to "Russell's paradox," which had a major impact on set theory. Russell's paradox showed that the following statement was self-contradictory (in spite of being provable by the axioms of set theory accepted at the time!): "There is a set A that contains all and only those sets that don't contain themselves."]

15. Everyone loves all lovers.
    Romeo loves Juliet.
    ∴ I love you.
    [Use Lxy, r, j, i, and u. This one is difficult.]

16. Not everyone loves everyone.
    ∴ Not everyone loves you.
    [Use Lxy and u.]

17. Every abortion kills some human fetus.
Every human fetus is a human being.
Anything that kills a human being is wrong.
∴ All abortions are wrong.
[Use Ax, Kxy (for "x kills y"), Fx (for "x is a human fetus"), Hx (for "x is a human being"), and Wx.]

18. All dogs are animals.
∴ All heads of dogs are heads of animals.
[Use Dx, Ax, and Hxy (for "x is a head of y"). Translate "x is a head of a dog" as "for some y, y is a dog and x is a head of y." Augustus DeMorgan in the nineteenth century claimed that this was a valid argument that traditional logic couldn't validate. This led to the logic of relations.]

19. [To prevent these last two from being very awkward, let's unofficially use time variables (t, t', t'', and so on) and time constants (u, u', u'', and so on).]
If everything is contingent, then there is some time at which everything fails to exist.
If there is some time at which everything fails to exist, then there is nothing in existence now.
There is something in existence now.
Everything that isn't contingent is necessary.
∴ There is a necessary being.
[Use Cx for "x is contingent," Ext for "x exists at time t," u for "now," and Nx for "x is necessary." This is part of St. Thomas Aquinas's third argument for the existence of God.]

20. For any contingent entity, there is some time at which it fails to exist.
∴ If everything is contingent, then there is some time at which everything fails to exist.
[Use Cx and Ext as in the previous argument. This is Aquinas's argument for the first premise of the previous argument. This one is difficult.]

5.5 DEFINITE DESCRIPTIONS

We'll call terms of the form "the so and so" *definite descriptions*, since they're meant to pick out a definite (single) person or thing. We've been translating definite descriptions using small constant letters. This is normally all right, but sometimes it leads to problems. Bertrand Russell suggested this alternative analysis of definite descriptions:

The A exists.
There is exactly one A.
For some x: x is A, and no one besides x is A.
$(\exists x)(Ax \cdot \sim(\exists y)(\sim y = x \cdot Ay))$

The A is B.
There is exactly one A, and it is B.
For some x: x is A, no one besides x is A, and x is B.
$(\exists x)((Ax \cdot \sim(\exists y)(\sim y = x \cdot Ay)) \cdot Bx)$

Russell claimed that his account helps to expose certain philosophical confusions. I'll explain one example. Then I'll give you some examples to work through (as an exercise). Consider this statement, which we might symbolize as "Bp":

> 1. The present king of France is bald.

Is "Bp" true or false? If "Bp" is true, then there is a present king of France. But, since France isn't a monarchy, there is no present king of France. So "Bp" isn't true. So "Bp" must be false. So it must be that the present king of France *isn't* bald ("~Bp"). But if the present king of France isn't bald, then he must have hair. But this doesn't seem right either.

On Russell's view, "The present king of France is bald" means this:

> 2. There is exactly one present king of France and this person is bald.

Statement 2 is false, since there is no present king of France. Now the falsity of statement 2 amounts to the truth of 3 [*not* the truth of 4]:

> 3. It's *not* the case that there is exactly one present king of France and this person is bald.

> 4. There is exactly one present king of France, and this person *isn't* bald.

The falsity of statement 2 *doesn't* entail the truth of statement 4. English is misleading here. If we deny statement 1, then we seem to be affirming "The present king of France *isn't* bald"—in other words, "The present king of France has hair." We resolve the problem by moving from the confusions of ordinary English to a more precise form of expression. Statement 1, understood precisely, means the same as statement 2. And denying 2 amounts to affirming 3 [*not* 4]. This in no way commits us to the existence of a hairy king over modern France.

Some logicians prefer a different approach to this puzzle. Some say that if there is no present king of France then "The present king of France is bald" is neither true nor false. Some think, too, that it's perfectly fine to translate any term of the form "the so and so" by a QC small constant letter—but only if the term refers to a single entity.

5.5A Exercises

You are to criticize each argument from the point of view of Russell's theory. First paraphrase out the word "the" in each argument. Then say what is wrong with the argument. (Maybe it's invalid or has a false premise or is question-begging.)

0. The perfect island either has palm trees, or it doesn't.
    If the perfect island has palm trees, then it exists.
    If the perfect island doesn't have palm trees, then it exists.
∴ The perfect island exists.

ANSWER: On Russell's view, "The perfect island doesn't have palm trees" is ambiguous. It might mean "There *is* exactly one perfect island and it *doesn't* have palm trees." This would make the first premise false, since this premise would then mean "There is exactly one perfect island and it has palm trees, or there is exactly one perfect island and it doesn't have palm trees." Alternatively, "The perfect island doesn't have palm trees" might mean that the following whole statement is false: "There is exactly one perfect island and it has palm trees." This meaning would make the third premise false, since this premise would mean that the falsity of "There is exactly one perfect island and it has palm trees" would imply the existence of exactly one perfect island. So in either case the argument has a false premise.

1. "The round square does not exist" is a true statement about the round square.
   If there is a true statement about something, then that something has to exist.
   ∴ The round square exists.
   But the round square isn't a real thing.
   ∴ Some things that exist aren't real things.
   [A philosopher named Meinong argued in this way that the realm of *existing* things extends much further than just the realm of *real* things. Russell was convinced of this view for a while. Later, Russell came to see the view as foolish and tried to expose the error of the reasoning by using his theory of descriptions.]
2. The perfect being has all positive perfections.
   Existence is a positive perfection.
   ∴ The perfect being (God) has existence.
   [René Descartes argued this way. He claimed that the first premise was true by definition. Just as a triangle by definition has three angles, so also the perfect being (God) by definition is the being with all positive perfections. He thought that the second premise (which claims that it's better to be than not to be) is also a necessary truth. So Descartes thought that the existence of the perfect being (God) is a matter of logical necessity. Russell also used his theory of descriptions against this argument.]
3. If the most perfect being conceivable doesn't exist, then this book (which *does* exist) is greater than the most perfect being conceivable.
   "This book is greater than the most perfect being conceivable" is self-contradictory and hence false.
   ∴ The most perfect being conceivable (God) exists.
   [St. Anselm's ontological argument for the existence of God went something like this. Russell attacked the argument using his theory of descriptions.]

6

Modal Logic

6.1 NECESSITY AND POSSIBILITY

Our *modal calculus* (MC) is a language for testing arguments involving necessity and possibility. MC builds on PC and QC and adds two further modal symbols:

SYMBOL	MEANING	NAME
$\square$	necessary	box
$\diamond$	possible	diamond

$\square$A   = It's *necessary* that A.
$\diamond$A   = It's *possible* that A.

We'll use "necessary" to mean "logically necessary." A *logically necessary* statement is one that's self-contradictory to deny. Alternatively, a *logically necessary* statement is one whose truth is based purely on logic, the meaning of concepts, and the necessary connections between properties.

Here are five examples of logically necessary statements:

1. $2 = 2$.
2. If it's raining, then it's raining.
3. All unmarried men are unmarried.
4. All bachelors are unmarried.
5. If this is entirely green, then it isn't entirely red.

We could prove statements 1, 2, and 3 using previous systems. Assuming one of these statements to be false (but not assuming further premises), we could derive a contradiction. Statement 4 is true by definition. Since "bachelor" means "unmarried man," statement 4 means the same as 3. Statement 5 is true because the property of being entirely green necessarily excludes the property of being entirely red.

People sometimes use "necessary" and "must" in senses other than "logically necessary." Here are some examples:

> "It's *necessary* to eat."

[This means that, if you don't eat, then you'll die. It doesn't mean that "You don't eat" is self-contradictory.]

> "You *must* keep your promises."

[This means that you ought to keep your promises. It doesn't mean that "You don't keep promises" is self-contradictory.]

> "It's *necessary* that unsupported objects fall."

[This means that, by a true law of nature, unsupported objects fall. It doesn't mean that "Unsupported objects don't fall" is self-contradictory.]

This chapter focuses on *logical necessity*. Later we'll develop similar systems for some other kinds of necessity.

Properly speaking, what is necessary is not a *sentence* but rather a *claim* or *idea* that we might use a sentence to express. In English, " = " could have been used to mean "≠." Then the sentence "2 = 2" would have expressed something false. But the claim that we *now* make using this sentence is a necessary truth. There's a certain idea that "2 = 2" (given current usage) expresses. And this idea itself has to be true (however we might express it).

We'll use these five terms as synonyms (although some philosophers introduce distinctions among them):

- necessary truth (or statement)
- logically necessary truth (or statement)
- logical truth (or statement)
- conceptual truth (or statement)
- analytic truth (or statement)

We'll also use "must be true" to mean "is necessary."

We'll use "possible" to mean "logically possible." A *logically possible* statement is one that isn't self-contradictory. Here are two examples:

> "It's raining."

> "I ran a mile in two minutes this morning."

The second example isn't *physically* possible. But it's *logically* possible, since it involves no self-contradiction. We'll use "A *could* be true," "A is *consistent*," and "A *isn't self-contradictory*" as different ways to say "A is *possible*."

In previous chapters, we sometimes used words like "must" and "can't" loosely—to emphasize, for example, the firmness of a claim. Now we'll take such words more strictly. We'll translate them using "□" and "◇."

6.2 MODAL FORMULAS AND POSSIBLE WORLDS

MC wffs are typographical strings constructable using the PC and QC rules, plus this new rule:

> The result of prefixing any wff with "□" or "◇" is itself a wff.

We'll begin with MC wffs that don't involve the QC symbols.

By our new rule, we don't add parentheses when we prefix a wff by "□" or "◇." So these four *aren't* wffs:

$$□(A) \qquad (□A) \qquad ◇(A) \qquad (◇A)$$

The correct forms are "□A" and "◇A." Similarly, these four aren't wffs:

$$\mathord{\sim}□(A) \qquad \mathord{\sim}(□A) \qquad \mathord{\sim}◇(A) \qquad \mathord{\sim}(◇A)$$

The correct forms are "∼□A" and "∼◇A."

Necessity and possibility are *modes* of truth (hence the term *modal logic*). "□A" is stronger than "A is true," and means "A *has to be true*." "◇A" is weaker than "A is true," and means "A *could be true*." This box compares the three claims:

□A	A	◇A
It's necessary that A.	A is true.	It's possible that A.
A is true in all possible worlds.	A is true in the actual world.	A is true in some possible worlds.
Strongest	Middle	Weakest

The box speaks of "possible worlds." A *possible world* is a consistent and complete description of how things might have been or might, in fact, be. Picture a possible world as a long (perhaps infinitely long) novel. The novel is *consistent*, in that its statements don't entail any self-contradictions. The novel describes a set of possible situations that are all possible together. The novel is *complete*, in that every possible fact or its denial is included. The novel may or may not be true. The *actual world* is the novel (or possible world) that's true—the complete description of how things, in fact, are.

A *necessary* statement (for example, "2 = 2") is one that's *true in all possible worlds*. A *true* statement (for example, "Gensler is a Michigan fan") is one that's *true in the actual world*. A *possible* statement (for example, "Gensler is an Ohio State fan") is one that's *true in some possible world*. A possible statement may or may not be true in the actual world.

There are two ways to express "impossible" in MC:

> A is *impossible*.
> = ~◇A [It *isn't possible* that A.]
> = □~A [It's *necessarily false* that A.]

An impossible statement (for example, "2 ≠ 2") is one that's false in every possible world.

MC expresses *entails* as "necessary if-then":

> A *entails* B.
> = It's necessary that if A then B.
> = □(A ⊃ B)

"A entails B" claims that "If A then B" is true in every possible world. "Entails" here means "logically entails" and is stronger than just "if-then" or "implies." This following claim is true (since I never teach logic on Thursdays):

> (T ⊃ ~L)   "*If* this is Thursday, *then* I don't
>   teach logic today."

But it just *happens* to be true. We could easily describe a possible world where it's Thursday and I'm teaching logic. So this claim is *false*:

> □(T ⊃ ~L)   " 'This is Thursday' (*logically*) *entails*
>   'I don't teach logic today.' "

"Entails" applies only when it's impossible for the first part to be true but the second false—as in this example:

> □(R ⊃ P)   " 'There's rain' (*logically*)
>   *entails* 'There's precipitation.' "

Every possible world that has rain has precipitation. Some logicians translate "□(A ⊃ B)" as "A *strictly implies* B" and "(A ⊃ B)" as "A *materially implies* B."

Here are some additional forms:

> A *doesn't entail* B.
> = It isn't necessary that if A then B.

= It isn't true in every possible world that if A then B.
= ~□(A ⊃ B)

A is *consistent* (*compatible*) with B.
= "A and B" is possible.
= In some possible world, A and B are both true.
= ◇(A · B)

A is *inconsistent* (*incompatible*) with B.
= "A and B" is impossible.
= In no possible world are A and B both true.
= ~◇(A · B)

A is a *contingent* (*synthetic*) *truth*.
= A is true but not-A is possible.
= A is true in the actual world but false in some possible world.
= (A · ◇~A)

A is a *contingent* (*synthetic*) *statement*.
= A is possible and not-A is possible.
= A is true in some possible world and false in some possible world.
= (◇A · ◇~A)

"Necessary" includes "possible" but excludes "contingent." "2 = 2" is necessary. It's also possible (= "not self-contradictory"). But it isn't contingent. A contingent statement is a possible statement whose denial is also possible. This chart may be helpful:

	Necessary	"2 = 2"	
			TRUE
Possible		"Gensler likes Michigan"	
	Contingent		
		"Gensler likes Ohio State"	
			FALSE
Impossible		"2 ≠ 2"	

6.3 MODAL TRANSLATIONS

When translating from English to MC, in general place the modal operator to mirror the English word order:

not necessary = ~□
necessary not = □~

"If," "both," and "either" often translate as "(":

necessary if = □(
if necessary = (□

Translate each "necessary" or "possible" into a separate box or diamond:

If A is necessary and B is possible, then C is possible.
= ((□A · ◇B) ⊃ ◇C)

    Certain English modal forms are ambiguous. You should translate these into *two* MC wffs, and say that the English could mean one or the other. Here's a common ambiguous form:

If A is true, then it's necessary that B.

This form could have either of these two meanings:

It's necessary that if A then B.	□(A ⊃ B)

If A is true, then B taken by itself is necessary.	(A ⊃ □B)

For example, "If there's rain, then it's necessary that there's precipitation" on the first interpretation would mean this:

It's logically necessary that if there's rain then there's precipitation.

□(R ⊃ P)

This *box-outside* form is true. Rain and precipitation are both contingent, but there's a necessary relation between the two. "If there's rain, then there's precipitation" is true in every possible world. The second interpretation is:

If there's rain, then "There's precipitation" (taken by itself) is logically necessary.

(R ⊃ □P)

This *box-inside* form is false. It says that, if "There's rain" is true in the *actual world*, then "There's precipitation" is true in *every possible world*. It's raining in Chicago, as I write this. So the antecedent "R" is true. But "There's precipitation" *isn't* logically necessary. Dry worlds are imaginable. So the consequent "□P" is false. Then the *box-inside* form is false—since the antecedent is true, but the consequent is false.

The medievals called the *box-outside* form "the necessity of the *consequence*." Here the connection between the parts is necessary. They called the *box-inside* form "the necessity of the *consequent*." Here the second part alone is necessary. I'll use the phrase "taken by itself" to indicate the *box-inside* form. So "If A is true, then B *taken by itself* is necessary" is "(A ⊃ □B)."

Here's another ambiguous form:

> If A is true, then it's impossible that B.

Again, this could have two meanings:

□(A ⊃ ∼B)
= It's necessary that if A then not-B.
= "If A then not-B" is true in all possible worlds.

(A ⊃ □∼B)
= (A ⊃ ∼◇B)
= If A is true, then B taken by itself is impossible.
= If A is true in the actual world, then B is false in every possible world.

The *box-outside* form asserts a *relative* necessity. What is necessary is the *connection* between the two parts—between A being true and B being false. But the *box-inside* form asserts an *inherent* necessity. Given that A is true, the falsity of B then is *inherently* necessary.

There's a similar ambiguity if we use "must" or "couldn't" in the second part of an *if-then* instead of "necessary" or "impossible." So these four forms are ambiguous:

- "*If* A is true, *then* B *must* be true."
- "*If* A is true, *then* it's *necessary* that B."
- "*If* A is true, *then* B *couldn't* be true."
- "*If* A is true, *then* it's *impossible* that B."

Later you'll see the importance of this ambiguity for various philosophical disputes. For now, just remember to translate an ambiguous *if-then* two ways.

6.3A Translation Exercises

Using these equivalences, translate each sentence into MC:

G = There's a God = God exists
E = There's evil = Evil exists
M = There's matter = Matter exists
R = There's rain
P = There's precipitation

Be sure to translate the ambiguous forms both ways.

0. "God exists and evil doesn't exist" entails "There's no matter."
ANSWER: □((G · ~E) ⊃ ~M)

1. It's (logically) necessary that God exists.
2. "There's a God" is self-contradictory.
3. It isn't necessary that there's matter.
4. It's necessary that there's no matter.
5. "There's rain" entails "There's precipitation."
6. "There's precipitation" doesn't entail "There's rain."
7. "There's no precipitation" entails "There's no rain."
8. If there's rain, then there must be rain.
9. If rain is necessary, then precipitation is necessary.
10. God exists.
11. It isn't possible that there's evil.
12. It's possible that there's no evil.
13. If there's rain, then it's possible that there's rain.
14. If there's rain, then there's matter.
15. "There's matter" is compatible with "There's evil."
16. "There's a God" is inconsistent with "There's evil."
17. Necessarily, if there's a God then there's no evil.
18. If there must be matter, then there's evil.
19. Necessarily, if there's a God then "There's evil" (taken by itself) is self-contradictory.
20. If there's a God, then there can't be evil.
21. It's necessary that either there's a God or there's matter.
22. Either it's necessary that there's a God or it's necessary that there's matter.
23. If precipitation is impossible, then rain is impossible.
24. "There's matter" is a contingent statement.
25. "There's rain" is a contingent truth.
26. "There's a God" isn't a contingent statement.
27. It's necessary that matter is possible.
28. If there's a God, then it must be that there's a God.
29. If there's a God, then "There's a God" (taken by itself) is necessary.
30. Necessarily it's the case that if there's a God then "There's a God" (taken by itself) is necessary.

6.4 PROOFS

Modal proofs work much like quantificational proofs. But we need to include possible worlds and four new inference rules. Consider this English version of a modal proof:

    1.   Necessarily, if there's matter then there's evil.
    2.   It's necessary that there's matter.
    [∴ It's necessary that there's evil.

3.   ⌈asm: It isn't necessary that there's evil.
4. | ∴ It's possible that there's no evil. [from 3]
5. | ∴ In some world-W: There's no evil. [from 4]
6. | ∴ In world-W: There's matter. [from 2]
7. | ∴ In world-W: If there's matter then there's evil. [from 1]
8.   ⌊∴ In world-W: There's evil. [from 6 and 7, contradicting 5]
9. ∴ It's necessary that there's evil. [from 3, 5, and 8]

Line 5 uses the principle that whatever is *possible* is true in *some* possible world. And lines 6 and 7 use the principle that whatever is *necessary* is true in *every* possible world. We get a contradiction in world-W. World-W contains "There's no evil" (line 5) and "There's evil" (line 8). So our argument is valid.

We must expand our proof mechanisms to allow such proofs. We define a *world prefix* to be a string of zero or more instances of "W." So " " (zero instances), "W," "WW," and so on are world prefixes. These represent possible worlds, with the blank world prefix (" ") representing the actual world.

A *derived step* from now on is a line consisting of a world prefix and then "∴" and then a wff. So these are now derived steps:

    ∴ P   ("So P is true in the actual world")
 W ∴ P   ("So P is true in world-W")
WW ∴ P   ("So P is true in world-WW")

Old derived steps with no W's before "∴" are still "derived steps" by the new definition. They use the blank world prefix. An *assumption* from now on is a line consisting of a world prefix and then "asm:" and then a wff. So these are now assumptions:

    asm: P   ("Assume that P is true in the actual world")
 W asm: P   ("Assume that P is true in world-W")
WW asm: P   ("Assume that P is true in world-WW")

Again, assumptions with no W's before "asm:" are still "assumptions" by the new definition. We'll often use derived steps with W's, but we'll seldom use assumptions with W's.

Unless specified otherwise, we can use an inference rule only within a given world. If we have "(A ⊃ B)" and "A" in the same world, then we can infer "B"— but only in this same world. We'll now take "→" in the inference rules to mean that, given each wff on the left as the wff of not-blocked-off lines *using the same world prefix*, we may infer a derived step consisting of *this same world prefix* (*unless otherwise noted*) followed by "∴" followed by any of the wffs on the right. "↔" means that this also holds for the right-to-left direction.

We'll use the S- and I-rules and RAA in MC proofs. We have to reword RAA to include world prefixes:

> *RAA* Suppose that some pair of not-blocked-off lines
> *using the same world prefix* have contradictory
> wffs. Then block off all the lines from the
> last not-blocked-off assumption on down. Infer
> a step consisting of *this assumption's world*
> *prefix followed by* "∴" followed by a contradictory
> of the wff of that assumption.

To apply RAA, lines with the *same* world prefix must have contradictory wffs. Having
"W ∴ A" and "WW ∴ ~A" isn't enough. "A" may well be true in one world but
false in another. But "WW ∴ A" and "WW ∴ ~A" provide a genuine contradiction.
The line derived using RAA must have the same world prefix as the assumption. If
"W asm: A" leads to a contradiction in any world at all, then by RAA we may derive
"W ∴ ~A."

The MC inference rules are much like the QC rules. M1 and M2 are *reverse*
squiggle rules. These hold regardless of what pair of contradictory wffs replaces
"A"/"~A":

M1 $\sim\Diamond A$ $\leftrightarrow$ $\Box\sim A$
M2 $\sim\Box A$ $\leftrightarrow$ $\Diamond\sim A$

By M1, "It isn't possible that $2 \neq 2$" = "It's necessarily false that $2 \neq 2$." By M2,
"It isn't necessary that there's evil" = "It's possible that there's no evil." Use these
rules only within the same world. We'll usually use them from left to right, to move
the modal operator to the outside:

So we usually go from "not possible" to "necessarily false," and from "not necessary"
to "possibly false."

M3 is the *box-dropping rule*. It holds regardless of what wff replaces "A":

M3 $\Box A$ $\rightarrow$ A   Here we can use *any* world prefixes (the
                                      same or different) in either line.

If $2 = 2$ is *necessary*, then it's true in any possible world we choose (including the
actual world). From "$\Box A$" we can derive "∴ A," "W ∴ A," "WW ∴ A," and so
forth:

> $\Box A$
> ―――――――――――
> W ∴ A   (use **any** string of W's, or no W's at all)

The box must *begin* the wff. If a squiggle or left-hand parentheses begins the wff, we
can't use M3:

$$\begin{array}{l} \sim\Box A \\ W \therefore \sim A \qquad [wrong!] \end{array}$$

Here we'd reverse the squiggle to go from ''$\sim\Box A$'' to ''$\Diamond\sim A$,'' and then use the diamond dropping rule. This one is also wrong:

$$\begin{array}{l} (\Box A \supset B) \\ W \therefore (A \supset B) \qquad [wrong!] \end{array}$$

We'd have to use the PC rules here. If we had ''$\Box A$,'' we could derive ''B.'' If we had ''$\sim B$,'' we could derive ''$\sim\Box A$.'' If we get stuck we might assume ''$\sim\Box A$.'' Only unwrap an outer (initial) box.

M4 is the *diamond-dropping rule*. It holds regardless of what wff replaces ''A'':

$$M4 \quad \Diamond A \quad \rightarrow \quad A \qquad \text{The derived line must use a world prefix}$$
consisting of a *new* string of W's (one
not occurring in earlier lines). The
deriving line can use any world prefix.

If ''There's no evil'' is possible, then it's true in some possible world. We may give this world an arbitrary, *and hence new*, name. The diamond must begin the wff and we must use a *new* world prefix:

$$\begin{array}{|l|} \hline \Diamond A \\ \hline W \therefore A \quad \text{(use a \textbf{new} string of W's)} \\ \hline \end{array}$$

Compare these two inferences:

$$\begin{array}{ll} \qquad\qquad \Diamond B & \qquad\qquad \Diamond B \\ {[wrong!]} \therefore B & W \therefore B \qquad [right!] \end{array}$$

The ''wrong'' inference uses the blank world prefix, which is always old (used in previous steps). It wrongly concludes that B is true from the premise that B is possible. The ''right'' inference uses world prefix ''W,'' which is new (not used in previous steps). Here's another pair:

$$\begin{array}{llll} & \qquad \Diamond A & \qquad \Diamond A & \\ & \qquad \Diamond B & \qquad \Diamond B & \\ {[right!]} & W \therefore A & \quad W \therefore A & [right!] \\ {[wrong!]} & W \therefore B & \quad WW \therefore B & [right!] \end{array}$$

The ''wrong'' inference here concludes that, since A and B are individually possible, they must be true in the *same* possible world. The ''right'' inference merely concludes that each is true in *some* (not necessarily the same) possible world. Use a new and different world prefix whenever you drop a diamond. A world prefix is ''new'' if it's never been used before in the proof. A world prefix once used becomes ''old.''

I call M4 the *Christopher Columbus rule*, after the one who said "It's *possible* to go to a *new* world!" This corny name might help you remember to use a *new* world when dropping "◇."

This modal proof is the symbolic version of the English proof given earlier:

1.	☐(M ⊃ E)	VALID
2.	☐M	
	[∴ ☐E	
⋆ **3.**	⌈asm: ~☐E	
⋆ **4.**	│∴ ◇~E [from 3]	
5.	│W ∴ ~E [from 4]	
6.	│W ∴ M [from 2]	
⋆ **7.**	│W ∴ (M ⊃ E) [from 1]	
8.	⌊W ∴ E [from 6 and 7, contradicting 5]	
9.	∴ ☐E [from 3, 5, and 8]	

After the assumption, we reverse the squiggle to move the operator to the outside (line 4). We drop the diamond, using new world-W (line 5). We drop the boxes, using world-W (lines 6 and 7). Then we apply the PC rules to get a contradiction (lines 5 and 8). RAA gives us the original conclusion (line 9).

We can star (and then ignore) a line when we use the reverse-squiggle or diamond-dropping rules on it. But we can't star a line when we use the box-dropping rule on it. We sometimes have to drop the box again using another world.

I suggest this strategy for getting a proof or refutation:

1. Block off the conclusion. Add "asm:" followed by its simpler contradictory to form the next line.

2. Use the S- and I-rules and RAA whenever you can. Use the same starring procedures as in the PC proofs.

3. *Reverse squiggles.* Apply M1 or M2 to each not-blocked-off line starting with "~◇" or "~☐." This will give you lines with initial modal operators. Star (and ignore) the lines used.

4. *Drop each initial diamond.* Apply M4 to each not-blocked-off line starting with "◇." Use a *new* world for each new line. Star (and ignore) the lines used. [Be sure to drop diamonds *before* dropping boxes.]

5. *Drop each initial box.* Apply M3 to each not-blocked-off line starting with "☐." Drop each "☐" *once for each old world*. (Include the actual world if it contains wffs without modal operators.) Don't star the lines used. You might have to use them again if more worlds pop into existence later in the argument.

6. Make additional assumptions on wffs of the form "(A ⊃ B)," "(A ∨ B)," or "~(A · B)." Do this only as a last resort, in order to use and star these wffs.

Eventually we'll reach a proof or refutation. We won't see invalid arguments or refutations until Section 6.5.

Here's another example of an MC proof:

Necessarily, if you don't decide then you decide.
∴ Necessarily you decide.

    1.   □(~D ⊃ D)   VALID
        [∴ □D
* **2.**   ⌜asm: ~□D
* **3.** | ∴ ◇~D [from 2]
  4. | W ∴ ~D [from 3]
* **5.** | W ∴ (~D ⊃ D) [from 1]
  6.   ⌞W ∴ D [from 4 and 5, contradicting 4]
  7. ∴ □D [from 2, 4, and 6]

After the assumption, we reverse a squiggle to get 3. We drop a diamond using new world-W to get 4 and drop a box using world-W to get 5. We get a contradiction between 4 and 6. Thus we prove the argument valid.

6.4A Exercises

Each of these is *valid*. Give a formal proof for each one.

0. □(A ⊃ B)
   ◇~B
∴ ◇~A

    ANSWER: 1.   □(A ⊃ B)   VALID
              * 2.   ◇~B
                 [∴ ◇~A
              * 3.   ⌜asm: ~◇~A
                4. | ∴ □A [from 3]
                5. | W ∴ ~B [from 2]
              * 6. | W ∴ (A ⊃ B) [from 1]
                7. | W ∴ A [from 4]
                8.   ⌞W ∴ B [from 6 and 7, contradicting 5]
                9. ∴ ◇B [from 3, 5, and 8]

    The jump from 3 to 4 might be confusing. We could reverse the squiggle in 3 in either of two ways:

$$\frac{\sim \Diamond \sim A}{\therefore\ \Box \sim \sim A} \qquad\qquad \frac{\sim \Diamond \sim A}{\therefore\ \Box A}$$

    On the left, we simply replace "~◇" with "□~" and end up with "~~" in the middle. Later we'd have to apply the double negation rule S-1. On the right, we collapse "~~" right away. This shortens the proof.

1. ◇(A · B)   **2.** A
∴ ◇A   ∴ ◇A
3. □(A ⊃ B)   **4.** ~◇(A · B)
   ◇A   ◇A
∴ ◇B   ∴ ~□B
5. ~◇(A · ~B)   **6.** □A
∴ □(A ⊃ B)   ∴ ◇A

7. □A
  ~□B
∴ ~□(A ⊃ B)
9. (◇A ∨ ◇B)
∴ ◇(A ∨ B)

8. (A ⊃ □B)
  ~B
∴ ◇~A
10. □(A ⊃ B)
∴ (□A ⊃ □B)

6.4B Exercises

Each of these arguments is *valid*. For each one, first read it over and see whether it's valid intuitively. Then translate it into MC and prove it valid.

0. "This is a square" entails "This is composed of straight lines."
"This is a circle" entails "This isn't composed of straight lines."
∴ "This is a square and also a circle" is logically self-contradictory.
[Use S, L, and C.]

ANSWER: 1.   □(S ⊃ L)   VALID
          2.   □(C ⊃ ~L)
            [∴ ~◇(S · C)
      * 3.   ┌asm: ◇(S · C)
      * 4.   │ W ∴ (S · C) [from 3]
      * 5.   │ W ∴ (S ⊃ L) [from 1]
      * 6.   │ W ∴ (C ⊃ ~L) [from 2]
        7.   │ W ∴ S [from 4]
        8.   │ W ∴ C [from 4]
        9.   │ W ∴ L [from 5 and 7]
       10.   └ W ∴ ~L [from 6 and 8, contradicting 9]
       11.   ∴ ~◇(S · C) [from 3, 9, and 10]

1. "You knowingly testify falsely because of threats to your life" entails "You lie."
"You knowingly testify falsely because of threats to your life" is consistent with "You don't intend to deceive."
∴ "You lie" doesn't entail "You intend to deceive."
[Use T, L, and I. From Tom Carson.]
2. Necessarily, if you don't decide then you decide not to decide.
Necessarily, if you decide not to decide then you decide.
∴ Necessarily, if you don't decide then you decide.
[Use D for "You decide" and N for "You decide not to decide." From Jean-Paul Sartre ("Not to decide is to decide"—"We are condemned to be free").]
3. "You do what you want" is consistent with "Your act is determined."
"You do what you want" entails "Your act is free."
∴ "Your act is free" is compatible with "Your act is determined."
[Use W, D, and F.]
4. There's a God.
There's evil in the world.
∴ "There's a God" is logically compatible with "There's evil in the world."
[Use G and E. Most people who think G and E incompatible wouldn't accept the first premise.]
5. "There's a God" is logically compatible with T.
T logically entails "There's evil in the world."

∴ "There's a God" is logically compatible with "There's evil in the world."

[Use G, T, and E. Here T is a theodicy (an explanation of why God might permit evil) which is consistent with G and entails E. T, for example, might be "God's goal in creating the world involves the significant use of human freedom to bring a half-completed world toward its fulfillment; moral evil results from the abuse of human freedom and physical evil from the half-completed state of the world." This T combines the "evil from free will" view of Augustine with the "evil needed for greater good" view of Iraneus. This argument (but not the particular T) is from Alvin Plantinga.]

6. God is omnipotent.

"You freely *always* do the right thing" is logically possible.

If "You freely *always* do the right thing" is logically possible and God is omnipotent, then it's possible for God to bring it about that you freely *always* do the right thing.

∴ It's possible for God to bring it about that you freely *always* do the right thing.

[Use O, F, and B. J. L. Mackie argued this way. He thought God had a third option besides making robots who always act rightly and making free beings who sometimes act wrongly. God could have made free beings who always act rightly.]

7. "God brings it about that you do A" is inconsistent with "You freely do A."

"God brings it about that you freely do A" entails "God brings it about that you do A."

"God brings it about that you freely do A" entails "You freely do A."

∴ It's impossible for God to bring it about that you freely do A.

[Use B, F, and X. This attacks the conclusion of the last argument. Be careful in your translation.]

8. "This is red, and there's a blue light that makes red things look violet to normal observers" entails "Normal observers won't sense redness."

"This is red, and there's a blue light that makes red things look violet to normal observers" is logically consistent.

∴ "This is red" doesn't entail "Normal observers will sense redness."

[Use R, B, and N. From Roderick Chisholm.]

9. "There are no truths" entails "It's true that there are no truths."

"It's true that there are no truths" entails "There are truths."

∴ "There are truths" is logically necessary.

[Use T for "There are truths" and X for "It's true that there are no truths."]

10. If truth is a correspondence with the mind, then "There are truths" entails "There are minds."

"There are truths" is logically necessary.

"There are minds" isn't logically necessary.

∴ Truth isn't a correspondence with the mind.

[Use C, T, and M. This attacks the correspondence theory of truth. The second premise is from the previous argument.]

6.5 REFUTATIONS

Applying our proof strategy on an *invalid* argument leads to a refutation:

Necessarily, if there's matter then there's evil.

It's *possible* that there's matter.

∴ It's *necessary* that there's evil.

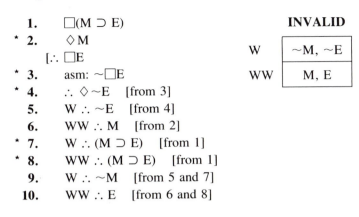

Here we have two diamonds to drop (lines 2 and 4). We use a new and different world each time (lines 5 and 6). We then drop the necessary thing (line 1) into both worlds (lines 7 and 8). We use the formulas in world-W to give us 9, and those in world-WW to give us 10. We reach no contradiction. (This requires contradictory wffs in the *same* world.) So we gather the simple pieces to provide a refutation.

A *refutation* in MC is a galaxy of possible worlds in which the premises are true but the conclusion is false. Here our refutation is a galaxy of two possible worlds. Both letters are false in world-W but true in world-WW. The first premise is true in this galaxy, since every world with matter has evil. The second premise is true, since some world has matter. But the conclusion is false, since there are worlds without evil. Our refutation shows the argument to be invalid.

Evaluate wffs starting with boxes or diamonds using these rules:

□A is true, if and only if A is true in *every* world of the galaxy.	◇A is true, if and only if A is true in *some* world of the galaxy.

Here the "□(M ⊃ E)" premise is true, since what it claims to be necessary is true in all possible worlds:

(M ⊃ E) is true in W, because (0 ⊃ 0) = 1.
(M ⊃ E) is true in WW, because (1 ⊃ 1) = 1.

And the "◇M" premise is true, since what it claims to be possible is true in at least one world:

M is true in WW.

The "□E" conclusion is false, since what it claims to be necessary isn't true in all the worlds:

E is false in W.

There's a galaxy of possible worlds making the premises true and the conclusion false. So the argument is invalid.

Suppose we were evaluating "(◇M ⊃ ~□E)" in this galaxy. We'd first evaluate the parts starting with modal operators. As before, ◇M = 1 and □E = 0. So "(◇M ⊃ ~□E)" is "(1 ⊃ ~0)," or "1."

Sometimes our galaxy contains the actual world. Consider this invalid argument and its refutation:

Necessarily, if there's matter then there's evil.
There's matter.
∴ It's necessary that there's evil.

1.	□(M ⊃ E)	**INVALID**		
2.	M			
	[∴ □E			M, E
* **3.**	asm: ~□E	W		~M, ~E
* **4.**	∴ ◇~E [from 3]			
5.	W ∴ ~E [from 4]			
* **6.**	W ∴ (M ⊃ E) [from 1]			
* **7.**	∴ (M ⊃ E) [from 1]			
8.	W ∴ ~M [from 5 and 6]			
9.	∴ E [from 2 and 7]			

This starts out as before. But here we put the necessary thing into old world-W (line 6) and *also* into the actual world (line 7). We put it into the actual world because we have wff "M" without modal operators in the actual world. Then we derive line 8 from the wffs in world-W, and line 9 from those in the actual world. We reach no contradiction. Instead we reach a refutation consisting of two possible worlds. Both letters are true in the actual world but false in world-W. In this galaxy, the first premise is true, since what it claims to be necessary is true in all the possible worlds (the actual world and world-W). The second premise is true in the galaxy, since "M" is true in the actual world. We evaluate a wff with no modal operators using its truth value in the actual world. The conclusion is false, since what it claims to be necessary isn't true in world-W. So in some galaxy of possible worlds the premises are true but the conclusion false. So the argument is invalid.

6.5A Exercises

Say whether valid (then give a proof) or invalid (then give a refutation). The first five are all invalid.

0. ◇A
◇B
∴ ◇(A · B)

W	A, ~B
WW	B, ~A

ANSWER: * 1.   ◇A
        * 2.   ◇B
               [∴ ◇(A · B)
        * 3.   asm: ~◇(A · B)
          4.   ∴ □~(A · B)   [from 3]
          5.   W ∴ A   [from 1]
          6.   WW ∴ B   [from 2]
        * 7.   W ∴ ~(A · B)   [from 4]
        * 8.   WW ∴ ~(A · B)   [from 4]
          9.   W ∴ ~B   [from 5 and 7]
          10.   WW ∴ ~A   [from 6 and 8]

Two things (for example, winning and losing) might each be possible separately without thereby being possible together.

1. ◇A
∴ □A

2. □(A ⊃ ~B)
B
∴ □~A

3. ◇A
~□B
∴ ~□(A ⊃ B)

4. □((A · B) ⊃ C)
◇A
◇B
∴ ◇C

5. □(A ∨ B)
∴ (□A ∨ □B)

6. (◇A ⊃ □B)
∴ □(A ⊃ B)

7. □((A · B) ⊃ C)
□A
∴ □(B ⊃ C)

8. □(C ⊃ (A ∨ B))
(~A · ◇~B)
∴ ◇~C

9. ◇A
∴ □◇A

10. (□A ⊃ □B)
∴ □(A ⊃ B)

6.5B Exercises

For each argument, first evaluate intuitively and then translate into MC and work it out. Say whether each is valid (then give a proof) or invalid (then give a refutation). Watch for ambiguous sentences such as "If A, then it must be that B." Translate ambiguous arguments into two different symbolic versions and test each for validity. Often one version is invalid while the other has false premises. The argument may be plausible only because of the ambiguity.

0. If I run, then it's necessary that I move.
I run.
∴ "I move" is a necessary truth (in other words, "I move" is true in all possible worlds).
[Use R and M. Boethius used this argument in the Middle Ages to explain the difference between the necessity of the *consequence* and the necessity of the *consequent*. This is the same as the *box-outside* versus *box-inside* distinction. Boxes hadn't yet been invented in Boethius's time.]

ANSWER: The first premise here is ambiguous. It could have either of these two meanings:

☐(R ⊃ M)

= It's necessary that if I run then I move.

= In all possible worlds "If I run, then I move" is true.

(R ⊃ ☐M)

= If I run, then "I move" (taken by itself) is necessary.

= If "I run" is true in the actual world, then "I move" is true in all possible worlds.

If we take the first premise as a *box-outside* form, then the argument is invalid:

1.	☐(R ⊃ M)	**INVALID**
2.	R	
	[∴ ☐M]	
* 3.	asm: ~☐M	
* 4.	∴ ◇~M [from 3]	
5.	W ∴ ~M [from 4]	
* 6.	∴ (R ⊃ M) [from 1]	
* 7.	W ∴ (R ⊃ M) [from 1]	
8.	∴ M [from 2 and 6]	
9.	W ∴ ~R [from 5 and 7]	

	R, M
W	~R, ~M

Our refutation has an actual world (in which I both run and move) and a possible world-W (in which I neither run nor move). Here the first premise is true, since in every world in which I run, I also move. The second premise is true, since in the actual world I run. But the conclusion is false, since I don't move in world-W. So the argument on this first interpretation is invalid. But if we take the first premise as a *box-inside* form, then the argument is valid (and already a formal proof, since the conclusion follows using one of our I-rules):

1. (R ⊃ ☐M)   VALID
2. R
3. ∴ ☐M   [from 1 and 2]

Here the first premise is *false*. It's false in the galaxy of possible worlds used to refute the first form of the argument, since then "R" is true and "☐M" is false. So neither form of the argument establishes the conclusion. The first form is invalid, while the second form has a false premise.

1. If truth is utility, then "P is true" entails "P is useful to believe."
   "P is true but not useful to believe" is consistent.
∴ Truth isn't utility.
   [Use U, T, and B.]
2. You have knowledge.
   "You're mistaken" is logically possible.
∴ "You have knowledge and you're mistaken" is logically possible.
   [Use H for "You *have* knowledge" and A for "You *are* mistaken."]

3. Necessarily, if this will be then this will be.

∴ If this will be, then it's necessary (in and of itself) that this will be.

[Use B. This illustrates the two senses of "Que será será"—"Whatever will be will be." The first sense is a truth of logic while the second is a form of fatalism.]

4. It's necessary that either I'll do it or I won't do it.

∴ Either it's necessary that I'll do it, or it's necessary that I won't do it.

[Use D for "I'll do it." Aristotle and the stoic Chrysippus mentioned this argument for fatalism. Chrysippus thought it was like arguing "Everything is either A or non-A; hence either everything is A or everything is non-A."]

5. Necessarily, if God brings it about that A then A.

A is a self-contradiction.

∴ It's impossible for God to bring it about that A.

[Use A and B, where B is for "God brings it about that A."]

6. If this is experienced, then this must be thought about.

"This is thought about" entails "This is put into the categories of judgments."

∴ If it's possible for this to be experienced, then it's possible for this to be put into the categories of judgments.

[Use E, T, and C. This one, on the more plausible interpretation, is from Immanuel Kant.]

7. Necessarily, if you mistakenly think that you exist then you don't exist.

Necessarily, if you mistakenly think that you exist then you exist.

∴ "You mistakenly think that you exist" is impossible.

[Use M and E. This relates to Descartes's "I think, therefore I am" ("Cogito ergo sum").]

8. It's necessarily true that if you're morally responsible for your actions then you're free.

It's necessarily true that if your actions are uncaused then you aren't morally responsible for your actions.

∴ "You're free" doesn't entail "Your actions are uncaused."

[Use R, F, and U. From A. J. Ayer. By "uncaused" he meant "not determined."]

9. Necessarily, if formula F has an all-1 truth table then F is true.

∴ If formula F has an all-1 truth table, then F (taken by itself) is necessary.

[Use A and T. This illustrates the *box-outside* versus *box-inside* distinction.]

10. If "One's conscious life won't continue forever" entails "Life has no meaning," then "One's conscious life will continue forever" entails "Life is eternally absurd."

Necessarily, if life is eternally absurd then life has no meaning.

∴ If it's possible for life to have a meaning, then "One's conscious life won't continue forever" doesn't entail "Life has no meaning."

[Use C, M, and A.]

11. "I seem to see a chair" doesn't entail "There's some actual chair that I seem to see."

If people sometimes directly perceive material objects, then "I seem to see a chair and there's some actual chair that I seem to see" is consistent.

∴ People never directly perceive material objects.

[Use S, A, and D. This attacks direct realism.]

12. If you act on your decision to study, then it's necessary that you'll study.

If it's necessary that you'll study, then you aren't free.

∴ If you act on your decision to study, then you aren't free.

[Use A, S, and F. This is an implausible attack on free will.]

13. "There's a God" is logically incompatible with "There's evil in the world."

There's evil in the world.

∴ "There's a God" is self-contradictory.
[Use G and E.]

14. If you have money, then it couldn't be that you're broke.
It could be that you're broke.
∴ You don't have money.
[Use H for "You *have* money" and A for "You *are* broke." Is this argument just a valid instance of *modus tollens*: "(P ⊃ Q), ~Q ∴ ~P"?]

15. If you have knowledge, then it couldn't be that you're mistaken.
It could be that you're mistaken.
∴ You don't have knowledge.
[Use H for "You *have* knowledge" and A for "You *are* mistaken." This is a classic and seemingly powerful argument for skepticism. Given that we could always be mistaken, it seems to follow that we never have knowledge. But the previous argument is far less plausible.]

16. "Every married bachelor is married" is a necessary truth.
"Every married bachelor is married" entails "Some married bachelor is married."
"Some married bachelor is married" entails "Some bachelor is married."
∴ "Some bachelor is married" is a necessary truth.
[Use A for "Every married bachelor is married," B for "Some married bachelor is married," and C for "Some bachelor is married." Many traditional logic books teach that "All S is P" entails "Some S is P," and thus seem committed to the second premise. The other two premises seem true. But the conclusion is clearly wrong.]

17. If thinking is just a chemical brain process, then "I think" entails "There's a chemical process in my brain."
"I think but I don't have a body" is logically consistent.
"There's a chemical process in my brain" entails "I have a body."
∴ Thinking isn't just a chemical brain process.
[Use J, T, C, and B. This attacks a form of materialism.]

18. If "I did that on purpose" entails "I made a prior purposeful decision to do that," then there's an infinite regress of purposeful decisions to decide.
It's impossible for there to be an infinite regress of purposeful decisions to decide.
∴ "I did that on purpose" is consistent with "I didn't make a prior purposeful decision to do that."
[Use D, P, and I. From Gilbert Ryle.]

19. God knew that you'd do it.
If God knew that you'd do it, then it was necessary that you'd do it.
If it was necessary that you'd do it, then you weren't free.
∴ You weren't free.
[Use K, D, and F. This argument is the focus of an ancient controversy. Would divine foreknowledge preclude human freedom? If it would, then should we reject human freedom (as did Luther)? Or should we reject divine foreknowledge (as does Charles Hartshorne)? Or perhaps (as the medieval thinkers Boethius, Thomas Aquinas, and William of Ockham all claimed) is the argument that divine foreknowledge precludes human freedom fallacious?]

20. It was always true that you'd do it.
If it was always true that you'd do it, then it was necessary that you'd do it.
If it was necessary that you'd do it, then you weren't free.
∴ You weren't free.
[Use T (for "It was always true that you'd do it"; don't use a box here), D, and F. This

argument is much like the former one. Are statements about future contingencies (for example, "I'll brush my teeth tomorrow") true or false before they take place? Should we do truth tables for such statements in the normal way, assigning them "1" or "0"? Does this somehow preclude human freedom? If so, should we then reject human freedom? Or should we claim that statements about future contingencies aren't "1" or "0" but must instead have some third truth value (for example, "½")? Or is the argument perhaps fallacious?]

6.6 S5 AND GALACTIC TRAVEL

Logicians seldom disagree on whether arguments are valid. But they disagree on some modal arguments. Conflicting systems reflect these disagreements. Now we'll study four such systems which are called (for historical reasons that we won't go into) *S5*, *S4*, *B*, and *T*.

These two questions are controversial:

1. What are rules for galactic travel? Suppose that "□A" is true in a world. Does it follow that "A" is true in every other world?
2. What rules govern strings like "◇◇," "□□," "◇□," and "□◇"? For example, does "◇◇A" entail "◇A"? Suppose that it's *possible* that it's *possible* that A. Does it follow that A itself is *possible*?

The two questions connect closely. If we answer one, we implicitly answer the other. We'll start with the first question.

The four systems (S5, S4, B, and T) all accept the reverse squiggle rules (M1 and M2) and the diamond dropping rule (M4). But they differ on how to formulate the box dropping rule (M3). In Section 6.4, we worded this rule as follows:

> *M3*   □A → A   Here we can use *any* world prefixes (the same or different) in either line.

This rule accords with system S5. It allows unlimited galactic travel—from *any* world to *any* world. If "□A" is true in any world at all, we can put "A" in the same world or in any other world. Something is a necessary truth in one world, if and only if it's true in all possible worlds. This implies that all worlds have the same necessary truths.

All four systems let us go from "□A" to "A" in the same world. But the weaker systems S4, B, and T restrict galactic travel between worlds. We can't always go from "□A" in one world to "A" in another world. Traveling between worlds requires a suitable "travel ticket."

We get travel tickets when we drop diamonds. Let "W1" and "W2" stand for any world prefixes. Suppose we go from "◇A" in world-W1 to "A" in a *new* world-W2. Then we get a free travel ticket from W1 to W2. To record this, we'll write "W1 → W2":

W1 → W2
We have a ticket to move from world-W1 to world-W2.

Here's an example. Suppose we are doing a proof with the wffs "◇◇A and "◇B." We'd get these travel tickets (where "#" stands for the actual world):

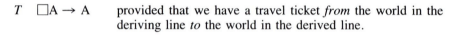

```
*  1.      ◇◇A                        TRAVEL TICKETS
*  2.      ◇B
      .........................................................
* 11.      W ∴ ◇A    [from 1]          # → W
  12.      WW ∴ A    [from 11]         W → WW
  13.      WWW ∴ B    [from 2]         # → WWW
```

Dropping a diamond gives us a free travel ticket from the world in the "from" step to the world in the "to" step. In line 11, for example, we get ticket "# → W"— because we've moved *from* "◇◇A" in the actual world ("#") *to* "◇A" in world-W.

In all four systems, tickets are reusable; we can use "W1 → W2" any number of times. But the rules for using tickets vary. Some systems let us combine tickets or use them in both direction. System T allows neither. With T, we can only use one ticket at a time, and we can only use it in the direction of the arrow. We can travel from W1 to W2 (when we drop boxes), if and only if we have a ticket going *from* W1 *to* W2:

> *T*   □A → A   provided that we have a travel ticket *from* the world in the deriving line *to* the world in the derived line.

So "W1 → W2" only lets us go from W1 to W2. T is the minimal "TICKET system."

The "BOTH directions" system B lets us use one ticket at a time—but we can use it in both directions:

> *B*   □A → A   provided that we have a ticket (in either direction) *between* the worlds in the two lines.

So "W1 → W2" lets us go from W1 to W2—or from W2 to W1.

The "SERIES of tickets" systems S4 and S5 let us combine a series of tickets. The weaker S4 still makes us follow the direction of the arrow:

> *S4*   □A → A   provided that we have a ticket or *series* of tickets *from* the world in the deriving line *to* the world in the derived line.

Given "W1 → W2" and "W2 → W3," we can go from W1 to W3.

The stronger S5 lets us combine tickets and use them in both direction:

> *S5*   □A → A   provided that we have a ticket or *series* of tickets *between* the worlds in the two lines. (The direction of the tickets doesn't matter.)

In previous sections, we used M3 as the rule for dropping boxes. M3 allowed unlimited travel between worlds—so that if we had "□A" in any world, then we could put

"A" in the same world or in any other world. S5 and M3 are equivalent in that they lead to the same arguments being provable.

This table sketches how the four systems let us use tickets:

SYSTEM	Can we use tickets in BOTH direction?	Can we combine a SERIES of tickets?
T	NO	NO
B	YES	NO
S4	NO	YES
S5	YES	YES

When we drop boxes, system T lets us use only one ticket at a time and only in the direction of the arrow. Suppose that we have these three travel tickets:

$\# \to$ W	W $\to$ WW	$\# \to$ WWW

Then system T only lets us travel from # to W, from W to WW, or from # to WWW. The other three systems allow these movements—and more. System B lets us use single tickets in both directions, so we can also go from W to #, from WW to W, and from WWW to #. System S4 lets us use a series of tickets in the direction of the arrow; this lets us go from # to WW (using "$\# \to$ W" and "W $\to$ WW" together). System S5 lets us travel from any world to any other world.

S5 is the most liberal, allowing the greatest number of arguments to be proved valid. So S5 is the *strongest* system. T is the *weakest* system, allowing the fewest proofs. B and S4 are *intermediate* systems, and each allows some proofs that the other doesn't.

The four systems give the same result for most arguments. But a few arguments are valid in one system but invalid in another. These arguments all involve wffs that apply a modal operator to a wff already containing a modal operator.

This argument is valid in S4 or S5 but invalid in T or B:

1. $\Box$A
   [∴ $\Box\Box$A
* 2. ⌈asm: ~$\Box\Box$A
* 3. | ∴ $\Diamond$~$\Box$A   [from 2]
* 4. | W ∴ ~ $\Box$A   [from 3] $\# \to$ W
* 5. | W ∴ $\Diamond$~A   [from 4]
6. | WW ∴ ~A   [from 5]   W $\to$ WW
7. ⌊WW ∴ A   [from 1, contradicting 6]   *NEED S4 OR S5*
8. ∴ $\Box\Box$A   [from 2, 6, and 7]

Step 7 requires that we combine a series of tickets in the direction of the arrow. Tickets "$\# \to$ W" and "W $\to$ WW" then let us move from "$\Box$A" in the actual world (#) to "A" in world-WW.

This next one is valid in B or S5 but invalid in T or S4:

   1.   A
        [∴ □◇A
* **2.** ⌐asm: ~□◇A
* **3.** │ ∴ ◇~◇A   [from 2]
* **4.** │ W ∴ ~ ◇A   [from 3]   # → W
   5. │ W ∴ □~A   [from 4]
   6. └ ∴ ~A   [from 5, contradicting 1]   *NEED B OR S5*
   7. ∴ □◇A   [from 2, 1, and 6]

Step 6 requires using a single ticket in the opposite direction of the arrow. Ticket "# → W" then lets us move from "□~A" in world-W to "~A" in the actual world (#).

This last one is valid in S5 but invalid in T or B or S4:

* **1.**   ◇A
        [∴ □◇A
* **2.** ⌐asm: ~□◇A
* **3.** │ ∴ ◇~◇A   [from 2]
* **4.** │ W ∴ ~◇A   [from 3]   # → W
   5. │ W ∴ □~A   [from 4]
   6. │ WW ∴ A   [from 1]   # → WW
   7. └ WW ∴ ~A   [from 5, contradicting 6]   *NEED S5*
   8. ∴ □◇A   [from 2, 6, and 7]

Step 7 requires that we combine a series of tickets and use them in both directions. Tickets "# → W" and "# → WW" then let us move from "□~A" in world-W to "~A" in world-WW.

S5 is the simplest system in several ways:

1. We can formulate S5 more simply. The box dropping rule doesn't have to mention travel tickets. We need only say that, if we have "□A" in *any* world, then we can put "A" in *any* world (the same or a different one).
2. S5 expresses some simple intuitions about necessity and possibility. "□A" is true in *any* world, if and only if "A" is true in *all* worlds. And "◇A" is true in *any* world, if and only if "A" is true in *some* world. So if "□A" or "◇A" is true in *any* world, then it's true in *all* worlds. What is necessary or possible doesn't vary from world to world.
3. S5 lets us simplify any string of modal operators. "□□A" is logically equivalent to "□A," "◇◇A," to "◇A," "□◇A" to "◇A," and "◇□A" to "□A." For any string of boxes or diamonds, we can substitute the last element of the string. So we don't have to worry about the difference between "□□◇ ◇□◇ ◇A" and "◇A." Here the second is just a simpler version of the first.

Which is the best system? This depends on what we take the box and diamond to mean. If we interpret the box and diamond to be about the *logical necessity and possibility of ideas*, then S5 is the best system. If an idea (for example, the claim that 2 = 2) is logically necessary, then it couldn't have been other than logically necessary. So if A is logically necessary, then it's logically necessary that A is logically necessary ["($\Box$A $\supset$ $\Box\Box$A)"]. Similarly, if an idea is logically possible, then it's logically necessary that it's logically possible ["($\Diamond$A $\supset$ $\Box\Diamond$A)"]. Of the four systems, only S5 accepts both these formulas. S5 contains truths that the other systems don't. All this presupposes that we use the box to talk about the *logical necessity of ideas*.

We could interpret the modal operators in other ways. We might take the box to be about the *logical necessity of sentences*. Now the *sentence* "2 = 2" just happens to express a necessary truth; it wouldn't have expressed one if English had used " = " to mean "≠." So the *sentence* is necessary, but it isn't necessary that it's necessary; this makes "($\Box$A $\supset$ $\Box\Box$A)" false. [But the *idea* that "2 = 2" *now* expresses is both necessary and necessarily necessary—and a change in how we use language wouldn't make this *idea* false.] So whether S5 is the best system can depend on whether we take the box to be about the necessity of *ideas* or of *sentences*.

There are still other ways to take the term "necessary" (see Section 6.1). Calling something "necessary" in a given context might mean that it's "physically necessary," or "proved," or "known," or "obligatory," or something else. Some logicians advocate weak systems like T because such weak systems hold for practically *any* sense of "necessary." Such logicians, however, might use S5 for arguments involving the logical necessity of ideas. I have some sympathy with this view. But most of the modal arguments that I'm interested in are about the logical necessity of ideas. So I find it convenient to use S5 as the standard system of modal logic. But we might have to switch to weaker systems for arguments about other sorts of necessity.

Here we've considered the four main modal systems. We could invent other systems—for example, ones in which we can combine travel tickets only in groups of three. Logicians develop such systems, not to help us in analyzing real arguments, but rather to explore interesting formal structures.*

6.6A Exercises

Each of these arguments is *valid* in system S5. Give a formal proof for each one. Also say if the argument is valid in systems T, B, or S4.

0. ~$\Box$A
∴ $\Box$~$\Box$A

ANSWER: * 1.   ~$\Box$A   VALID
                 [∴ $\Box$~$\Box$A
    * 2.   ⌐asm: ~$\Box$~$\Box$A
    * 3.   │ ∴ $\Diamond\Box$A   [from 2]
    * 4.   │ ∴ $\Diamond$~A   [from 1]
      5.   │ W ∴ $\Box$A   [from 3]   # → W
      6.   │ WW ∴ ~A   [from 4]   # → WW
      7.   └WW ∴ A   [from 5, contradicting 6]   *NEED S5*
      8.   ∴ $\Box$~$\Box$A   [from 2, 6, and 7]

---

* For more on such systems, consult G. E. Hughes and M. J. Cresswell, *An Introduction to Modal Logic* (London: Methuen, 1968).

Step 7 requires combining a series of tickets and using them in both directions. This demands system S5.

1. ◇□A
∴ A

2. ◇A
∴ ◇◇A

3. ◇◇A
∴ ◇A

4. ◇□A
∴ □A

5. (□A ⊃ □B)
∴ □(□A ⊃ □B)

6. □(A ⊃ B)
∴ □(□A ⊃ □B)

7. (◇A ⊃ □B)
∴ □(A ⊃ □B)

8. □(A ⊃ □B)
∴ (◇A ⊃ □B)

9. ◇□◇A
∴ ◇A

10. ◇A
∴ ◇□◇A

11. □A
∴ □(B ⊃ □A)

12. □◇□◇A
∴ □◇A

13. □◇A
∴ □◇□◇A

14. □(A ⊃ □B)
◇A
∴ □B

15. □A
∴ □□□A

6.6B Exercises

Each of these arguments is *valid* in system S5. For each one, first read it over and see if it's valid intuitively. Then translate it into MC and give a formal proof. Also say if the argument is valid in systems T, B, or S4.

0. Necessarily it's the case that if there's a necessary being then "There's a necessary being" (taken by itself) is a necessary truth.
"There's a necessary being" is logically possible.
∴ There's a necessary being.
[Use N for "There's a (logically) necessary being." A *logically necessary being* is a being which, if it exists at all, exists of logical necessity. This makes the first premise true by definition. This argument is Charles Hartshorne's version of St. Anselm's second ontological argument for the existence of God (identified with the necessary being).]

ANSWER: 1.   □(N ⊃ □N)   VALID
     * 2.   ◇N
       [∴ N
     3.   ⌐asm: ~N
     4.   W ∴ N   [from 2]   # → W
     * 5.   W ∴ (N ⊃ □N)   [from 1]
     6.   W ∴ □N   [from 4 and 5]
     7.   ∴ N   [from 6, contradicting 3]   *NEED S5 OR B*
     8. ∴ N   [from 3 and 7]

Step 7 requires that we use a single ticket in the opposite direction of the arrow. This demands systems S5 or B.

The validity of this nonintuitive argument depends on which system of modal logic is correct. Some philosophers defend the argument (often after defending S5 or B). Others argue that, since the argument is invalid, systems S5 and B, which would make it valid, must be wrong. Others reject one of the premises. Still others deny the theological import of the conclusion. They say that a necessary being could be a prime number or the world and need not be God.

Another version uses "God" in place of "necessary being":

Necessarily it's the case that if there's a God then "There's a God" (taken by itself) is a necessary truth. [God couldn't exist contingently.]

"There's a God" is logically possible.

∴ There's a God.

1. "There's a necessary being" isn't a contingent proposition.

"There's a necessary being" is logically possible.

∴ There's a necessary being.

[Use N. This reformulation of the Anselm-Hartshorne argument is clearly valid. Does it depend on S5 or B?]

2. Prove that the first premise of argument 0 is logically equivalent to the first premise of argument 1. You can prove that two statements are logically equivalent by first deducing the second from the first, and then deducing the first from the second. In which systems does this equivalence hold?

3. Necessarily it's the case that if there's a necessary being then "There's a necessary being" (taken by itself) is a necessary truth.

"There's *no* necessary being" is logically possible.

∴ There's *no* necessary being.

[Use N. Some object that the first premise of the Anselm-Hartshorne argument just as easily leads to the opposite conclusion.]

4. It's necessary that 2 + 2 = 4.

It's possible that no language ever existed.

If necessary truths hold because of language conventions, then "It's necessary that 2 + 2 = 4" entails "Some language has sometime existed."

∴ Necessary truths don't hold because of language conventions.

[Use T, L, and N. This attacks the linguistic theory of logical necessity.]

5. Necessarily, if "possible" means "imagined to be true" and it's possible that it's raining, then there are imaginers.

It isn't logically necessary that there are imaginers.

It's possible that it's raining.

∴ It isn't logically necessary that "possible" means "imagined to be true."

[Use M, R, and I. From Paul Moser and Arnold VanderNat.]

6.7 QUANTIFIED MODAL TRANSLATIONS

In this section and the next, we'll develop a quantified modal calculus (QMC). We'll call this system our "naïve QMC," since it ignores certain problems. Later we'll bring in refinements and have a "sophisticated QMC." My basic understanding of quantified modal logic follows that of Alvin Plantinga's *The Nature of Necessity* (London: Oxford University Press, 1974).*

* For related discussions, see A. N. Prior's article on "Logic, Modal" in the *Encyclopedia of Philosophy* and Kenneth Konyndyk's *Introductory Modal Logic* (Notre Dame, Indiana: Notre Dame Press, 1986).

Many QMC translations follow the patterns that we've already learned. For example, we translate "everyone" into a universal quantifier following the English word order:

> It's possible for *everyone* to be above average. (FALSE)
> = It's possible that, for all x, x is above average.
> = ◇(x)Ax

But "anyone" (regardless of where it occurs) translates into a universal quantifier at the *beginning* of the wff:

> It's possible for *anyone* to be above average. (TRUE)
> = For all x, it's possible that x is above average.
> = (x)◇Ax

QMC can express the difference between *necessary* and *contingent* properties. Numbers have both kinds of properties. The number 8, for example, has the necessary properties of *being even* and of *being one greater than seven*. These are properties that 8 couldn't have lacked. But 8 also has contingent properties, such as *being one of my favorite numbers* and *being less than the number of chapters in this book*. These are properties that 8 has but might have lacked. We can symbolize "necessary property" and "contingent property" as follows:

> □Fx = F is a *necessary* (*essential*) *property* of x.
> = x has the property of being necessarily F.
> = x has the necessary property of being F.
> = x is necessarily F.
> = In all possible worlds, x would be F.

> (Fx · ◇~Fx) = F is a *contingent* (*accidental*) *property* of x.
> = x is F—but could have lacked F.
> = In the actual world x is F; but in some possible world x isn't F.

Human beings have mostly contingent properties. Socrates had contingent properties such as *having a beard* and *being a philosopher*. These are contingent, because Socrates could (without self-contradiction) have been a clean-shaven nonphilosopher. But Socrates also had some necessary properties, such as *being self-identical* and *not being a square circle*. These two are properties that every being has of necessity.

Are there properties that some beings have of necessity but that some other beings lack? *Aristotelian essentialism* is the controversial view that there are such properties. Plantinga, in support of this view, suggests that Socrates had these properties (that some other beings lack) of necessity:

- not being a prime number
- being snubnosed in W (a specific possible world)

- being a person
- being capable of conscious activity
- being identical with Socrates

The last property differs from that of *being named "Socrates."*

Plantinga explains "necessary property" as follows. Suppose that "A" *names* a being and "F" *names* a property. Then the entity named by "A" has the property named by "F" necessarily, if and only if the proposition expressed by "A is non-F" is logically impossible. Then to say that Socrates necessarily has the property of not being a prime number is to say that the proposition "Socrates is a prime number" (with the name "Socrates" referring to the person Socrates) is logically impossible. It's important that we use *names* (for example, "Socrates") here and not *definite descriptions* (for example, "the entity I'm thinking about").

We've talked before about the ambiguity between *box-outside* and *box-inside* forms. Here's an ambiguous QMC statement:

> All persons are necessarily persons.

The statement could mean either of these:

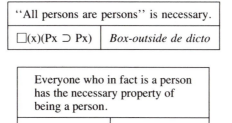

"All persons are persons" is necessary.	
□(x)(Px ⊃ Px)	*Box-outside de dicto*

Everyone who in fact is a person has the necessary property of being a person.	
(x)(Px ⊃ □Px)	*Box-inside de re*

The medievals distinguished necessity *de dicto* (" of the saying") from necessity *de re* ("of the thing"). The form in the middle box attributes necessity to the *proposition* (or saying) "All persons are persons." The claim here is trivially true. In the last box, the necessary property of being a person is attributed to each *person*. This last claim is more controversial. If this claim is true, you couldn't have been a nonperson. Then it's self-contradictory to imagine yourself existing but not being a person. This would exclude, for example, the possibility of your being reincarnated as an unconscious doorknob.

Watch for this ambiguous form:

> All A's are necessarily B's.

When you translate it, say that it could mean either of these:

□(x)(Ax ⊃ Bx)
= It's necessary that all A's are B's.

$(x)(Ax \supset \Box\, Bx)$
= All A's have the necessary property of being B's.

As before, various philosophical fallacies result from confusing the two forms.

6.7A Exercises

Translate each sentence into MC. Be sure to translate the ambiguous forms both ways.

0. It's necessary that all mathematicians have the property of being necessarily rational.
ANSWER: $\Box(x)(Mx \supset \Box Rx)$

This has boxes outside and inside. The first box symbolizes *de dicto* necessity ("It's necessary that . . . .") while the second symbolizes *de re* necessity ("have the property of being necessarily . . . .").

1. All bachelors are necessarily unmarried.
2. It's necessary that all bachelors are unmarried.
3. All bachelors have the necessary property of being unmarried.
4. Being Greek is a contingent property of Socrates.
5. Being identical to Socrates is a necessary property of Socrates.
6. Being named "Socrates" is a contingent property of Socrates.
7. Mathematicians are necessarily rational.
8. Mathematicians aren't necessarily two-legged.
9. Everyone that we observe to be sitting is necessarily sitting.
10. All numbers have the property of being necessarily abstract entities.
11. It's necessary that all the living beings in this room are persons.
12. All the living beings in this room are necessarily persons.
13. All the living beings in this room have the necessary property of being persons.
14. It's contingently true that all the living beings in this room are persons.
15. All the living beings in this room have as a contingent property that they are persons.
16. Any contingent claim could be true. [Use Cx for "x is a contingent claim" and Tx for "x is true."]
17. "All contingent claims are true" is possible.
18. It's necessary that everything is self-identical.
19. Everything is such that it has the property of being necessarily self-identical.
20. Nine is necessarily odd.
21. There's some being that has the necessary property of being unsurpassed in greatness.
22. God has the necessary property of being unsurpassed in greatness.
23. It's possible for someone to be unsurpassed in greatness.
24. It's possible for anyone to be unsurpassed in greatness.
25. It's possible for everyone to be unsurpassed in greatness.
26. All mathematical statements that are true are necessarily true. [Use Mx and Tx.]
27. Necessarily, everything that's red has the necessary property of being colored.
28. It's possible for every being to go out of existence.
29. It's possible for any being to go out of existence.
30. It's possible for someone who isn't standing to stand. [Look for two possible meanings.]

6.8 QUANTIFIED MODAL PROOFS

Our naïve QMC incorporates all the previous inference rules. Proofs and refutations go much like before. Here's an easy proof (for simplicity, we'll translate "x is self-identical" as "Sx" instead of "x = x"):

It's necessary that everything is self-identical.
∴ Every entity has the property of being necessarily self-identical.

It's necessary that everything is S.
∴ For each x, it's necessary that x is S.

1.	□(x)Sx	VALID

[∴ (x)□Sx

* **2.** ⌐asm: ~(x)□Sx
* **3.** │∴ (∃x)~□Sx [from 2]
* **4.** │∴ ~□Sa [from 3]
* **5.** │∴ ◇~Sa [from 4]
  6. │W ∴ ~Sa [from 5]
  7. │W ∴ (x)Sx [from 1]
  8. └W ∴ Sa [from 7, contradicting 6]
  9. ∴ (x)□Sx [from 2, 6, and 8]

These steps use rules that we've already learned.

QMC arguments sometimes lead into infinite loops (see Section 5.4). Consider this invalid argument:

Anyone could be above average.
∴ Everyone could be above average.

For each x, it's possible that x is above average.
∴ It's possible that everyone is above average.

If we try to prove this mechanically, we are led into an endless loop:

1. (x)◇Ax

[∴ ◇(x)Ax

* **2.** asm: ~◇(x)Ax
  3. ∴ □~(x)Ax [from 2]
* **4.** ∴ ◇Aa [from 1]
  5. W ∴ Aa [from 4]
* **6.** W ∴ ~(x)Ax [from 3]
* **7.** W ∴ (∃x)~Ax [from 6]
  8. W ∴ ~Ab [from 7]

These steps keep getting repeated with new constants and worlds.

```
 * 9.       ∴ ◇Ab [from 1]
   10.    WW ∴ Ab [from 9]
* 11.    WW ∴ ~(x)Ax [from 3]
* 12.    WW ∴ (∃x)~Ax [from 11]
   13.    WW ∴ ~Ac [from 12]
   14.      ∴ ◇Ac [from 1]
........................................
```

This is
the first
repetition.

We can intuitively come up with a refutation involving two possible worlds, each with entities a and b:

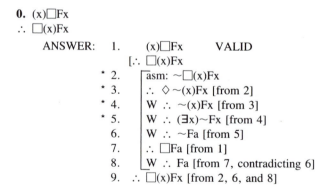

```
                              a, b
                     ┌──────────────┐
         W           │  Aa, ~Ab     │
                     ├──────────────┤
        WW           │  Ab, ~Aa     │
                     └──────────────┘
```

In this galaxy, the premise is true—since each entity is above average in some world or other. But the conclusion is false—since in no world is *every* entity above average. For now, we'll assume in our refutations that every world contains the same entities (and at least one such entity).

6.8A Exercises

Say whether valid (then give a proof) or invalid (then give a refutation). Assume our naïve QMC.

0. (x)□Fx
∴ □(x)Fx

```
ANSWER:   1.      (x)□Fx          VALID
                 [∴ □(x)Fx
        * 2.    ┌ asm: ~□(x)Fx
        * 3.    │ ∴ ◇~(x)Fx [from 2]
        * 4.    │ W ∴ ~(x)Fx [from 3]
        * 5.    │ W ∴ (∃x)~Fx [from 4]
          6.    │ W ∴ ~Fa [from 5]
          7.    │ ∴ □Fa [from 1]
          8.    └ W ∴ Fa [from 7, contradicting 6]
          9.   ∴ □(x)Fx [from 2, 6, and 8]
```

This is called the "Barcan inference," after Ruth Barcan Marcus. It was much disputed in the early days of quantified modal logic. It's doubtful that our naïve QMC gives the right results for this argument and for many similar arguments. We'll discuss this in Section 6.9.

1. (∃x)□Fx
∴ □(∃x)Fx
3. ∴ □(∃x)x = a

2. a = b
∴ (□Fa ⊃ □Fb)
4. ∴ (∃x)□x = a

5. □(x)Fx
∴ (x)□Fx
7. ∴ (x)□x = x
9. ◊(∃x)Fx
∴ (∃x)◊Fx
11. □(x)(Fx ⊃ Gx)
∴ (x)(Fx ⊃ □Gx)
13. ∴ (x)(y)(x = y ⊃ □x = y)
15. ~a = b
∴ □~a = b

6. ◊(x)Fx
∴ (x)◊Fx
8. ∴ □(x)x = x
10. (∃x)◊Fx
∴ ~(∃x)Fx
12. (◊(x)Fx ⊃ (x)◊Fx)
∴ ((∃x)~Fx ⊃ □(∃x)~Fx)
14. □(x)(Fx ⊃ Gx)
□Fa
∴ □Ga

6.8B Exercises

For each argument, first evaluate intuitively and then translate into our naïve QMC and work it out. Say whether valid (then give a proof) or invalid (then give a refutation). Look for ambiguous sentences like "All A's are necessarily B's." Translate ambiguous arguments into two different symbolic versions and test each for validity.

0. All mathematicians are necessarily rational.
   Paul is a mathematician.
   ∴ Paul is necessarily rational.
   [Use Mx, Rx, and p.]

ANSWER: The first premise is ambiguous. If we take it in the *de dicto* sense (to mean "It's necessary that all mathematicians are rational"), then the argument is invalid:

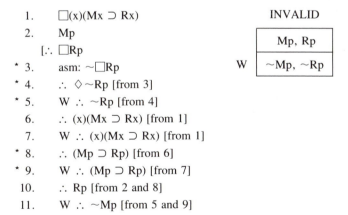

1.	□(x)(Mx ⊃ Rx)	INVALID
2.	Mp	
	[∴ □Rp]	
* 3.	asm: ~□Rp	W
* 4.	∴ ◊~Rp [from 3]	
* 5.	W ∴ ~Rp [from 4]	
6.	∴ (x)(Mx ⊃ Rx) [from 1]	
7.	W ∴ (x)(Mx ⊃ Rx) [from 1]	
* 8.	∴ (Mp ⊃ Rp) [from 6]	
* 9.	W ∴ (Mp ⊃ Rp) [from 7]	
10.	∴ Rp [from 2 and 8]	
11.	W ∴ ~Mp [from 5 and 9]	

Table (right side):

Mp, Rp
~Mp, ~Rp

In our refutation, Paul is a mathematician and rational in the actual world, but neither a mathematician nor rational in possible world-W. Paul might be a frog in world-W.

The argument is valid if we take the first premise *de re* (to mean "Each mathematician has the property of being necessarily rational"):

1.   (x)(Mx ⊃ □Rx)   VALID
2.   Mp
    [∴ □Rp
* 3.   ⌈asm: ~□Rp
* 4. | ∴ (Mp ⊃ □Rp) [from 1]
5.   ⌊∴ □Rp [from 2 and 4, contradicting 3]
6.   ∴ □Rp [from 3 and 5]

1. "Whoever doesn't have a beard isn't me" is a necessary truth.
∴ "I don't have a beard" is self-contradictory.
[Use Bx and i. The next example argues for the premise.]
2. I have a beard.
∴ "Whoever doesn't have a beard isn't me" is a necessary truth.
[Use Bx and i. G. E. Moore criticized this kind of argument, which he saw as essential to the idealistic metaphysics of his day. The conclusion is a fancy way to say "It's a necessary truth that I have a beard." If this form of reasoning works, then all my properties are necessary.]
3. Aristotle isn't identical to Plato.
If some being has the property of being necessarily identical to Plato but not all beings have the property of being necessarily identical to Plato, then some beings have necessary properties that other beings lack.
If some beings have necessary properties that other beings lack, then Aristotelian essentialism is true.
∴ Aristotelian essentialism is true.
[Use a, p, S (for "Some beings have necessary properties that other beings lack"), and E. This argument is essentially from Alvin Plantinga.]
4. All bachelors are necessarily unmarried.
You're a bachelor.
∴ "You are unmarried" is a necessary truth.
[Use Bx, Ux, and u.]
5. Necessarily there exists something unsurpassed in greatness.
∴ There exists something which necessarily is unsurpassed in greatness.
[Use Ux for "x is unsurpassed in greatness."]
6. The number that I'm thinking of isn't necessarily even.
6 = the number that I'm thinking of.
∴ 6 isn't necessarily even.
[Use n, E, and s. Does our naïve QMC correctly decide whether this argument is valid?]
7. "I'm a thinking being, and there are no material objects" is logically possible.
Everything that's a material object has the necessary property of being a material object.
∴ I'm not a material object.
[Use Tx, Mx, and i. From Alvin Plantinga.]
8. Every human is necessarily rational.
Every living being in this room is human.
∴ Every living being in this room is necessarily rational.
[Use Hx, Rx, and Lx. Aristotle, the first logician and the first to combine quantification with modality, used this example.]

9. It isn't necessary that all cyclists are rational.
Paul is a cyclist.
Paul is rational.
∴ Paul is contingently rational.
[Use Cx, Rx, and r.]

10. "Socrates has a little pain in his toe but doesn't show pain behavior" is logically consistent.
Necessarily it's the case that everyone who has a little pain in his toe is in pain.
∴ "All in pain show pain behavior" isn't a necessary truth.
[Use s, Tx for "x has a little pain in his toe," Bx for "x shows pain behavior," and Px for "x is in pain." This attacks a semantic behaviorism that analyzes the meaning of "pain" in behavioral terms.]

11. If Q (the question "Why is there something and not nothing?") is a meaningful question, then it's possible that there's an answer for Q.
Necessarily, every answer for Q refers to an existent that explains the existence of other things.
Necessarily, nothing that refers to an existent that explains the existence of other things is an answer to Q.
∴ Q isn't a meaningful question.
[Use M, Ax, and Rx.]

12. The number of apostles is 12.
12 is necessarily greater than 8.
∴ The number of apostles is necessarily greater than 8.
[Use n, t, e, and Gxy. Does our naïve QMC correctly decide whether this argument is valid?]

13. Something exists in the understanding than which nothing could possibly be greater. [In other words, there's some x such that x exists in the understanding, and it isn't possible that there be something greater than x.]
Anything that exists in the reality is greater than anything that doesn't exist in reality.
Socrates exists in reality.
∴ Something exists in reality than which nothing could possibly be greater. [In other words, there's some x such that x exists in reality, and it isn't possible that there be something greater than x.]
[Use Ux for "x exists in the understanding," Rx for "x exists in reality," Gxy for "x is greater than y," and s for "Socrates." Use a universe of discourse of *possible beings*— including fictional or mythological beings like Santa Claus in addition to actual beings. (Is this legitimate?) This is a form of St. Anselm's first ontological argument for the existence of God.]

14. "There's a being having unsurpassable greatness" is logically possible.
"Whatever has unsurpassable greatness has maximal excellence in every possible world" is necessarily true.
Whatever has maximal excellence is omnipotent, omniscient, and morally perfect.
∴ There's a being that is omnipotent, omniscient, and morally perfect.
[Use Ux, Mx, and Ox. Translate "x has maximal excellence in every possible world" as "necessarily, x has maximal excellence." This is Alvin Plantinga's ontological argument for the existence of God. It combines elements from Anselm's first and second arguments (the previous example and example 0 of Section 6.6B). Plantinga regards the second and third premises as true by definition. He thinks the first premise is reasonable, but controversial.]

15. All (well-formed) cyclists are necessarily two-legged.
Paul is a (well-formed) cyclist.
∴ Paul is necessarily two-legged.
[Use Cx, Tx, and p.]

6.9 SOPHISTICATED QUANTIFIED MODAL LOGIC

Our naïve QMC has two main problems. The first one is that it mishandles definite descriptions (terms of the form "the so and so").

So far, we've normally translated definite descriptions using small letters. For example, we've translated "the number that I'm thinking of" as "n." But this can get us into trouble in modal contexts.

Consider this ambiguous English statement:

The number that I'm thinking of is necessarily odd.
"□On" in naïve QMC

The ambiguous statement could mean either of these:

"There's just one number that I'm thinking of, and this number is odd" is necessary.
$\Box(\exists x)((Tx \cdot \sim(\exists y)(\sim x = y \cdot Ty)) \cdot Ox)$

There's just one number that I'm thinking of, and this number has the property of being necessarily odd.
$(\exists x)((Tx \cdot \sim(\exists y)(\sim x = y \cdot Ty)) \cdot \Box Ox)$

The *box-outside* form is false, since I could be thinking of no number, or more than one number, or an even number. But the *box-inside* form might be true. Suppose I'm thinking of just the number 7. Now 7 is odd. But 7 doesn't just happen to be odd. Being odd is a *necessary* property of the number 7. So if 7 is the number that I'm thinking of, then the *box-inside* form is true.

"□On" is as ambiguous as the English statement. So in our sophisticated QMC we won't symbolize definite descriptions using small letters. Rather, we'll use Russell's analysis (see Section 5.5)—as we did in the above boxes.

Our new translation policy blocks the proof of some invalid arguments. Here's an example:

8 = the number that I'm thinking of.
It's necessary that (8 = 8).
∴ It's necessary that (8 = the number that I'm thinking of).

e = n   VALID in naïve QMC
□e = e
∴ □e = n

Both premises are true, but the conclusion is false. So the argument is invalid. But yet it's provable in naïve QMC, since the conclusion follows from the premises by the equals-may-substitute-for-equals rule Q6 (Section 5.2).

Sophisticated QMC disallows the translation. The proof fails if we analyze definite descriptions using Russell's approach:

Exactly one being x is a number that I'm thinking of and x = 8.
It's necessary that (8 = 8).
∴ It's necessary that (exactly one being x is a number that I'm thinking of and x = 8).

(∃x)((Nx · ~(∃z)(Nz · ~z = x)) · x = e)   INVALID
□e = e   in sophisticated QMC
∴ □(∃x)((Nx · ~(∃z)(Nz · ~z = x)) · x = e)

If you don't believe me that the proof fails, try it!*

The second problem is that our naïve QMC implicitly assumes that the same objects exist in all possible worlds. This leads to implausible results.

Our naïve QMC makes Gensler (and everyone else) into a logically necessary being:

In every possible world, there exists a being who is Gensler.
□(∃x)x = g

This claim is false, since there are impoverished possible worlds without Gensler. But yet we can prove the claim using naïve QMC:

	[∴ □(∃x)x = g VALID in naïve QMC
* 1.	⌈asm: ~□(∃x)x = g
* 2.	│∴ ◇~(∃x)x = g [from 1]
* 3.	│W ∴ ~(∃x)x = g [from 2]
* 4.	│W ∴ (x)~x = g [from 3]
5.	│W ∴ ~g = g [from 4]
6.	⌊W ∴ g = g [from Q5]
7.	∴ □(∃x)x = g [from 1, 5, and 6]

* The following is another argument that's provable in our naïve QMC but unprovable if we use Russell's analysis of "the F":

"The F is the F" is necessary.   □f = f
∴ There's something that has   ∴ (∃x)□x = f
  the necessary property
  of being the F.

There are two ways out of this problem. One way is to change the interpretation of "(∃x)." The provable "□(∃x)x = g" is false because we take "(∃x)" to mean "for some *actually existing being* x." Some suggest that we instead take "(∃x)" to mean "for some *possible being* x." Then our provable formula would mean:

In every possible world, there's a *possible being* who is Gensler.

Some think that there's a possible being Gensler in every world. In some of these worlds Gensler actually exists, and in others he doesn't. We need an existence predicate "Ex" to distinguish between possible beings that exist and those that don't. We'd use the formula "(∃x)~Ex" to say that there are possible beings that don't exist.

This view is paradoxical, since it says that *there are nonexistent beings.* Alvin Plantinga (in *The Nature of Necessity*) defends the opposite view, which he calls "actualism." *Actualism* holds that to be a being and to exist is the same thing. There neither are nor could have been nonexistent beings. Of course there could be beings other than those that now exist. But this doesn't mean that *there now are beings that don't exist.* Actualism denies the latter claim.

Since I favor actualism, I don't want to posit nonexistent beings. I'll continue to take "(∃x)" to mean "for some *actually existing being*." On this reading, "□(∃x)x = g" means "It's necessary that there's an actually existing being who is Gensler." This is false—since I might not have existed. So we must reject some step of the proposed proof.

The faulty step seems to be the derivation of 5 from 4:

4   W ∴ (x)~x = g
5   W ∴ ~g = g [from 4]

The step presumes that this inference is valid:

For each actually existing being x,
x is distinct from Gensler.   (x)~x = g
∴ Gensler is distinct from Gensler.   ∴ ~g = g

This inference isn't valid—unless we presuppose the premise that Gensler is an actually existing being.

Since we reject the inference here, we must move to a *free logic*—one free of the assumption that individual constants like "g" always refer to actually existing beings. This involves changing our rule for dropping universal quantifiers (Section 4.3):

Old Q3   (x)Fx   →   Fa

[Everything is F   →   a is F.]

Suppose that every existing being is F; a might not be an existing being, and so might not be F. So we need this modified rule:

New Q3   (x)Fx, (∃x)x = a   →   Fa

[Everything is F, a exists   →   a is F.]

Here we symbolize "a exists" ("a is an existing being") by "(∃x)x = a" ("For some existing being x, x is identical to a").

We'll also strengthen our rule for dropping existential quantifiers:*

New Q4   (∃x)Fx   →   Fa, (∃x)x = a

[Something is F   →   a is F, a exists.]

[As in the old Q4, "a"'s replacement must be a *new* constant (one not occurring in earlier lines) and there must be a not-blocked-off assumption.]

Q3 is weakened, but Q4 is strengthened. The resulting system can prove *almost* everything that we could prove before; but now the proofs are longer. The main effect is to block a few proofs. We can no longer prove that Gensler exists in all possible worlds. And our proof of the Barcan inference (example 0 of Section 6.8A) now fails:

Each being that actually exists has the necessary property of being F.
∴ In every possible world, every existing being is F.

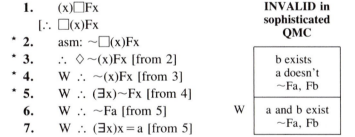

1.	(x)□Fx			**INVALID in**
	[∴ □(x)Fx			**sophisticated**
* 2.	asm: ~□(x)Fx			**QMC**
* 3.	∴ ◇~(x)Fx [from 2]			b exists
* 4.	W ∴ ~(x)Fx [from 3]			a doesn't
* 5.	W ∴ (∃x)~Fx [from 4]			~Fa, Fb
6.	W ∴ ~Fa [from 5]		W	a and b exist
7.	W ∴ (∃x)x = a [from 5]			~Fa, Fb

Our new rule for dropping existential quantifiers tells us that a exists in world-W (line 7). But we don't know whether a exists in the actual world. So we can't conclude "□Fa" from "(x)□Fx" in line 1. Using our naïve QMC, we could conclude "□Fa"— and then put "Fa" in world-W to contradict line 6. But now the step is blocked, and the proof fails.

We don't automatically get a refutation, but we can invent one on our own. In our refutation, we list which entities exist in which worlds. We'll use "a exists" for the equivalent "(∃x)x = a." Here "Each being that actually exists has the necessary

---

* Our "normal strategy" for doing proofs says to drop each unstarred initial existential quantifier and write in a new letter. But we won't follow this strategy for the existential wffs [like "(∃x)x = a"] that we get using Q4 and Q7. The "normal strategy" would lead into an infinite loop! Rather, we'll just keep these new existential wffs around to use with Q3 in dropping universal quantifiers.

property of being F" is true—since entity-b is the only being that actually exists and in every world it is F. But "In every possible world, every existing being is F" is false—since in world-W there is an existing being that isn't F.

Let's take a more abstract example. Suppose that only abstract objects existed (numbers, sets, propositions, etc.) and that these had the necessary property of being abstract. Then "Each being that actually exists has the necessary property of being abstract" would be true. But "In every possible world, every existing being is abstract" could still be false—if there were possible worlds containing concrete entities.

Or suppose that God had created nothing—and that uncreated beings had the necessary property of being uncreated. Then "Each being that actually exists has the necessary property of being uncreated" would be true. But "In every possible world, every existing being is uncreated" could still be false—since God might conceive of possible worlds containing created beings. Thus the Barcan inference fails.

Our new approach allows different worlds to have different existing entities. Gensler might exist in one world but not in another. We shouldn't picture existing in different worlds as anything spooky. It's just a way of talking about different possibilities. I might not have existed. We can tell consistent stories where my parents didn't meet and where I never came into existence. If these stories had been true, then I wouldn't have existed. So I don't exist in these stories (although I might exist in other stories). Existing in a possible world is much like existing in a story. (A "possible world" is a technical analogue of a "consistent story.") "I exist in world-W" just means "If world-W had been actual, then I would have existed." We'll also allow possible worlds with no entities. In such worlds, all wffs starting with existential quantifiers are false and all those starting with universal quantifiers are true.

Should we allow this as a possible world when we do our refutations?

W	a doesn't exist
	Fa

This seems peculiar—a doesn't exist but yet a has the property of being F! This seems impossible. Only existing beings have properties. In a consistent story where Gensler *doesn't* exist, Gensler couldn't be a logician or a backpacker. So if "a exists" isn't true in a possible world, then "a has property F" isn't true in that world either.

We'll put this idea into an inference rule:

$$Q7 \qquad Fa \quad \rightarrow \quad (\exists x)x = a$$

$$[a \text{ has property } F \quad \rightarrow \quad a \text{ exists}]$$

[Rule Q7 holds regardless of what constant replaces "a," what variable replaces "x," and what wff containing only a capital letter and "a" and perhaps other small letters (but nothing else) replaces "Fa."] By Q7, "Descartes thinks" entails "Descartes exists." Conversely, the falsity of "Descartes exists" entails the falsity of "Descartes thinks." Rule Q7 expresses that it's a necessary truth that only existing objects have properties. Plantinga calls this view "*serious actualism.*" Actualists who reject Q7 are frivolous.

It's important that the following *isn't* an instance of Q7 (since the wff substituting for "Fa" in Q7 can't contain "~"):

$\sim Fa \quad \rightarrow \quad (\exists x)x = a \qquad$ [wrong!]

This case is confusing because "a isn't F" in English can have two different senses.

Consider the sentence "Descartes *doesn't* think." Here's one possible meaning (the internal negation sense):

Descartes has the property of being a nonthinker.
= Descartes is an existing being who doesn't think.
= $(\exists x)(x = d \cdot \sim Td)$

This claim is *de re*. It affirms something of an existing being Descartes. Taken this way, "Descartes doesn't think" entails "Descartes exists." Here's the other possible meaning (the external negation sense):

It's false that Descartes has the property of being a thinker.
= It isn't the case that Descartes is an existing being who thinks.
= $\sim(\exists x)(x = d \cdot Td)$
= $\sim Td$

This claim is *de dicto*. It says that a proposition is false. Taken this way, "Descartes doesn't think" doesn't attribute a property to an existing object. And it doesn't entail "Descartes exists."

One might object to Q7 on the grounds that Santa Claus has properties (such as *being fat*)—but doesn't exist. But various stories predicate conflicting properties to Santa (regarding, for example, when he delivers presents). Does Santa have contradictory properties? Or is one Santa story uniquely "true"? What would that mean? When we say "Santa is fat," we mean that in such and such a story (or possible world) there's a being called Santa who is fat. We shouldn't think of Santa as a nonexisting being in our actual world who has properties such as being fat. Rather, what exists in our actual world is stories about there being someone with certain properties—and children who believe these stories. So Santa needn't make us give up Q7.

We need to modify our definition of "necessary property." Consider this definition:

F is a *necessary property* of a.
= In all possible worlds, a is F.
= $\Box Fa$

Let's grant that Socrates has properties only in worlds where he exists—and that there are worlds where he doesn't exist. Then there are worlds where Socrates has no properties. It follows that there aren't any properties that Socrates has in all worlds. By our definition, Socrates would have no necessary properties.

Socrates might still have some necessary *combinations* of properties. Perhaps it's true in all worlds that, *if* Socrates exists, *then* Socrates is a person. This suggests a looser definition of "necessary property":

F is a *necessary property* of a.

= In all possible world where a exists, a is F.

= $\Box((\exists x)x = a \supset Fa)$

This looser definition reflects more clearly what philosophers mean when they speak of "necessary properties." It lets us claim that Socrates has the property of being necessarily a person. This claim would mean that Socrates is a person in every possible world where he exists. Equivalently, in no possible world does Socrates exist as anything other than a person. Here's an analogous definition of "contingent property":

F is a *contingent property* of a.

= a is F; but in some possible world where a exists, a isn't F.

= $(Fa \cdot \Diamond((\exists x)x = a \cdot \sim Fa))$

These refinements overcome some problems. But they make our system harder to use. We don't need the refinements for most QMC arguments. So we'll keep the naïve QMC of earlier sections as our "official system." This is the system that we'll build on in the following chapters. The sophisticated QMC will be our back-up system. We'll keep it around for the few cases where the naïve approach gives questionable results.

6.9A Exercises

Say whether each argument is valid (then give a proof) or invalid (then give a refutation). Assume the sophisticated QMC sketched in this section.

0. $\Diamond(x)Fx$
∴ $(x)\Diamond Fx$ ANSWER:

It could have been that every existing being is F.
∴ Each being that actually exists could have been F.

1.	$\Diamond(x)Fx$		INVALID

	[∴ $(x)\Diamond Fx$		
* 2.	asm: $\sim(x)\Diamond Fx$		a and b exist
* 3.	∴ $(\exists x)\sim\Diamond Fx$ [from 2]		$\sim Fa$, Fb
* 4.	∴ $\sim\Diamond Fa$ [from 3]	W	b exists
5.	∴ $(\exists x)x = a$ [from 3]		a doesn't exist
6.	∴ $\Box\sim Fa$ [from 4]		$\sim Fa$, Fb
7.	W ∴ $(x)Fx$ [from 1]		
8.	∴ $\sim Fa$ [from 6]		
9.	W ∴ $\sim Fa$ [from 6]		

We can't infer "W ∴ Fa" from "W ∴ (x)Fx" (step 7), since we don't know if a exists in world-W. (This is OK in the naïve QMC and would make the argument valid.) We invent our own refutation. Here "It could have been that every existing being is F" is true—since in world-W all existing beings are F. But "Each being that actually exists could have been F" is false—since entity-a actually exists but couldn't have been F (since it is F in no possible world).

The medieval thinker Jean Buridan (after whom this one is called the "Buridan inference") gave this example. "It could have been that every being is God" is true—since we can imagine a possible world where only God exists. But "Each being that actually exists could have been God" is false—since you or I couldn't have been God.

Also, "It could have been that every existing being is distinct from Gensler" is true—since we can imagine an impoverished world where Gensler didn't exist. But "Each being that actually exists could have been distinct from Gensler" is false—since Gensler couldn't have been distinct from Gensler.

1. ∴ (∃x)x = a

2. ∴ (x)□x = x

3. (∃x)□Fx
∴ □(∃x)Fx

4. (x)Fx
∴ ~(x)~Fx

5. (∃x)◇Fx
∴ ◇(∃x)Fx

6. ◇(∃x)Fx
∴ (∃x)◇Fx

7. ~(∃x)x = a
∴ ~Fa

8. (∃x)(Fx ∨ ~Fx)
∴ (∃x)□(Fx ∨ ~Fx)

9. (∃x)◇~Fx
∴ ◇(∃x)~Fx

10. (∃x)◇((∃y)y = x · ~Fx)
∴ ◇(∃x)~Fx

11. ∴ (∃x)x = x

12. ∴ (∃x)□(∃y)y = x

13. □a = a
∴ (∃x)□x = a

14. (∃x)x = a
∴ (∃x)□x = a

15. □(x)Fx
∴ (x)□((∃y)y = x ⊃ Fx)

16. □(x)~Fx
∴ (x)□~Fx

17. □Fa
∴ (∃x)□Fx

18. □(x)(Fx ⊃ Gx)
□((∃x)x = a ⊃ Fa)
∴ □((∃x)x = a ⊃ Ga)

19. (x)~Fx
∴ ~Fa

20. Criticize this argument:

It's necessary that either Descartes thinks or Descartes doesn't think.

It's necessary that if Descartes thinks then Descartes exists.

It's necessary that if Descartes doesn't think then Descartes exists.

∴ It's necessary that Descartes exists.

7

Imperative and Deontic Logic

7.1 IMPERATIVE TRANSLATIONS

Our *deontic calculus* (DC) builds on previous systems. It deals with deontic (normative) notions like "ought" and with imperatives like "Do this!" We'll start with imperatives. We'll mostly follow the approach of Hector-Neri Castañeda.*

We need two rules for constructing imperative wffs:

1. Any underlined capital letter is a wff.

2. Any capital letter followed by one or more small letters, one small letter of which is underlined, is a wff.

Underlining turns indicatives into imperatives:

INDICATIVE	IMPERATIVE
You're doing A.	Do A!
A	$\underline{A}$
Au	A$\underline{u}$

Here are examples that build on PC:

$$\underline{A} = \text{Do (or be) A.}$$
$$= \text{Let it be the case that A.}$$
$$= \text{Would that act A be done.}$$

* See Castañeda's "Imperative Reasonings," *Philosophy and Phenomenological Research* 21 (1960), 21–49; "Outline of a Theory on the General Logical Structure of the Language of Action," *Theoria* 26 (1960), 151–82; "Actions, Imperatives, and Obligations," *Proceedings of the Aristotelian Society* 68 (1967–68), 25–48; and "On the Semantics of the Ought-to-do," *Synthese* 21 (1970), 448–68.

$\sim\underline{A}$ = Don't do A.

$(\underline{A} \lor \underline{B})$ = Do A or B.

$\sim(\underline{A} \lor \underline{B})$ = Don't do either A or B.

$(\underline{A} \cdot \underline{B})$ = Do A and B.

$\sim(\underline{A} \cdot \underline{B})$ = Don't both do A and do B.

Don't combine doing A with doing B.

$\sim(\underline{A} \cdot \sim\underline{B})$ = Don't both do A and not do B.

Don't combine doing A with not doing B.

Don't do A *without* doing B.

Be sure to underline imperative parts but not factual parts:

$(A \cdot B)$ = You're doing A and you're doing B.

$(A \cdot \underline{B})$ = You're doing A, but do B.

$(\underline{A} \cdot \underline{B})$ = Do A and B.

$(A \supset B)$ = If you're doing A, then you're doing B.

$(A \supset \underline{B})$ = If you (in fact) are doing A, then do B.

$(\underline{A} \supset B)$ = Do A, only if you (in fact) are doing B.

Since English doesn't permit an imperative after "if," we can't read "$(A \supset B)$" as "If do A, then you're doing B." But we can read it as the equivalent "Do A, only if you're doing B." This means the same thing as "$(\sim B \supset \sim \underline{A})$" ("If you aren't doing B, then don't do A").

There's a subtle difference between these two:

$(A \supset \sim\underline{B})$ = If you (in fact) are doing A, then don't do B.

$\sim(\underline{A} \cdot B)$ = Don't both do A and do B.

= Don't combine doing A with doing B.

"A" is underlined in the second but not the first; otherwise, the two wffs would be equivalent. The *if-then* "$(A \supset \sim B)$" says that, if A is *true*, then we aren't to do B. But the *don't-combine* "$\sim(\underline{A} \cdot B)$" forbids a combination—doing A and also doing B. If we are doing A, it doesn't follow that we aren't to do B. Maybe instead we are to stop doing A. We'll see more on this distinction later.

Here are examples that build on QC:

$A\underline{x}$ = x, do (or be) A.

= Let it be the case that x does A.

= Would that x do A.

$A\underline{xy}$ = x, do A to y.

= Let it be the case that x does A to y.

= Would that x do A to y.

We underline the letter referring to the agent. Here are examples using quantifiers:

$$(x)Ax = \text{Everyone does A.}$$
$$(x)A\underline{x} = \text{Let everyone do A.}$$
$$(\exists x)A\underline{x} = \text{Let someone do A.}$$

$$(x)(Ax \supset B\underline{x}) = \text{Let everyone who (in fact) is doing A do B.}$$
$$(\exists x)(Ax \cdot B\underline{x}) = \text{Let someone who (in fact) is doing A do B.}$$
$$(\exists x)(A\underline{x} \cdot B\underline{x}) = \text{Let someone do both A and B.}$$

Notice which letters are underlined.

7.1A Exercises

Translate each of these into an imperative wff. Take "you" here as a singular "you."

0. If the cocoa is about to boil, remove it from the heat. [Use B and R.]
ANSWER: $(B \supset \underline{R})$

1. Leave or shut up. [Use L and S.]
2. If you don't leave, then shut up.
3. Do A, only if you want to do A. [Use A for "You do A" and W for "You want to do A."]
4. Do A, only if you want to do A. [This time use Ax for "x does A," Wx for "x wants to do A," and u for "you."]
5. Don't combine accelerating with braking. [Use A and B.]
6. If you accelerate, then don't brake.
7. If you brake, then don't accelerate.
8. If you believe that you ought to do A, then do A. [Use B for "You believe that you ought to do A" and A for "You do A."]
9. If you don't do A, then don't believe that you ought to do A.
10. Don't combine believing that you ought to do A with not doing A.
11. If you have a cold, then take Dristan. [Use Cx, Dx, and u.]
12. Let everyone who has a cold take Dristan.
13. If everyone does A, then do A yourself. [Use Ax and u.]
14. Gensler, rob Jones! [Use Rxy, g, and j.]
15. If Jones hits you, then hit Jones! [Use Hxy, j, and u.]
16. If you're shifting, then pedal. [Use S and P.]
17. Don't combine shifting with not pedaling.
18. If you believe that A is wrong, then don't do A. [Use B for "You believe that A is wrong" and A for "You do A."]
19. Don't combine believing that A is wrong with doing A.
20. "Become a teacher!" entails "If studying is needed to become a teacher, then study!" [Use B for "You become a teacher," N for "Studying is needed to become a teacher," and S for "You study."]
21. Would that someone be sick and also well. [Use Sx and Wx.]
22. Would that someone who is sick be well.

23. Would that someone be sick who is well.
24. Be with someone you love. [Use Wxy, Lxy, and u.]
25. Love someone you're with.

7.2 IMPERATIVE ARGUMENTS

Imperative proofs work much like indicative proofs. We need no new inference rules. But we must treat "A" and "<u>A</u>" as different wffs. So "A" and "~<u>A</u>" aren't contradictories. It's consistent to say "You're now doing A—but don't!"

Here's a valid imperative argument:

If you're accelerating, then don't brake.
You're accelerating.
∴ Don't brake.

$(A ⊃ ~\underline{B})$ VALID
A
∴ ~<u>B</u>

The conclusion follows by our rule *modus ponens*.

But there's a problem here. We defined *valid* using the words *true* and *false* (Section 1.2):

> A *valid* argument is one in which it would be contradictory for the premises to be **true** but the conclusion **false.**

But "Don't brake" and other imperatives aren't true or false. So how can the *valid/invalid* distinction apply to imperative arguments?

We need a broader definition of *valid* that applies to indicative and imperative arguments equally well. This new definition (which avoids the words *true* and *false*) does the job:

> A *valid* argument is one in which the conjunction of the premises with the contradictory of the conclusion is inconsistent.

Then to say that our argument is *valid* means that this combination is inconsistent:

> "If you're accelerating, then don't brake; you're accelerating; brake."

The combination *is* inconsistent. So our argument is *valid* in this new sense.*

---

* We could equivalently define a *valid* argument as one in which every set of imperatives and indicatives consistent with the premises is also consistent with the conclusion.

This next argument is just like the first one except that it uses a *don't-combine* premise:

Don't combine accelerating with braking.
You're accelerating.
∴ Don't brake.

$\sim(\underline{A} \cdot \underline{B})$ INVALID
A
∴ $\sim\underline{B}$

The first premise forbids the accelerating-and-braking combination. Suppose we are accelerating. It doesn't follow that we shouldn't brake. Maybe we need to brake (and stop accelerating) to avoid hitting another car! So the argument is invalid. It's consistent to conjoin the premises with the contradictory of the conclusion:

> "Don't combine accelerating with braking;
> you're accelerating; (there's a car right in
> front of you—stop accelerating!); brake."

The part in parentheses helps us see that it's consistent to endorse the premises and yet reject the conclusion.
We'd work the problem out this way:

*1. $\sim(\underline{A} \cdot \underline{B}) = 1$ INVALID
2.   A = 1
     [∴ $\sim\underline{B} = 0$
3.   asm: $\underline{B}$
4.   ∴ $\sim\underline{A}$ [from 1 and 3]

> A, $\sim\underline{A}$, $\underline{B}$
>
> [A = 1, $\underline{A}$ = 0, $\underline{B}$ = 1]

> You're accelerating.
> Don't accelerate!
> Brake!

We quickly get a refutation—a set of assignments of 1 and 0 to the letters making the premises 1 but the conclusion 0.
But there's a problem here. Our refutation assigns *false* to "Accelerate!"—so A = 0. But "Accelerate!" and other imperatives aren't true or false. So what could "$\underline{A}$ = 0" mean?
We'll read "1" as "correct" and "0" as "incorrect." Applied to indicatives, these mean *true* or *false*. Applied to imperatives, these mean that the action prescribed is *correct* or *incorrect* relative to some standard that divides the actions prescribed by the imperative letters into correct and incorrect actions. The standard could be of different sorts, based on such things as morality, the law, or traffic safety goals. Generally we won't specify the standard.
Suppose we have an argument containing just indicative and imperative letters, "$\sim$," and propositional connectives. The argument is valid, if and only if, relative

to *every* assignment of ''1'' or ''0'' to the indicative and imperative letters, if the premises are ''1,'' then so is the conclusion. Equivalently, the argument is valid if, and only if, any possible facts and standards for correct actions which make the premises correct would also make the conclusion correct.

So our refutation amounts to this:

A = 1	''You're accelerating'' is true.
$\underline{A}$ = 0	Accelerating is incorrect.
$\underline{B}$ = 1	Braking is correct.

Our argument is invalid because the premises could be correct while the conclusion wasn't correct.

Our two arguments were the same, except that one used the *if-then* ''(A ⊃ $\underline{B}$),'' while the other used the *don't-combine* ''~($\underline{A}$ · $\underline{B}$).'' Since one argument was valid and the other was invalid, these two forms aren't equivalent.

Consider these three forms:

(~A ⊃ ~ $\underline{B}$)	If you don't do A, then don't believe that you ought to do A.
(B ⊃ $\underline{A}$)	If you believe that you ought to do A, then do A.
~($\underline{B}$ · ~$\underline{A}$)	Don't combine believing that you ought to do A with not doing A.

The first two are mixtures of indicatives and imperatives, saying that, if you're (in fact) doing one thing, then you aren't to do another. The last one is a pure imperative of the *don't-combine* form, forbidding the accelerating-and-braking combination.

Imagine that you find yourself accelerating and braking (making ''A'' and ''B'' both true)—thus wearing down your brakes and wasting gasoline. Then you're violating all three imperatives. But the three differ on what you're to do next. The first form tells you not to brake, and the second tells you not to acelerate. But the last form leaves it open whether you're to stop accelerating or stop braking. Maybe you need to brake (and stop accelerating) to avoid hitting another car. Or maybe you need to accelerate (and stop braking) to pass another car. The *don't-combine* form doesn't tell a person in this forbidden combination exactly what to do.

These distinctions are crucial for understanding consistency imperatives. Consider these three forms (using A for ''You do A'' and B for ''You believe that A is wrong''):

(A ⊃ ~ $\underline{B}$)	If you do A, then don't believe that A is wrong.
(B ⊃ ~$\underline{A}$)	If you believe that A is wrong, then don't do A.
~($\underline{B}$ · $\underline{A}$)	Don't combine believing that A is wrong with doing A.

Imagine that you find yourself *believing that A is wrong* and yet *doing A*. Then you're violating all three imperatives. But the three differ on what you're to do next. The first form tells you to stop believing that A is wrong, and the second tells you to stop doing A. Which of these is better advice depends on the situation. Maybe your belief in the wrongness of A is correct and well-founded, and you need to stop doing A. The first form is faulty, since it would tell you to change your belief. Or maybe your belief is incorrect. Maybe you believe, irrationally, that it's wrong to treat dark-skinned people fairly; but you treat them fairly anyway. Then you should give up your belief. The second form is faulty, since it would tell you to stop treating these people fairly. So the two *if-then* forms sometimes give wrong advice.

The third form is a *don't-combine* imperative forbidding this combination:

BELIEVING THAT A IS WRONG	+	DOING A

This combination always has a faulty element. If your believing is correct, then your doing is faulty. If your doing is correct, then your believing is faulty. If you *believe that A is wrong* and yet *do A*, then you're inconsistent. How should you regain consistency? Should you reject the belief, or reject the action? We've seen that that depends on the situation. The *don't-combine* form forbids an inconsistency, but it doesn't tell a person in this forbidden combination exactly what to do.

Here's an analogous, but more complex, trio:

$(\sim A \supset \sim \underline{B})$	If you don't do A, then don't believe that you ought to do A.
$(B \supset \underline{A})$	If you believe that you ought to do A, then do A.
$\sim(\underline{B} \cdot \sim\underline{A})$	Don't combine believing that you ought to do A with not doing A.

"Follow your conscience" is often seen as equivalent to the second form. But this form can lead to absurdities (see example 10 of section 7.2B). The *don't-combine* form is better. It simply forbids an inconsistent combination. Even when we add premises describing a person's beliefs or actions, this form doesn't prescribe or forbid individual actions.

7.2A Exercises

Say whether valid (give a proof) or invalid (give a refutation).

0. $(A \supset \sim\underline{B})$
  $(\sim A \supset \sim \underline{C})$
∴ $\sim(\underline{B} \cdot \underline{C})$ ANSWER:

    * 1. $(A \supset \sim\underline{B})$ INVALID
    * 2. $(\sim A \supset \sim\underline{C})$
        $[\therefore \sim(\underline{B} \cdot \underline{C})]$

```
  * 3.    ⎡asm: (B · C)
    4.    ⎢∴ B    [from 3]
    5.    ⎢∴ C    [from 3]
    6.    ⎢∴ ~A    [from 1 and 4]
    7.    ⎣∴ ~C    [from 2 and 6, contradicting 5]
    8.    ∴ ~ (B · C)    [from 3, 5, and 7]
```

1. (A ⊃ B)   **2.** ~A
∴ (~B ⊃ ~A)   ∴ ~(A · B)
3. (A ⊃ B)   **4.** ~(A · ~B)
∴ ~(A · ~B)   ∴ (A ⊃ B)
5. ~ ◇(A · B)   **6.** (x)(Fx ⊃ Gx)
~(C · ~A)   Fa
∴ ~(C · B)   ∴ Ga
7. (x)~(Fx · Gx)   **8.** (x)(Fx ⊃ Gx)
(x)(Hx ⊃ Fx)   (x)(Gx ⊃ Hx)
∴ (x)(Gx ⊃ ~Hx)   ∴ (x)(Fx ⊃ Hx)
9. (~A ∨ ~B)   **10.** ~(A · ~B)
∴ ~(A · B)   ∴ (~A ∨ B)

7.2B Exercises

For each argument, first evaluate intuitively and then translate into DC and work it out.
Say whether valid (give a proof) or invalid (give a refutation).

0. Get 100 percent.
If you get 100 percent, then celebrate.
∴ Celebrate.
[Use G and C.]

```
ANSWER:    1.      G                    INVALID
         * 2.      (G ⊃ C
              [∴ C                      ┌─────────────────┐
           3.      asm: ~C              │  G, ~G, ~C      │
           4.      ∴ ~G                 └─────────────────┘
```

Don't celebrate yet. Maybe you flunked the test! Our argument is clearly invalid.
Two proposed criteria of validity would judge our argument to be valid. The *obedience
view* says that an imperative argument is valid if doing what the premises prescribe necessarily
involves doing what the conclusion prescribes. This is fulfilled in the present case. If you do
what the first premise prescribes, then you get 100 percent; if you then do what the second
premise prescribes, you'll celebrate. So the obedience view says that our argument is valid. So
the obedience view is wrong.

The *threat view* analyzes the imperative "Do A!" as "Either you will do A or else S will happen"—where "S" ("the sanction") represents some unspecified bad thing. So "A" is taken to mean "(A ∨ S)." But if we replace "G̲" with "(G ∨ S)" and "C̲" with "(C ∨ S)," then our argument becomes valid. So the threat view says that our argument is valid. So the threat view is wrong.

1. Either bunt or hit a sacrifice fly.
   Don't bunt.
   ∴ Hit a sacrifice fly.
   [Use B and H.]
2. If the cocoa is about to boil, remove it from the heat.
   If the cocoa is steaming, it's about to boil.
   ∴ If the cocoa is steaming, remove it from the heat.
   [Use B, R, and S.]
3. Don't eat cake.
   If you don't eat cake, then give yourself a gold star.
   ∴ Give yourself a gold star.
   [Use E and G.]
4. If you're shifting, then pedal.
   You're shifting.
   ∴ Pedal.
   [Use S and P. It's bad for a 10-speed bicycle if you shift without pedaling.]
5. Don't combine shifting with not pedaling.
   You're shifting.
   ∴ Pedal.
   [Use S and P.]
6. Don't shift.
   ∴ Don't combine shifting with not pedaling.
   [Use S and P.]
7. If he's in the street, wear your gun.
   Don't wear your gun.
   ∴ He isn't in the street.
   [Use S and G. This has imperative premises and an indicative conclusion. It's valid, in that the conjunction of the premises with the denial of the conclusion is inconsistent.]
8. Back up all your floppy disks.
   This is one of your floppy disks.
   ∴ Back this up.
   [Use Bxy, Dx, u, and t.]
9. If you take logic, then you'll make logic mistakes.
   Take logic.
   ∴ Make logic mistakes.
   [Use T and M.]
10. If you believe that you ought to commit mass murder, then commit mass murder.
    You believe that you ought to commit mass murder.
    ∴ Commit mass murder.
    [Use B and C. Suppose we take "Always follow your conscience" to mean "If you believe

that you ought to do A, then do A.'' Then this principle can tell us to do evil things. Would the corresponding *don't-combine* form also tell us to do evil things? See the next example.]

11. Don't combine believing that you ought to commit mass murder with not committing murder.
You believe that you ought to commit mass murder.

∴ Commit mass murder.
[Use B and C.]

12. Don't both drive and watch the scenery.
Drive!

∴ Don't watch the scenery.
[Use D and W.]

13. Get a soda.
If you get a soda, then pay a dollar.

∴ Pay a dollar.
[Use G and P.]

14. ∴ Either do A or don't do A.
[This (vacuous) imperative tautology is analogous to the logical truth ''You're doing A or you aren't doing A.'']

15. Don't combine believing that A is wrong with doing A.

∴ Either don't believe that A is wrong, or don't do A.
[Use B and A.]

16. Mail it.

∴ Mail it or burn it.
[Use M and B. This one has been used to discredit imperative logic. The argument is valid, since this is inconsistent: ''Mail it; don't either mail it or burn it.'' Strictly taken, ''Mail it or burn it'' doesn't entail ''You may burn it.'' It's consistent to follow ''Mail it or burn it'' with ''Don't burn it.'' See example 1.]

17. Don't combine having this end with not taking this means.
Don't take this means.

∴ Don't have this end.
[Use E and M.]

18. Lie to your friend only if you want people to lie to you under such circumstances.
You don't want people to lie to you under such circumstances.

∴ Don't lie to your friend.
[Use L and W. The first premise is a simplified version of Kant's formula of universal law. We'll see a more sophisticated version in Chapter 9.]

19. Let every incumbent who will be honest be endorsed.

∴ Let every incumbent who won't be endorsed not be honest.
[Use Hx, Ex, and universe of discourse of incumbents.]

20. Studying is needed to become a teacher.
''Become a teacher'' entails ''Do what is needed to become a teacher.''
''Do what is needed to become a teacher'' entails ''If studying is needed to become a teacher, then study.''

∴ Either study or don't become a teacher!
[Use N for ''Studying is needed to become a teacher,'' B for ''You become a teacher,'' D for ''You do what is needed to become a teacher,'' and S for ''You study.'' This example shows that we can deduce some *complex* imperatives from purely descriptive premises.]

7.3 DEONTIC TRANSLATIONS

Our deontic calculus will use "O" for the all-things-considered sense of "ought" that we normally use in discussing moral issues. The following passage contrasts this sense with the *prima facie* ("at first glance") sense:

> Insofar as I promised to take my children to the zoo, I *ought* to do this (*prima facie duty*). But insofar as my wife needs me to drive her to the hospital, I *ought* to do this instead (*prima facie duty*). Since this duty to my wife is more urgent, in the final analysis I *ought* to drive my wife to the hospital (*all-things-considered duty*).

Both senses of "ought" in this paragraph are *normative*, in that they're used to evaluate, and not just to describe or categorize. Sometimes we use "ought" *descriptively*. We might say that something "ought to be done," meaning that it's commanded by company policy or would promote one's self-interest, but not intending to express a positive evaluation of the action. These other senses of "ought" may follow different logical patterns.

So we'll use "O" for the all-things-considered sense of "ought" that we normally use when discussing moral issues. Now philosophers disagree on how to understand and justify such ought-judgments. They dispute questions like these:

- What do such ought-judgments mean?
- Are they objectively true or false?
- Can we argue rationally about ethical principles?
- How can we justify an ethical judgment?

We'll largely avoid these questions here. Deontic logic doesn't presuppose any particular answer to them. My way of speaking will sometimes presume that ought-judgments are true or false (since I think they are). But someone who thinks otherwise could rephrase my remarks to avoid this presumption.

Our DC uses two deontic operators: "O" for "ought" and "R" for "all right." "O" and "R" work much like "$\Box$" and "$\Diamond$." Deontic operators attach to *imperative* wffs, just as modal operators attach to *indicative* wffs.

Here are examples that build on PC:

O$\underline{A}$	= It ought to be that A.
	= Act A is obligatory (required, mandatory, a duty).
R$\underline{A}$	= It would be all right that A.
	= Act A is right (permissible, OK).
~R$\underline{A}$	= O~$\underline{A}$ = It would be wrong that A.
	= Act A isn't right.
	= Act A ought not to be done.
~O$\underline{A}$	= R~$\underline{A}$ = It isn't obligatory that A.
	= It's all right that A be omitted.
O($\underline{A} \cdot \underline{B}$)	= It ought to be that A and B.

$R(\underline{A} \vee \underline{B})$ = It's all right that A or B.

$(A \supset O{\sim}\underline{B})$ = If you do A, then you ought not to do B.

$(B \supset O{\sim}\underline{A})$ = If you do B, then you ought not to do A.

$O{\sim}(\underline{A} \cdot \underline{B})$ = You ought not to combine doing A with doing B.

$(A \supset O\underline{B})$ = If you do A, then you ought to do B.

$({\sim}B \supset O{\sim}\underline{A})$ = If you don't do B, then you ought not to do A.

$O{\sim}(\underline{A} \cdot {\sim}\underline{B})$ = You ought not to combine doing A with not doing B.

                 = You ought not to do A *without* doing B.

Note the *if-then* and *don't-combine* deontic forms in the last two groups. Here are examples that build on QC:

$OA\underline{x}$ = x ought to do A.

$OA\underline{x}y$ = x ought to do A to y.

$RA\underline{x}$ = It would be all right for x to do A.

$RA\underline{x}y$ = It would be all right for x to do A to y.

$O(x)A\underline{x}$ = It's obligatory that everyone do A.

${\sim}O(x)A\underline{x}$ = It isn't obligatory that everyone do A.

$O{\sim}(x)A\underline{x}$ = It's obligatory that not everyone do A.

$O(x){\sim}A\underline{x}$ = It's obligatory that everyone refrain from doing A.

Compare these two:

$O(\exists x)A\underline{x}$ = It's obligatory that someone or other answer the phone.

$(\exists x)OA\underline{x}$ = There's someone (some specific person) who has an obligation to answer the phone.

The first might be true while the second is false. It could be obligatory that someone or other answer the phone—while no specific person has the obligation to answer it. The danger in such cases is that everyone will say, "Let the other person answer it!" We sometimes need to assign duties to insure that things get done.

Consider these three wffs:

$O(\exists x)(K\underline{x} \cdot R\underline{x})$ = It's obligatory that some who kill repent.

$O(\exists x)(K\underline{x} \cdot R\underline{x})$ = It's obligatory that some kill who repent.

$O(\exists x)(K\underline{x} \cdot R\underline{x})$ = It's obligatory that some both kill and repent.

The three are very different. The underlined letters show what parts are obligatory: repenting, killing, or killing-and-repenting. If we attached "O" to indicatives, our formulas couldn't distinguish the forms. All three would translate the same:

O(∃x)(Kx · Rx)

Because of such examples, we need to attach "O" to imperative wffs, not to indicative ones.*

So "O" and "R" attach to imperative wffs. "A" is an imperative wff, and so "OA" is a wff. But many cases are unclear. Is "◇A̅" an imperative wff? Is "O◇A̲" a wff? The answer is NO to both questions. We have to give rules to cover such cases.

Our rules will distinguish *descriptive*, *imperative*, and *deontic* wffs. Here are examples of each:

DESCRIPTIVE	IMPERATIVE	DEONTIC (NORMATIVE)	
You're doing A.	Do A!	You ought to do A.	It's all right for you to do A.
A	A̲	OA	RA
Au	A̲u	OA̅u	RA̅u

To distinguishing each kind of wff, we'll rework all our old rules and add some new ones:

1. Any not-underlined capital letter not immediately followed by a small letter is a *descriptive* wff.
2. A string consisting of a not-underlined capital letter followed by one or more not-underlined small letters is a *descriptive* wff.
3. Any underlined capital letter not immediately followed by a small letter is an *imperative* wff.
4. A string consisting of a not-underlined capital letter followed by one or more small letters, one small letter of which is underlined, is an *imperative* wff.
5. The result of prefixing any wff with "~" is a wff and is *descriptive*, *imperative*, or *deontic*, depending on what the original wff was.
6. The result of joining any two wffs by "·" or "∨" or "⊃" or "≡" and enclosing the result in parentheses is a wff. It's *descriptive* if both original wffs were descriptive. It's *imperative* if both were imperative or if one was imperative and the other descriptive or deontic. It's *deontic* if both were deontic or if one was deontic and the other descriptive.
7. Suppose we have a wff containing a given variable but not containing any quantifier using that variable. The result of prefixing this wff with a quantifier using that variable is a wff. It's *descriptive*, *imperative*, or *deontic* depending on what the original wff was.
8. The result of joining any two not-underlined small letters by " = " is a *descriptive* wff.

---

* Couldn't we distinguish the three as "(∃x)(Kx · ORx)," "(∃x)(OKx · Rx)," and "(∃x)O(Kx · Rx)"? The problem with this is that putting "(∃x)" outside the "O" changes the meaning. See the previous paragraph.

9. The result of prefixing any wff with "□" or "◇" is a *descriptive* wff.

10. The result of prefixing any imperative wff with "O" or "R" is a *deontic* wff.

In applying these rules, we'll consider letters with primes to be distinct letters.

7.3A Exercises

Translate each of these into a DC wff.

0. "You ought to do A" entails "It's possible for you to do A."

ANSWER: □ (O<u>A</u> ⊃ ◇A)

Here we shouldn't underline "A" in "◇A." "◇<u>A</u>" means "The imperative 'Do A!' is logically consistent," while "◇A" means "It's possible that you do A."

1. If you're accelerating, then you ought not to brake. [Use A and B.]
2. If you're braking, then you ought not to accelerate.
3. You ought not to combine accelerating with braking.
4. If A is wrong, then don't do A.
5. Do A, only if A is permissible.
6. "Do A" entails "A is permissible."
7. Act A is morally indifferent (morally optional).
8. If A is permissible and B is permissible, then A-and-B is permissible.
9. If it's all right for you to believe that, then you ought not to act that way. [Use B and A.]
10. It isn't your duty to do A, but it's your duty not to do A.
11. "Everyone does A" doesn't entail "It would be all right for me to do A." [Use Ax and i.]
12. If you believe that you ought to do A, then you ought to do A. [Use B for "You believe that you ought to do A" and A for "You do A."]
13. You ought not to combine believing that you ought to do A with not doing A.
14. If it's all right for x to do A to y, then it's all right for y to do A to x. [Use Axy.]
15. It's your duty to do A, only if it's possible for you to do A.
16. It's obligatory that the state send only guilty persons to prison. [Use Gx, Sxy, and s.]
17. If you ought to do A, then you'll (in fact) do A.
18. If it isn't possible for everyone to do A, then you ought not to do A. [Use Ax and u.]
19. It's obligatory that someone do A.
20. There's someone who is obligated to do A.
21. If you ought not to do A, then it's obligatory that no one do A.
22. It's all right for someone to do A.
23. It's all right for everyone to do A.
24. If it's all right for you to do A, then it's all right for anyone to do A.
25. It isn't all right for anyone to do A.
26. It's permissible that everyone who isn't sinful be thankful. [Use Sx and Tx.]
27. It's permissible that everyone who isn't thankful be sinful.
28. It's obligatory that Smith help someone or other whom Jones is beating up. [Use Hxy, Bxy, s, and j.]
29. It isn't a duty for everyone to do A.
30. It isn't a duty for anyone to do A.

7.4 DEONTIC ARGUMENTS

We'll now add six inference rules for the deontic operators. The first four are much like the QC and MC rules. We have two reverse squiggle rules, a rule for dropping "O," and a rule for dropping "R."

D1 and D2 are *reverse squiggle rules*. These hold regardless of what pair of contradictory imperative wffs replaces "$\underline{A}$"/"$\sim\underline{A}$":

D1 $\sim R\underline{A} \leftrightarrow O\sim\underline{A}$

D2 $\sim O\underline{A} \leftrightarrow R\sim\underline{A}$

By D1, "A isn't permissible" = "It's obligatory that A not be done." Each means "A is wrong." By D2, "A isn't obligatory" = "It's permissible not to do A." Use these rules only within the same world. We'll usually use them from left to right, to move the deontic operator to the outside:

$\sim R\underline{A}$
$\therefore\ O\sim\underline{A}$

$\sim O\underline{A}$
$\therefore\ R\sim\underline{A}$

Before we go to our next two rules, we need to expand our worlds. From now on, a *possible world* will be a complete and consistent set of indicatives *and imperatives*. We'll define "deontic world" so that these equivalences hold:

> OA = It's obligatory that A.
>   = The imperative "$\underline{A}$" is in *all* deontic worlds.

> R$\underline{A}$ = It's permissible that A.
>   = The imperative "$\underline{A}$" is in *some* deontic worlds.

A *deontic world* is a possible world in which the imperatives prescribe some set of actions jointly permitted by the deontic judgments in the world. A deontic world that depends on the actual world contains:

- the same descriptive and deontic judgments that the actual world contains, and
- imperatives prescribing some set of actions jointly permitted by these deontic judgments.

Later we'll refine these notions further.

We redefine a *world prefix* to be a string of zero or more instances of the letters "W" or "D." As before, world prefixes represent possible worlds. "D," "DD," and so on now represent deontic worlds. We can use these in derived steps and assumptions, such as:

D $\therefore$ $\underline{A}$ ("So $\underline{A}$ is in deontic world-D")

DD asm: $\underline{A}$ ("Assume that $\underline{A}$ is in deontic world-DD")

Given all this, we can explain the rules for dropping deontic operators.

    D3 is the *"O" dropping rule*. It holds regardless of what imperative wff replaces "A̲":

    D3   OA̲ → A̲   The world prefix of the derived step must either be the same as that of the earlier step or else consist of the world prefix of the earlier step plus a string of one or more D's.

If act A is obligatory, then A̲ is in *all* deontic worlds. So if we have "OA̲" in the actual world, then we can derive "∴ A̲," "D ∴ A̲," "DD ∴ A̲," and so on:

OA̲	(Use **any** string
——	of D's—or no
D ∴ A̲	D's at all)

"O" must begin the wff. If "~" or "(" begins the wff, we can't use D3.

    D3 lets us go from "OA̲" in a world to "A̲" in the same world. This accords with "Hare's Law" (named after R. M. Hare):

HARE'S LAW	An ought-judgment entails the corresponding imperative: "You
□(OA̲ ⊃ A̲)	ought to do A" entails "Do A."

Hare's Law equivalently claims that "You ought to—but don't" is inconsistent. The law fails for some weaker senses of "ought." There's no inconsistency in either of these:

"You ought (that is, ought according to company policy) to do this—but don't!"

"You have a duty (a *prima facie* duty that other considerations might override) to keep your promise. But you need to drive this sick person to the hospital. So don't keep your promise."

But Hare's Law seems to hold for the all-things-considered sense of "ought" that we normally use when discussing moral issues. This seems to be inconsistent:

"All things considered, you ought to do A. But don't do A!"

Hare's Law is also called "prescriptivity." Not all philosophers accept it. Those who reject it would want to specify that in applying D3 the world prefix of the derived step can't be the same as that of the earlier step.

    D4 is the *"R" dropping rule*. It holds regardless of what imperative wff replaces "A̲":

    D4   RA → A   The world prefix of the derived step must be like that of the earlier step, except that it ends in a *new* string of one or more D's (a string not occurring in earlier lines).

If act A is permissible, then A is in *some* deontic world. We may give this world an arbitrary *and hence new* name. "R" must begin the wff and we must use a *new* string of D's:

RA	(Use a
———	**new** string
D ∴ A	of D's)

    Here's a simple deontic proof using these rules:

  1.   O~(A · B)   VALID
  2.   OA
     [∴ O~B
* **3.**   ⌈asm: ~O~B
* **4.**   │ ∴ RB [from 3]
  5.   │ D ∴ B [from 4]
  6.   │ D ∴ A [from 2]
* **7.**   │ D ∴ ~(A · B) [from 1]
  8.   ⌊D ∴ ~B [from 6 and 7, contradicting 5]
  9. ∴ O~B [from 3, 5, and 8]

This is just like an MC proof, except for the underlining and the use of "O," "R," and "D" in place of "□," "◇," and "W." As before, we can star (and then ignore) a line when we use a reverse squiggle or "R" dropping rule on it.

    D3 and D4 apply easily to wffs in the actual world:

       RA
       OB
    ————
   D ∴ A
   D ∴ B

But suppose the wffs above the line were in world-W. Then the world prefix of the derived steps would contain "W" (the world prefix of the earlier steps) followed by "D":

   W ∴ RA
   W ∴ OB
   ————————
  WD ∴ A
  WD ∴ B

The provisos on D3 and D4 about world prefixes allow these moves.

World-WD is a deontic world that *depends on* possible world-W. Let "W1" represent any world. Then a deontic world that *depends on* a world-W1 is a possible world that contains:

* the same descriptive and deontic judgments that W1 contains, and
* imperatives prescribing some set of actions jointly permitted by these deontic judgments.

So world-D (that depends on the actual world) contains the same descriptive and deontic judgments that the actual world contains. And world-WD (that depends on world-W) contains the same descriptive and deontic judgments that world-W contains. Then:

> "O$\underline{A}$" is *true in W1*, if and only if the imperative
> "$\underline{A}$" is in *every* deontic world that *depends on W1*.

> "R$\underline{A}$" is *true in W1*, if and only if the imperative
> "$\underline{A}$" is in *some* deontic world that *depends on W1*.

The following proof uses the complex world prefix "WD" in steps 7 to 9:

$$[\therefore \ \Box(O(\underline{A} \cdot \underline{B}) \supset O\underline{A}) \qquad \text{VALID}$$

* **1.** ⌜ asm: $\sim\Box(O(\underline{A} \cdot \underline{B}) \supset O\underline{A})$
* **2.** │ $\therefore \ \Diamond \sim(O(\underline{A} \cdot \underline{B}) \supset O\underline{A})$ [from 1]
* **3.** │ W $\therefore \ \sim(O(\underline{A} \cdot \underline{B}) \supset O\underline{A})$ [from 2]
* **4.** │ W $\therefore \ O(\underline{A} \cdot \underline{B})$ [from 3]
* **5.** │ W $\therefore \ \sim O\underline{A}$ [from 3]
* **6.** │ W $\therefore \ R\sim\underline{A}$ [from 5]
* **7.** │ WD $\therefore \ \sim\underline{A}$ [from 6]
* **8.** │ WD $\therefore \ (\underline{A} \cdot \underline{B})$ [from 4]
* **9.** ⌞ WD $\therefore \ \underline{A}$ [from 8, contradicting 7]
* **10.** $\therefore \ \Box(O(\underline{A} \cdot \underline{B}) \supset O\underline{A})$ [from 1, 7, and 9]

The next two chapters will often use complex world prefixes such as "WD."

We need to explain two more inference rules. The *indicative transfer rule* D5 lets us transfer indicatives freely between a deontic world and whatever world it depends on. This rule holds because a deontic world and whatever world it depends on have the same indicative (descriptive or deontic) wffs. D5 holds regardless of what *descriptive or deontic* wff replaces "A":

> *D5*   A → A   Here the world prefixes of the derived and the deriving steps must be identical except that one ends in one or more D's.

We won't use D5 very often.

Our final inference rule is a version of Kant's Law (named for Immanuel Kant):

KANT'S LAW	"Ought" entails "can"—"You ought to do A" entails "It's possible for you to do A."
$\Box(O\underline{A} \supset \Diamond \underline{A})$	

Kant's Law equivalently claims that "You ought to do A, but A is impossible" is inconsistent. This law fails for some weaker senses of "ought." This is consistent:

> "You ought (that is, ought according to company policy) to do this—but it's impossible."

Company policy may command impossible things! But Kant's Law holds for the all-things-considered sense of "ought" that we normally use when discussing moral issues. This is inconsistent:

> "All things considered, you ought to do A.
> But it's impossible for you to do A!"

No one can have an all-things-considered moral obligation to do what is impossible.

Rule D6 is our version of Kant's Law:

> *D6* $O\underline{A} \rightarrow \Diamond \underline{A}$
>
> ["A is obligatory" $\rightarrow$ "A is logically possible."]

D6 holds regardless of what imperative wff replaces "$\underline{A}$" and what indicative wff replaces "A"—provided that the former is exactly like the latter except for underlining, and every wff out of which the former is constructed is an imperative wff.*

D6 gives us a weak version of Kant's Law. It tells us that "You *ought* to do A" entails "It's *logically possible* for you to do A." This is true as far as it goes, but it would be better to go further. "You *ought* to do A" also entails "You are *capable* of doing A." Conversely, if you are *incapable* of doing something (for example, running a two-minute mile), then you have *no obligation* to do it—even though it may be logically possible for you to do the thing in question.

But even our weak version of Kant's Law is useful for many arguments. And it won't hurt if in analyzing some arguments we interpret "$\Diamond$" in terms of what we are *capable* of doing (instead of what is *logically possible*). If we read "$\Diamond A\underline{x}$" as "x is *capable* of doing A," then we should read "$\Box A\underline{x}$" as "x is *incapable of not* doing A" ("$\sim\Diamond\sim A\underline{x}$").

---

* The proviso outlaws this kind of inference:

> It's obligatory that someone
>   who is lying not lie. $O(\exists x)(Lx \cdot \sim L\underline{x})$
> ∴ It's possible that someone   ∴ $\Diamond(\exists x)(Lx \cdot \sim L\underline{x})$
>   both lie and not lie.

Since the first "Lx" in the premise isn't an imperative wff, this (incorrect) derivation doesn't satisfy D6.

We've mentioned the first two of these four laws already:

Hare's Law:	An "ought" entails the corresponding imperative.
Kant's Law:	"Ought" entails "can."
Hume's Law:	You can't get an "ought" from an "is."
Poincaré's Law:	You can't get an imperative from an "is."

Now we'll briefly consider the last two.

Hume's Law (named for David Hume) claims that we can't validly deduce what we *ought* to do from premises that don't contain "ought" or similar notions. Hume's Law fails for some weak senses of "ought." Given a description of the company policy and of the situation, we can validly deduce what ought (that is, ought according to company policy) to be done. But Hume's Law seems to hold for the all-things-considered sense of "ought" that we normally use when discussing moral issues. It seems to me that we can't validly deduce from purely descriptive (nonmoral) premises a moral conclusion like "All things considered, I ought to return the money."*

Here's a more careful wording of Hume's Law:

HUME'S LAW $\sim\Box(B \supset O\underline{A})$	You can't get an "ought" from an "is"—more precisely: if B is a consistent nonevaluative statement and A represents any simple contingent action, then B doesn't entail "Act A ought to be done."

We need the complex wording because there are trivial cases where we clearly can deduce an "ought" from an "is." (See problems 3, 17, 23, 29, and 30 of Section 7.4A.) But these don't affect the main idea behind Hume's Law.

Poincaré's Law (named for the mathematician Jules Henri Poincaré) is similar. It claims that we can't validly deduce an imperative from indicative premises that don't contain "ought" or similar notions. Here's a more careful wording:

POINCARÉ'S LAW $\sim\Box(B \supset \underline{A})$	You can't get an imperative from an "is"—more precisely: if B is a consistent nonevaluative statement and A represents any simple contingent action, then B doesn't entail "Do act A."

---

* Some philosophers disagree and say we can deduce moral conclusions using only premises about what we like, or what is socially approved, or what accords with evolution, or what God commands. I can't argue against these views here.

Again, the qualifications block trivial objections. (See problem 20 of Section 7.2B.)
We won't build Hume's Law or Poincaré's Law into our system.

The DC proof strategy is much like that of MC. After assuming the opposite of
the conclusion, try to derive a contradiction using the old rules plus the DC rules.
Apply the reverse squiggle rules to put "O" and "R" at the beginning. Then drop
any initial "R," putting each permissible thing into a new and different deontic world.
Lastly drop each initial "O," putting each obligatory thing into each of the old deontic
worlds. Also put each obligatory thing into the actual world, if you have imperative
wffs without deontic operators in the actual world. You may have to use Kant's Law
if the argument contains modal operators. In a few cases you may have to use the
indicative transfer rule.

From now on, we won't do refutations on invalid arguments. Refutations get
messy when we mix various kinds of worlds.

7.4A Exercises

Say whether valid (then give a proof) or invalid (*no refutation necessary*). Remember
that many of the arguments with modal operators require us to use rule D6 (Kant's Law).

0. ∴ ~◊(O$\underline{A}$ · O~$\underline{A}$)

ANSWER: [∴ ~◊(O$\underline{A}$ · O~$\underline{A}$)   VALID
* 1.   ⌈ asm: ◊(O$\underline{A}$ · O~$\underline{A}$)
* 2.   │ W ∴ (O$\underline{A}$ · O~$\underline{A}$)   [from 1]
  3.   │ W ∴ O$\underline{A}$ [from 2]
  4.   │ W ∴ O~$\underline{A}$ [from 2]
  5.   │ W ∴ A   [from 3]
  6.   ⌊ W ∴ ~A   [from 4, contradicting 3]
  7. ∴ ~◊(O$\underline{A}$ · O~$\underline{A}$)   [from 1, 5, and 6]

The wff proved here says "It isn't logically possible that you ought to do A and also
ought not to do A." This principle is correct if we take "ought" in the all-things-considered
sense used in moral discussions. Morality can't make impossible demands on us. If we thought
otherwise (and some people do!), our lives would likely be filled with irrational guilt for not
fulfilling these (impossible) moral demands.

"~◊(O$\underline{A}$ · O~$\underline{A}$)" wouldn't be a correct principle if we took "O" to mean such things
as "ought according to company policy" or "*prima facie* ought." Inconsistent company policies
may require that we do A and also require that we omit doing A. And we often have a *prima
facie* duty to do A and another *prima facie* duty to omit doing A; see my example at the beginning
of Section 7.3.

1. O~$\underline{A}$
∴ O~($\underline{A}$ · $\underline{B}$)
3. b = c
∴ (OF$\underline{ab}$ ⊃ OF$\underline{ac}$)
5. O($\underline{A}$ ∨ $\underline{B}$)
(B ⊃ C)
∴ (~C ⊃ $\underline{A}$)
7. ∴ O(O$\underline{A}$ ⊃ $\underline{A}$)

2. (∃x)OA$\underline{x}$
∴ O(∃x)A$\underline{x}$
4. (A ⊃ O$\underline{B}$)
∴ O(A ⊃ $\underline{B}$)
6. O$\underline{A}$
O$\underline{B}$
∴ ◊(A · B)
8. ∴ O($\underline{A}$ ⊃ R$\underline{A}$)

9. O(x)(Fx ⊃ Gx)
   RFa
   ∴ Ga
11. (x)OFx
   ∴ O(x)Fx
13. O(A ∨ B)
   ∴ (~◇A ⊃ RB)
15. □(A ⊃ B)
   OA
   ∴ OB
17. A
   ∴ O(B ∨ ~B)
19. ∴ O(A ⊃ OA)
21. (OA ⊃ B)
   ∴ R(A · B)
23. A
   ~A
   ∴ OB
25. O(A ⊃ B)
   ∴ (A ⊃ OB)
27. ∴ O(~RA ⊃ ~A)
29. A
   ∴ (A ∨ OB)

10. OA
   OB
   ∴ O(A · B)
12. RA
   ∴ A
14. (A ⊃ OB)
   ∴ O(A ⊃ B)
16. OA
   RB
   ∴ R(A · B)
18. (x)RAx
   ∴ R(x)Ax
20. ∴ (RA ∨ R~A)
22. ~◇A
   ∴ R~A
24. O(x)(Fx ⊃ Gx)
   OFa
   ∴ OGa
26. O(x)Ax
   ∴ (x)OAx
28. ∴ ~(RA · R~A)
30. (A ∨ OB)
   ~A
   ∴ OB

Problems 3, 17, and 23 show how to deduce an "ought" from an "is." If "(A ∨ OB)" is an "ought," then 29 gives another example. If it's an "is," then 30 gives another example. Problem 19 of Section 7.4B gives another example. We formulated Hume's Law so that these examples don't refute it.

7.4B Exercises

For each argument, first evaluate intuitively and then translate into DC and work it out. Say whether valid (then give a proof) or invalid (*no refutation necessary*). To make many of these arguments more plausible, we should take "possible" (and the "◇") to express what is *capable of being done* (instead of what is *logically possible*).

0. You ought not to combine braking with accelerating.
   You ought to brake.
   ∴ You ought to brake and not accelerate.
   [Use B and A.]

   ANSWER: 1.   O~(B · A)   VALID
              2.   OB
                    [∴ O(B · ~A)
         * 3. ⌈ asm: ~O(B · ~A)
         * 4. │ ∴ R~(B · ~A)   [from 3]
         * 5. │ D ∴ ~(B · ~A)   [from 4]
         * 6. │ D ∴ ~(B · A)   [from 1]

7. | D ∴ B   [from 2]
8. | D ∴ A   [from 5 and 7]
9. | D ∴ ~A   [from 6 and 7, contradicting 8]
10. ∴ O(B · ~A)   [from 3, 8, and 9]

1. It isn't all right for you to drink and drive.
You ought to drive.
∴ Don't drink.
[Use K for "You drink" and V for "You drive."]

2. ∴ Either it's your duty to do A or it's your duty not to do A.
[The view called "rigorism" denies that there are morally neutral acts (acts permissible to do and also permissible not to do).]

3. I did A.
I ought not to have done A.
If I did A and it was possible for me not to have done A, then I have free will.
∴ I have free will.
[Use A and F. The argument is from Immanuel Kant, who argued that ethics requires belief in free will.]

4. ∴ If you ought to do A, then do A.

5. ∴ If you ought to do A, then you'll (in fact) do A.

6. It isn't possible for you to be perfect.
∴ It isn't your duty to be perfect.
[Use P.]

7. You ought not to combine drinking with driving.
You don't have a duty to drive.
∴ It's all right for you to drink.
[Use K and V.]

8. If it's all right for you to insult Jones, then it's all right for Jones to insult you.
∴ If Jones ought not to insult you, then don't you insult Jones!
[Use Ixy, u, and j. The premise here follows from the universalizability principle (that what's right for one person would be right for another in similar circumstances) plus the assumption that the circumstances are similar. The conclusion is a distant relative of the golden rule.]

9. ∴ Do A, only if it would be all right for you to do A.

10. ∴ "Do A" entails "A is permissible."

11. It's all right for someone to do A.
∴ It's all right for anyone to do A.
[Can you think of an example to show that this is invalid?]

12. If fatalism (the view that whatever happens could not have been otherwise) is true and I do A, then my doing A (taken by itself) is necessary.
∴ If fatalism is true and I do A, it's all right for me to do A.
[Use F and A.]

13. If it isn't right to do A, then it isn't right to promise to do A.
∴ Promise to do A, only if it's all right to do A.
[Use A and P.]

14. It's obligatory that someone or other answer the phone.
∴ There's some specific person who has an obligation to answer the phone.
[Use Ax.]

15. If it's all right for you to complain, then you ought to take action.
∴ You ought to either take action or else not complain.
   [Use C and T. This is the "Put up or shut up" argument.]
16. Jones ought to be happy in proportion to his moral virtue.
   Necessarily, if Jones is happy in proportion to his moral virtue, then Jones will be rewarded either in the present life or in an afterlife.
   It isn't possible for Jones to be rewarded in the present life.
   If it's possible for Jones to be rewarded in an afterlife, then there is a God.
∴ There is a God.
   [Use H for "Jones is happy in proportion to his moral virtue," P for "Jones will be rewarded in the present life," A for "Jones will be rewarded in an afterlife," and G for "There is a God." This is Kant's moral argument for the existence of God. To make the third premise plausible, we must interpret "possible" as "factually possible" (instead of "logically possible"). But does *"ought to be"* (the first premise uses this—and not the more usual *"ought to do"*) entail "is factually possible"?]
17. If it's right for you to litter, then it's wrong for you to preach concern for the environment.
∴ It isn't right for you to combine preaching concern for the environment with littering.
   [Use P and L.]
18. I ought to stay with my brother while he's sick in bed.
   It's impossible for me to combine these two things: staying with my brother while he's sick in bed and driving you to the airport.
∴ It's all right for me not to drive you to the airport.
   [Use S and D.]
19. Studying is needed to become a teacher.
   "Become a teacher" entails "Do what is needed to become a teacher."
   "Do what is needed to become a teacher" entails "If studying is needed to become a teacher, then study."
∴ You *ought* to either study or not become a teacher!
   [Use N for "Studying is needed to become a teacher," B for "You become a teacher," D for "You do what is needed to become a teacher," and S for "You study." This is the ought-version of the last example of Section 7.2B. It shows that we can deduce a *complex* ought-judgment from purely descriptive premises.]
20. It's permissible that you do A.
   It's permissible that you do B.
∴ It's permissible that you do A-and-B.
   [Can you think of an example to show that this is invalid?]
21. You ought to pay by check or pay by Mastercard.
   If your Mastercard is expired, then you ought not to pay by Mastercard.
∴ If your Mastercard is expired, then pay by check.
   [Use C, M, and E.]
22. If you ought to be better than everyone else, then it's obligatory that everyone be better than everyone else.
   "Everyone is better than everyone else" is self-contradictory.
∴ It's all right for you not to be better than everyone else.
   [Use Bx for "x is better than everyone else" and u.]
23. You're obliged to either pay cash or pay by check.
∴ Either you're obliged to pay cash or you're obliged to pay by check.
   [Use C and K.]

24. If the results of everyone stealing would be disastrous, then it wouldn't be all right for everyone to steal.

The results of everyone stealing would be disastrous.

∴ It wouldn't be all right for you to steal.

[Use D for "The results of everyone stealing would be disastrous," Sx for "x steals," and u.]

25. "Do A" is logically consistent with "It would be permissible to omit doing A."

∴ "A ought to be done" isn't logically equivalent to "Do A."

["P is logically equivalent to Q" means "Necessarily, P if and only if Q." This argument attacks an analysis of ought-judgments in terms of imperatives. The analysis is noncognitivist in that it claims that ought-judgments, like imperatives, aren't true or false.]

26. "Everyone breaks promises" is impossible.

∴ It's all right for there to be someone who doesn't break promises.

[Use Bx. Kant thought that universal promise-breaking would be impossible, since no one would make promises if everyone broke them. But he wanted to draw the stronger conclusion that it's always wrong to break promises. See problem 34.]

27. ∴ "It's obligatory to do A-*and*-B" logically entails "It's obligatory to do B."

[This is valid if "and" stands for logical conjunction. But "and" can mean other things. "You do that *and* I'll quit" means "*If* you do that, *then* I'll quit." Likewise, "You ought to take an apple *and* pay for it" might mean "You ought to take an apple—*and, if* you do take it, *then* you ought to pay for it" ["(O̲T̲ · (T ⊃ OP))"]. This by itself doesn't entail "You ought to pay for the apple." In doing these exercises, assume that "and" represents logical conjunction.]

28. It's all right for you to punish Judy for the accident, only if Judy ought to have stopped her car more quickly.

Judy couldn't have stopped her car more quickly.

∴ You ought not to punish Judy for the accident.

[Use P and S.]

29. You ought to help your neighbor.

It ought to be that, if you (in fact) help your neighbor, then you tell him that you'll help him.

You don't help your neighbor.

If you don't help your neighbor, then you ought not to tell him that you'll help him.

∴ You ought to tell your neighbor that you'll help him and you ought not to tell him that you'll help him.

[Use H and T. Roderick Chisholm pointed out that this clearly invalid argument was provable in many systems of deontic logic. Is it provable in our system?]

30. It's obligatory that Smith help someone or other whom Jones is beating up.

∴ It's obligatory that Jones beat up someone.

[Use Hxy, Bxy, s, and j. This "good Samaritan paradox" is provable in most deontic systems that attach "O" to indicative wffs. There are similar examples where the evil deed happens *after* the good deed. It might be obligatory that Smith warn someone or other that Jones *will* try to beat up. This doesn't entail that Jones ought to try to beat up someone.]

31. You ought to give to some charity or other.

∴ There is some particular charity that you ought to give to.

[Use Cx, Gxy, and u.]

32. If you take logic, then you'll make mistakes.

You ought not to make mistakes.

∴ You ought not to take logic.
   [Use T and M.]

33. If I ought to name you acting mayor because you served 20 years on the city council, then I ought to name Jennifer acting mayor because she served 20 years on the city council.
    I couldn't name both you and Jennifer as acting mayor.
∴ "I ought to name you acting mayor because you served 20 years on the city council" is false.
   [Use U and J.]

34. If it's all right for you to do A, then you ought to do A.
    If you ought to do A, then it's obligatory that everyone do A.
∴ If it's logically impossible that everyone do A, then you ought not to do A.
   [Use Ax and u. The premises here are doubtful, as is the conclusion. (The conclusion entails "If it's logically impossible that everyone become the first woman president, then you ought not to become the first woman president.") The conclusion is a relative of Immanuel Kant's formula of universal law. It's also a *"formal ethical principle"*—an ethical principle that we can formulate using the abstract notions of a system of logic but leaving unspecified the meaning of the individual, property, relational, and statement letters. We'll see formal ethical principles of greater plausibility in the next two chapters.]

35. If killing the innocent is wrong, then one ought not to intend to kill the innocent.
    If it's permissible to have a nuclear retaliation policy, then intending to kill the innocent is permissible.
∴ If killing the innocent is wrong, then it's wrong to have a nuclear retaliation policy.
   [Use K, I, and N. This is a common argument against a defense policy of retaliating for a nuclear attack.]

8

Belief Logic

8.1 BELIEF TRANSLATIONS

Our *belief calculus* (BC) builds on previous systems. It deals with rational believing and willing. Even though it adds only one symbol and two inference rules, it's a difficult system. We'll start with a simplified version of BC. Later, we'll give a more complex version with more qualifications.

We'll use ":" to construct descriptive and imperative belief formulas. Here's how to construct the former:

1. The result of writing a small letter and then ":" and then a wff is a *descriptive* wff.

For now, we'll put only indicatives after ":" and we'll translate ":" as "believes." Here are some simple examples:

$$u:A = \text{You believe that A.}$$
$$= \text{You believe that A is true.}$$
$$= \text{You endorse (assent to, say in your heart) A.}$$

$$u:\sim A = \text{You believe that A is false.}$$

$$\sim u:A = \text{You don't believe that A is true.}$$

The last one might be true because you believe that A is false or because you take no position either way. Here are some further examples:

$$(\sim u:A \cdot \sim u:\sim A) = \text{You take no position on A.}$$
$$= \text{You don't believe that A is true, and you don't believe that A is false.}$$

$$(\underline{u}:A \supset \sim\underline{u}:B) = \text{If you believe A, then you don't believe B.}$$
$$(\underline{u}:A \supset \underline{u}:B) = \text{If you believe A, then you believe B.}$$

$$\underline{u}:OA\underline{u} = \text{You believe that you ought to do A.}$$
$$\underline{u}:(x)OAx = \text{You believe that everyone ought to do A.}$$
$$(x)x:OA\underline{u} = \text{Everyone believes that you ought to do A.}$$

Here's the rule for constructing imperative belief formulas, together with some examples:

2. The result of writing an *underlined* small letter and then "`:`" and then a wff is an *imperative* wff.

$$\underline{u}:A = \text{Believe that A.}$$
$$= \text{Believe that A is true.}$$
$$= \text{Endorse (assent to, say in your heart) A.}$$

$$\underline{u}:\sim A = \text{Believe that A is false.}$$

$$\sim\underline{u}:A = \text{Don't believe that A is true.}$$

$$(\sim\underline{u}:A \cdot \sim\underline{u}:\sim A) = \text{Take no position on A.}$$
$$= \text{Don't believe that A is true, and don't believe that A is}$$
$$\text{false.}$$

$$\sim(\underline{u}:A \cdot \underline{u}:B) = \text{Don't combine believing A with believing B.}$$

$$\sim(\underline{u}:A \cdot \sim\underline{u}:B) = \text{Don't combine believing A with not believing B.}$$
$$= \text{Don't believe A without believing B.}$$

$$\underline{u}:OA\underline{u} = \text{Believe that you ought to do A.}$$
$$\underline{u}:(x)OAx = \text{Believe that everyone ought to do A.}$$
$$(x)\underline{x}:OA\underline{u} = \text{Let everyone believe that you ought to do A.}$$

8.1A Exercises

Using these letters, translate each sentence into a BC wff:

u = you (singular)　　　　G = There is a God

0. You believe that there is a God.　[You're a *theist*.]
ANSWER:　　　u:G

1. You believe that there is no God.　[You're an *atheist*.]
2. You take no position on whether there is a God.　[You're an *agnostic*.]
3. You don't believe that there is a God.　[You're a *nontheist*.]
4. You believe that "There is a God" is self-contradictory.

5. Necessarily, if you're a theist, then you aren't an atheist. [Is this statement true?]
6. Believe that there is a God!
7. Believe that there is a God, only if it's logically possible that there is a God.
8. If you believe A, then you don't believe not-A.
9. If you believe A, then don't believe not-A.
10. Don't combine believing A with believing not-A.

8.2 THE LOGIC OF CONSISTENT BELIEF

There are three approaches we might take to belief logic. On the first approach, belief logic deals with what belief formulas validly follow from what other belief formulas. We might try to prove inferences such as this one:

<blockquote>
You believe A.   u:A
∴ You don't believe not-A. ∴ ~u:~A
</blockquote>

But this inference is invalid. People can be confused and illogical. Students and politicians can assert A and assert not-A almost in the same breath. Given that someone believes A, we can deduce little or nothing about what else the person may believe. So this first approach is doomed from the start.

The other two approaches use the following technical notion of a (completely) consistent being:

<blockquote>
Let S be the nonempty set of propositions that x believes.
Then x is (completely) consistent, if and only if:
 1. set S is logically consistent, and
 2. x believes anything that follows logically from set S.
</blockquote>

This has to do with *consistent believing*. Later we'll broaden the definition to cover *consistent willing* as well.

The second approach to belief logic studies how people *would* believe *if they were consistent*. If we added the premise "You're completely consistent," then our previous argument would become valid:

<blockquote>
(You're completely consistent.)
You believe A.   u:A
∴ You don't believe not-A. ∴ ~u:~A
</blockquote>

A belief logic on this approach takes "You're completely consistent" as an implicit premise of its arguments. This premise is assumed, even though it's false, to help us explore what belief patterns a consistent being would follow. This second approach to belief logic is a good one.* But I prefer a third approach, in view of what I want to do later.

---

* Jaakko Hintikka uses roughly this second approach in his classic *Knowledge and Belief* (Ithaca, New York: Cornell University Press, 1962).

My approach is to construct a belief logic that generates consistency imperatives like this one:

∴ Don't combine believing A   ∴ ~(u:A · u:~A)
    with believing not-A.

Belief logic then tells us to avoid inconsistent combinations. The above premiseless argument isn't valid in the normal sense. But it becomes valid if we add the premise, "You *ought* to be consistent." We'll assume such an implicit premise in all our BC arguments. So when we call an argument "valid in BC" we'll mean that it's valid if we assume this additional premise. Here are two further arguments that are valid in BC in this extended sense:

A is inconsistent with B.   ~◇(A · B)
∴ Don't combine believing A.   ∴ ~(u:A · u:B)
    with believing B.

A logically entails B.   □(A ⊃ B)
∴ Don't combine believing A   ∴ ~(u:A · ~u:B)
    with not believing B.

BC forbids combinations of believing or not-believing that would eliminate us from the ranks of "completely consistent beings."

Let's formulate this last idea more precisely. Let an *atomic belief formula* be any wff of the form "x:A" or "~x:A" (where any small letter and wff may replace "x" and "A"). And let a *belief combination* be any conjunction of one or more atomic belief formulas using the same small letter (before ":"). Here are six belief combinations:

(x:A · x:~A)   x:A
(x:A · x:B)   (~x:A · ~x:B)
(x:A · ~x:B)   ((x:A · x:B) · x:C)

A belief combination is *inconsistent* if it would be impossible for a completely consistent being to believe in the way described. Our first belief combination is inconsistent:

(x:A · x:~A)

The others could also be inconsistent, depending on what the letters represent. Suppose that A is a self-contradiction. Then this belief combination is also inconsistent:

x:A

We want BC to forbid inconsistent belief combinations.

A *consistency imperative* is the result of writing "~" at the beginning of an inconsistent belief combination and making each conjunct into an imperative. Here's an example:

INCONSISTENT BELIEF COMBINATION	CONSISTENCY IMPERATIVE
(u:A · u:~A) You believe A and you believe not-A.	~(u:A · u:~A) Don't combine believing A with believing not-A!

We want BC to generate such consistency imperatives.

Similarly, if A is self-contradictory, then "x:A" ("x believes A") is an inconsistent belief combination and "~x:A" ("x, don't believe A!") is a consistency imperative. So we want BC to generate such arguments as this one:

$$\text{A is self-contradictory.} \qquad \sim \Diamond A$$
$$\therefore \text{ Don't believe A.} \qquad \therefore \sim \underline{u}{:}A$$

Recall our definition of a "consistent being":

> Let S be the nonempty set of propositions that x believes.
> Then x is (*completely*) *consistent*, if and only if:
>   1. set S is logically consistent, and
>   2. x believes anything that follows logically from set S.

Let's call any possible world that contains set S a *belief world of x*. In other words, a belief world of x is any possible world consistent with everything that x believes. Then x is a (completely) consistent being if these conditions are satisfied:

(a) There are belief worlds of x, and

(b) "x believes A" is true, if and only if A is in all of x's belief worlds.

By (a), some possible world contains the set S of propositions that x believes. So set S is logically consistent. This satisfies the first condition of our definition. Now any belief world of x, being a possible world, contains anything logically entailed by any set of propositions that it contains. So every belief world of x contains whatever follows from set S. So, by (b), x believes anything that follows from set S. This satisfies the second condition of our definition.

Suppose that we have a set of *imperative belief formulas* and we want to test whether x can consistently believe as directed by this set. To test this, we'll assume that x in believing this way could satisfy conditions (a) and (b). Then we'll see if this assumption is possible.

This leads to two inference rules. Following (a), we assume that x would have at least one belief world. We call the belief worlds of x "world-x," "world-xx," "world-xxx," and so forth. (These might be the same world.) By (b), if x is to believe A, then A would be in *all* of x's belief worlds. This gives us inference rule B+ to use on *positive imperative belief formulas*:

$$B+ \quad \boxed{\begin{array}{l} \underline{x}{:}A \\ \hline x \therefore A \end{array}} \quad \begin{array}{l}\text{(use \textbf{any} string} \\ \text{of one or} \\ \text{more x's)}\end{array}$$

> Assuming that x rationally believes
> as prescribed: if x is to believe A,
> then A is in *every* belief world of x.

By (b), if x *is not* to believe A, then *not all* belief worlds of x have A. Then *some* belief world of x has not-A. This gives us rule B − to use on *negative imperative belief formulas*:

$$B- \quad \boxed{\begin{array}{l} {\sim}\underline{x}{:}A \\ \hline x \therefore {\sim}A \end{array}} \quad \begin{array}{l}\text{(use a} \\ \textbf{new} \text{ string} \\ \text{of x's)}\end{array}$$

> Assuming that x rationally believes as
> prescribed: if x *is not* to believe A,
> then *not*-A is in *some* belief world of x.

Our proof strategy will go as follows:

- If you have *negative imperative belief formulas*, use rule B − on these first. Put the *denial* of each thing that *isn't* to be believed in a *new* belief world. You can star (and then ignore) a line when you use B − on it.
- If you have *positive imperative belief formulas*, use rule B + on these next. Put the thing to be believed in *each* old belief world that you have. (Use a single new belief world if you have no old belief worlds.) Don't star a line when you use B + on it.

Note that both rules operate only on *imperative* belief formulas. We'll formulate these rules more precisely later.

Here's an example of a proof:

∴ Don't combine believing A with believing not-A.

```
     [∴ ~(u:A · u:~A)              VALID
*1.   ⌐asm: (u:A · u:~A)
 2.   │∴ u:A    [from 1]
 3.   │∴ u:~A   [from 1]
 4.   │u ∴ A    [from 2]
 5.   │u ∴ ~A   [from 3, contradicting 4]
 6.  ∴ ~(u:~A · u:~A)   [from 1, 4, and 5]
```

After assuming the denial of what we want to prove, we break up the assumption using PC rules (steps 1 to 3). Then we use rule B + to get steps 4 and 5—and a contradiction. This shows that a completely consistent being couldn't believe in the manner prescribed. We derive our original conclusion using RAA: "Don't combine believing A with believing not-A."

Our proof doesn't show that the formula is logically necessary. Rather it shows that it's a consistency imperative that follows from an implicit "One *ought* to be consistent" premise. Having rules B + and B − is equivalent to assuming in every BC argument an implicit "One *ought* to be consistent" premise.

Here's another proof:

A logically entails B.
∴ Don't combine believing A with not believing B.

1.	☐(A ⊃ B)		VALID
	[∴ ~(u:A · ~u:B)		
*2.	asm: (u:A · ~u:B)		
3.	∴ u:A [from 2]		
*4.	∴ ~u:B [from 2]		
5.	u ∴ ~B [from 4]		
6.	u ∴ A [from 3]		
*7.	u ∴ (A ⊃ B) [from 1]		
8.	u ∴ ~A [from 5 and 7, contradicting 6]		
9.	∴ ~(u:A · ~u:B) [from 2, 6, and 8]		

This proof goes much like before. But now we first use rule B − on the *negative* belief formula in step 4 (and star this step) to get step 5. Then we use rule B + on the *positive* belief formula in step 3 to get step 6.

The *don't-combine* form of the conclusion here is important. Suppose that we have an argument where P (PREMISE) logically entails C (CONCLUSION). Compare these three forms:

(u:P ⊃ u:C)	If you believe the PREMISE, then believe the CONCLUSION.
(~u:C ⊃ ~u:P)	If you don't believe the CONCLUSION, then don't believe the PREMISE.
~(u:P · ~u:C)	Don't combine believing the PREMISE with not believing the CONCLUSION.

Suppose you believe the PREMISE but don't believe the (logically entailed) CONCLUSION. Then you're violating all three imperatives. But what should you do next? That depends. Maybe the PREMISE is solidly based and you should believe it and the entailed CONCLUSION. Or maybe the CONCLUSION is stupid and you should drop it and the PREMISE that entails it. The last imperative, with its *don't-combine* form, leaves open both possibilities. It simply forbids the inconsistent combination of believing the PREMISE but not believing the entailed CONCLUSION. It doesn't tell you what to do if you get into this forbidden combination. The first two imperatives tell you exactly what to do. The first tells you to stick with the CON-CLUSION—even though it may be stupid! The second tells you to drop the PREMISE—

even though it may be solidly based! So the *if-then* forms can tell you to do the wrong thing. The *don't-combine* form is superior. We'll use this form for our BC consistency imperatives.

Let's take another example. Assume that A is logically inconsistent with B. Compare these three forms:

(u:A ⊃ ~u:B)	If you believe A, then don't believe B.
(u:B ⊃ ~u:A)	If you believe B, then don't believe A.
~(u:A · u:B)	Don't combine believing A with believing B.

Suppose you believe A and also believe B, although the two are logically inconsistent. The first form tells you to drop B, while the second tells you to drop A. But the last form leaves it open whether you're to drop A or drop B. Which you should drop depends on the situation. BC's consistency imperatives have a *don't-combine* form. They just tell you to make your beliefs logically coherent with each other. But they don't say what beliefs to add or subtract to bring about this coherence.

We'll now formulate our inference rules B + and B − more precisely. We redefine a *world prefix* to be any string of zero or more instances of letters from the set <W, D, a, b, c, . . .>—where <a, b, c, . . .> is the set of small letters (including those with primes). This *positive belief rule* holds regardless of what wff replaces "A" and what small letter replaces "x":*

$B+$ x:A → A   The derived line's world prefix must consist of that of the earlier line plus a string of one or more x's. Also, neither line's world prefix can contain small letters or a "W."

This *negative belief rule* holds regardless of what pair of contradictory wffs replaces "A"/"~A" and what small letter replaces "x":

$B-$ ~x:A → ~A   The derived line's world prefix must consist of that of the earlier line plus a *new* string of x's (one not occurring in earlier lines). Also, neither line's world prefix can contain small letters or a "W."

B − is as close as BC gets to a reverse squiggle rule. Note that the outside "~" stays when you derive the new step.

This chart might be helpful for proofs involving operators of various types:

---

* The second sentence of the proviso on B + and B − blocks the proof of puzzling wffs that place one belief operator within another [for example, "z:~(x:A · x:~A)"] or claim logical necessity for consistency imperatives [for example, "□~(x:A · x:~A)"].

FIRST DROP THESE WEAK OPERATORS	THEN DROP THESE STRONG OPERATORS
~x: R ◊ (∃x) Use new worlds/constants. Star the line used.	x: O □ (x) Use old worlds/constants if you have them. Don't star the line used.

In addition, we should drop "x:" and "O" before dropping the very strong "□."
Other than that, the order of dropping operators within each of the two groups doesn't
matter much.*

In Section 5.2, we saw that the "Equals may substitute for equals" rule Q6 can
fail in arguments about beliefs. Consider this example:

> j:Dc   Jones believes that *Cicero* denounced Cataline.
>
> c = t   *Cicero* is the same person as *Tully*.
>
> ∴ j:Dt   Jones believes that *Tully* denounced Cataline.

In ancient Rome, Cicero (= Tully) denounced Cataline. Jones might know enough
history to believe that *Cicero* denounced Cataline, but not enough to believe that *Tully*
denounced Cataline. Then the premises would be true but the conclusion false. So the
argument is invalid. But yet the conclusion results from the premises using our "Equals
may substitute for equals" rule Q6:

> Q6   Fa, a = b → Fb

We have to qualify rule Q6 so it doesn't apply in belief contexts. From now on,
Q6 holds *only if no interchanged instance of the constants occurs within a wff immediately preceded by a small letter (underlined or not) followed by ":".*

8.2A Exercises

Say whether valid (give a proof) or invalid (no refutation necessary).

> 0. □(A ⊃ B)
> ∴ (u:A ⊃ u:B)

>   ANSWER: 1.   □(A ⊃ B)   INVALID
>   [∴ (u:A ⊃ u:B)
>   *   2.   asm: ~(u:A ⊃ u:B)
>   3.   ∴ u:A   [from 2]
>   *   4.   ∴ ~u:B   [from 2]
>   5.   u ∴ ~B   [from 4]
>   *   6.   u ∴ (A ⊃ B)   [from 1]
>   7.   u ∴ ~A   [from 5 and 6]

---

* Why doesn't BC follow the pattern of QC, MC, and DC in having four inference rules: two for
reversing squiggles and two for dropping operators? The main reason is that we don't have any convenient
word for "not believing false." Thus it would be difficult to work with a symbol that means this.

Rules B + and B − work only on imperative belief formulas. So we can't go from "u:A" in line 3 to "A" in belief world-u. The conclusion here has the bad *if-then* form. Suppose that A entails B and you believe A. It doesn't follow that you should believe B. Maybe you should reject A and also reject B.

1. ~◇(A · B)
∴ ~(u:A · u:B)
3. ~◇(A · B)
∴ (u:B ⊃ ~u:A)
5. □(A ⊃ B)
   u:A
∴ u:B
7. □(A ⊃ B)
   ~u:~A
∴ ~u:~B
9. ~◇(A · B)
∴ ~(u:A · ~u:~B)

2. ~◇(A · B)
∴ (u:A ⊃ ~u:B)
4. ~◇(A · B)
∴ (~u:A ∨ ~u:B)
6. □(A ⊃ B)
   u:A
∴ u:B
8. □(A ⊃ B)
   ~u:B
∴ u:~A
10. ~◇(A · B)
∴ ~(~u:~A · ~u:~B)

8.2B Exercises

For each argument, first evaluate intuitively and then translate into BC and work it out. Say whether valid (then give a proof) or invalid (no refutation necessary).

0. Believe A.
∴ Don't believe not-A.

ANSWER: 1.   u:A   VALID
             [∴ ~u:~A
          2. ⌈ asm: u:~A
          3. | u ∴ A   [from 1]
          4. ⌊ u ∴ ~A   [from 2, contradicting 3]
          5. ∴ ~u:~A   [from 2, 3, and 4]

The conclusion here isn't a consistency imperative. But the premise plus the provable consistency imperative "~(u:A · u:~A)" logically entails the conclusion.

1. A is logically inconsistent with B.
   You (in fact) believe A.
∴ Don't believe B.
2. ∴ Either believe A or believe not-A.
3. A is self-contradictory.
∴ Don't believe A.
4. ∴ Don't believe that A is true without believing that A is possible.
5. ∴ If you believe A, then you don't (in fact) believe not-A.
6. A logically entails B.
   Don't believe B.
∴ Don't believe A.
7. (A and B) entails C.
∴ Don't combine believing A and believing B, with not believing C.

8. A logically entails (B and C).
Don't believe that B is true.
∴ Believe that A is false.
9. ∴ If A is true, then believe A.
10. ∴ Believe A, only if A is logically possible.

8.3 BELIEVING AND WILLING

Now we'll expand BC to include *willing* as well as *believing*. We'll interpret ":" to mean "accepts." To accept an *indicative* is to *believe* it to be true:

x:A = x *accepts* (endorses, assents to, says in his or her
        heart) "A is true."
    = x *believes* that A.

Similarly, to accept an *imperative* is to *will* (in the fullest way) the corresponding action(s) to be done:

x:$\underline{A}$ = x *accepts* (endorses, assents to, says in his or her
        heart) "Let act A be done!"
    = x *wills* that act A be done.

In translating "x:$\underline{A}$," we'll often use terms more specific than "wills." We'll use terms like "acts," or "resolves to act," or "desires." Which of these fits depends on whether the imperative is present or future—and whether it applies to oneself or to another. Here are three examples:

> If "Ax" represents a present act, then:
>
> x:A$\underline{x}$ = x acts (in order) to do A.

[To accept a present tense imperative addressed to oneself (for example, "Gensler, do this now!" or "Let me do this now!") is to *act* (in order) to do the thing.]

> If "Ax" represents a future act, then:
>
> x:A$\underline{x}$ = x is resolved to do A.

[To accept a future tense imperative addressed to oneself (for example, "Gensler, do this later!" or "Let me do this later!") is to *be resolved* to do the thing.]

> If x ≠ y, then:
>
> x:A$\underline{y}$ = x wants y to do A.

[To accept an imperative addressed to another ("Jones, do this!" or "Would that Jones do this!") is to *want* (or desire) the thing to be done.]

Similarly, to accept "Would that I had done that!" is to wish that the thing had been done, or be sorry that it wasn't done.

There's a subtle difference between "x:A$\underline{x}$" and "Ax":

> x:A$\underline{x}$ = x acts (in order) to do A.
> = x says in his or her heart, "Would that I do A!"

> Ax = x does A.

"x:A$\underline{x}$" has to do with the imperative (or "plan") that x is acting on. Thus "x:A$\underline{x}$" is about what x *intends* to do. But "Ax" is about what x *actually does*. You might act (in order) to throw a touchdown pass, but fail. Then "u:T$\underline{u}$" is true, but "Tu" is false. Or you might throw an interception without wanting to. Then "Iu" is true, but "u:I$\underline{u}$" is false.

We can also give formulas telling someone to will something. Here are examples:

> $\underline{x:A}$ = x, accept (endorse, assent to, say in your heart) "Let act A be done!"!
> = x, will that act A be done!

> If "Ax" represents a present act, then:
> $\underline{x:Ax}$ = x, act (in order) to do A!

> If "Ax" represents a future act, then:
> $\underline{x:Ax}$ = x, be resolved to do A!

> If x $\neq$ y, then:
> $\underline{x:Ay}$ = x, want y to do A!

The difficult part in symbolizing such sentences is knowing what to underline. Study these rules and examples:

IF THE SENTENCE
is about *believing*, put an indicative after "`:`"
is about *willing*, put an imperative after "`:`"
describes what is believed or willed, then don't underline the letter before "`:`"
tells what to believe or will, then underline the letter before "`:`"

INDICATIVES	IMPERATIVES
u:A (You believe A.)	u̲:A (Believe A!)
u:A̲ (You will A.)	u̲:A̲ (Will A!)

Here are some further translations:

Alicia wants everyone to admire her.
= Alicia says in her heart, "Would that everyone admire Alicia!"
= a:(x)Ax̲a

If Roderick believes he ought to quit, then he'll quit.
= If Roderick says in his heart "I ought to quit," then he'll quit.
= (r:OQr̲ ⊃ Qr)

If you want to succeed, then don't think that it's impossible for you to succeed.
= If you say in your heart, "Succeed!" then don't say in your heart "It's impossible for me to succeed."
= (u:Su̲ ⊃ ~u:~ ◊ Su)

We noted in Section 7.3 that, if we prefixed "O" to indicatives (instead of to imperatives), then we wouldn't be able to make important distinctions. Something similar applies here. Consider these three wffs:

u:(∃x)(Kx · Rx̲) = You desire that some who kill *repent*.
                = You say in your heart "Would that some who kill *repent*."

u:(∃x)(Kx̲ · Rx) = You desire that some *kill* who repent.
                = You say in your heart "Would that some *kill* who repent."

u:(∃x)(Kx̲ · Rx̲) = You desire that some both *kill* and *repent*.
                = You say in your heart "Would that some *kill* and *repent*."

The three are very different. The underlining shows what parts are desired: repenting, or killing, or killing-and-repenting. If we attached "desire" to indicative formulas, then all three would translate the same way:

You desire that "(∃x)(Kx · Rx)" be true.
= You desire it to be the case that there is someone who both kills and repents.

Because of such examples, it's better to symbolize "desire" in terms of accepting an imperative.

Be careful in translating *if-then* and *don't-combine* forms (see Section 7.2):

If you believe that it's wrong for you to do A, then don't do A.
= (u:O~Au̲ ⊃ ~Au̲)

If you do A, then don't believe that it's wrong for you to do A.
= (Au̲ ⊃ ~u:O~Au̲)

Don't combine *believing* that it's wrong for you to do A with *acting* to do A.
= ~(u:O~Au · u:Au)

The last of these is provable in BC (see problem 0 of Section 8.3C). BC demands consistency between beliefs and actions. "Consistency" now is taken more broadly than before:

Let S be the nonempty set of *indicatives and imperatives* that x accepts.
Then x is *(completely) consistent*, if and only if:
1. set S is logically consistent, and
2. x accepts anything that follows logically from set S.

This definition covers consistent *believing and willing*.

Proofs go much the same as before. But now we have to be more careful about underlining.

8.3A Exercises

Translate each sentence into a BC wff. Use u for "you" and j for "Jones."

0. Don't act to do A without holding that A would be all right.
ANSWER: ~(u:Au · ~u:RAu)

1. You want Jones to sit down. [Use Sx for "x sits down."]
2. You think Jones ought to sit down.
3. Eat nothing! [Use Exy for "x eats y."]
4. You resolve to eat nothing.
5. You fall down, but you don't act (in order) to fall down. [Use Fx.]
6. You act to kick the goal, but you don't in fact kick the goal. [Use Kx.]
7. If you believe that you ought to do A, then do A.
8. Don't combine believing that you ought to do A with not acting to do A.
9. Don't resolve to eat nothing! [Use Exy.]
10. Don't combine resolving to eat nothing, with acting to eat this. [Use Exy and t.]
11. You're resolved that if killing were needed to save your family, then you wouldn't kill. [Use N and Kx.]
12. Do A, only if you want everyone to do A. (Act only as you'd want everyone to act.) [This is a crude version of Kant's formula of universal law. What are some objections to it?]
13. If x does A to you, then do A to x. (Treat others as they treat you.) [Use Axy. This principle, often confused with the golden rule, entails "Knock out x's eye if x knocks out your eye."]
14. If you do A to x, then x will do A to you. (People will treat you as you treat them.) [Use Axy. This one is also confused with the golden rule.]
15. If you want x to do A to you, then do A to x. (Treat others as you want to be treated.) [Use Axy. This is a crude version of the golden rule. What are some objections to it?]

8.3B Exercises

Say whether valid (give a proof) or invalid (no refutation necessary).

0. ∴ (u:O~A̲u̲ ⊃ ~u:A̲u̲)
ANSWER: [∴ (u:O~A̲u̲ ⊃ ~u:A̲u̲)   INVALID
* 1.   asm: ~(u:O~A̲u̲ ⊃ ~u:A̲u̲)
  2.   ∴ u:O~A̲u̲ [from 1]
  3.   ∴ u̲:A̲u̲ [from 1]
  4.   u ∴ A̲u̲ [from 3]

Here the wff doesn't have the correct *don't-combine* form. In our proof, we can't put "O~A̲u̲" in belief world-u, since "u:O~A̲u̲" in step 2 is an indicative.

This formula means, "If you believe that it's wrong for you to do A, then don't act to do A." But maybe you believe, irrationally, that it's wrong to treat dark-skinned people fairly. It doesn't follow that you aren't to act fairly toward dark-skinned people. Maybe you should act fairly toward everyone, and give up your belief to the contrary.

1. ∴ ~(u̲:A · u̲:~A̲)   **2.** ∴ u:(Ba ⊃ RBa)
3. ∴ (u:B̲a̲ ∨ u̲:~B̲a̲)   **4.** ∴ ~((u̲:(A ⊃ B) · u̲:A) · ~u̲:B)
5. u:(x)OA̲x̲   **6.** ~u:Au̲
∴ u̲:A̲u̲   ∴ ~u̲:OA̲u̲
7. ∴ u̲:(OA̲u̲ ⊃ A̲u̲)   **8.** ∴ (u̲:A̲u̲ ∨ ~u̲:OA̲u̲)
9. u̲:A̲u̲   **10.** ☐(A̲ ⊃ B̲)
∴ ~u̲:O~A   ∴ ~(u̲:OA̲ · ~u̲:B̲)

8.3C Exercises

For each argument, first evaluate intuitively and then translate into BC and work it out. Say whether valid (then give a proof) or invalid (no refutation necessary).

0. ∴ Don't combine *believing* that it's wrong for you to do A with *acting* to do A.
ANSWER: [∴ ~(u:O~A̲u̲ · u̲:A̲u̲)   VALID
* 1.   ┌ asm: (u:O~A̲u̲ · u̲:A̲u̲)
  2.   │ ∴ u:O~A̲u̲ [from 1]
  3.   │ ∴ u̲:A̲u̲ [from 1]
  4.   │ u ∴ O~A̲u̲ [from 2]
  5.   │ u ∴ A̲u̲ [from 3]
  6.   └ u ∴ ~A̲u̲ [from 4, contradicting 5]
  7.   ∴ ~(u:O~A̲u̲ · u̲:A̲u̲) [from 1, 5, and 6]

It's important to translate "acting to do A" as "u̲:A̲u̲" and not "A̲u̲." On the wrong translation, the formula forbids accidentally doing what one thinks is wrong. (There's no inconsistency in this, except maybe on an external level.) The correct version forbids this inconsistent combination: thinking that A is wrong and at the same time acting with the intention of doing A.

1. ∴ If you believe that you ought to do A, then do A.
[This *if-then* interpretation of the "Follow your conscience" principle is faulty. It tells people to commit mass murder, if they believe that they ought to do this.]

2. ∴ Don't combine *believing* that you ought to do A with *not acting* to do A.
[This *don't-combine* interpretation of "Follow your conscience" is better.]

3. ∴ Don't combine accepting "It's wrong for Jones to do A," with wanting Jones to do A.
[Use j.]

4. ∴ Don't *believe* that the state ought to execute all murderers, without *desiring* that if your friend be a murderer then the state execute your friend.
[Use s for "the state," Exy for "x executes y," Mx for "x is a murderer," f for "Your friend," and u.]

5. ∴ Don't act to do A without accepting that A is all right.

6. ∴ If you act to do A, then accept that act A is all right.

7. ∴ Don't combine resolving to eat nothing with acting to eat this.
[Use Exy and t.]

8. ∴ Don't act to do A without accepting that A is obligatory.

9. ∴ Don't combine believing that everyone ought to do A with not acting/resolving to do A yourself.
[This is BC's version of "Practice what you preach."]

10. Believe that you ought to do A.
∴ Act to do A.

11. "It's all right for you to do A" entails "It's obligatory that everyone do A."
∴ Don't combine acting to do A with not willing that everyone do A.
[The conclusion is a crude version of Kant's formula of universal law. The premise and conclusion are both questionable. To see this, substitute "become a doctor" for "do A" in both statements. We'll formulate a better version of Kant's formula in Chapter 9.]

12. "Attain this end" entails "If taking this means is needed to attain this end, then take this means."
∴ Don't combine (1) wanting to attain this end and (2) believing that taking this means is needed to attain this end with (3) not acting to take this means.
[Use E for "You attain this end," N for "Taking this means is needed to attain this end," M for "You take this means," and u. The conclusion is an ends-means consistency imperative.]

13. "Attain this end" entails "If taking this means is needed to attain this end, then take this means."
∴ If you want to attain this end and believe that taking this means is needed to attain this end, then act to take this means.
[Use E, N, M, and u.]

14. ∴ Don't *accept* "It's wrong for anyone to kill," without *it being the case that* if killing were needed to save your family, then you wouldn't kill.
[Use Kx and N. A draft board challenged a pacifist friend of mine, "If killing were needed to save your family, then would you kill?" My friend answered, "I don't know—I might lose control and kill (it's hard to predict what you'll do in a panic situation); but I *now* firmly hope and resolve that I wouldn't kill in such a situation." Maybe my friend didn't satisfy example 14; but he satisfied example 15. Are both provable in BC?]

15. ∴ Don't *accept* "It's wrong for anyone to kill," without *being resolved* that if killing were needed to save your family, then you wouldn't kill.
[Use Kx and N.]

8.4 DEONTIC BELIEF TRANSLATIONS

Beliefs can be "evident" or "reasonable" for a given person. As I shade my eyes from the bright sun, my belief that it's sunny is *evident*. It's very solidly grounded. As I hear the prediction of rain, my belief that it will rain is *reasonable*. My belief accords with reason, but isn't solid enough to be *evident*. "Evident" expresses a higher certitude than does "reasonable." We'll symbolize these notions as follows:

O$\underline{x}$:A = A is *evident* to x.
    = It's *obligatory* (rationally required) that x believe that A.
    = It wouldn't be reasonable for x to refrain from believing that A. ["~R~$\underline{u}$:A"]
    = Insofar as purely intellectual considerations are concerned (including x's experiences), x *ought* to believe that A.

R$\underline{x}$:A = A would be *reasonable* for x to believe.
    = It's *all right* (rationally permissible) that x believe that A.
    = x could reasonably (justifiably) believe that A.
    = Insofar as purely intellectual considerations are concerned (including x's experiences), it would be *all right* for x to believe that A.

Neither of these entails that x believes A. If we want to say that a proposition A *that x believes* is evident or reasonable, we'll use "(x:A · O$\underline{x}$:A)" or "(x:A · R$\underline{x}$:A)."

"Evident" and "reasonable" are relational. "It's raining" might be evident to a person outside, but not to someone in a windowless office. Sometimes I'll say that a belief is "evident" or "reasonable," but without saying *to whom*. When this happens, assume that I mean "evident" or "reasonable" *to you*.

Here are further translations:

~R$\underline{x}$:A   = It would be *unreasonable* for x *to believe* that A.
O~$\underline{x}$:A   = It's obligatory that x not believe that A.

R(~$\underline{x}$:A · ~$\underline{x}$:~A) = It would be *reasonable* for x *to take no position* on A.
    = It would be reasonable for x (1) not to believe that A is true and (2) not to believe that A is false.

O$\underline{x}$:(A ⊃ B)   = It's evident to x that if A then B.
(O$\underline{x}$:A ⊃ O$\underline{x}$:B)   = If it's evident to x that A, then it's evident to x that B.

O~($\underline{x}$:A · $\underline{x}$:~A)   = x ought not to combine believing A with believing not-A.

Remember that "O" and "R" attach only to *imperative* wffs. So "O$\underline{x}$:A" and "R$\underline{x}$:A" aren't correct wffs.

We can *almost* define "knowledge" in this simple way:

x knows that A = A is evident to x, A is true, and x believes A.
    knowledge = evident true belief
        xKA = (O$\underline{x}$:A · (A · x:A))

Knowing requires more than just having a true belief. If you guess right, you have a true belief—but not knowledge. To rate as knowledge, your true belief must be well-grounded. It must be more than just *reasonable* (*permitted* by the evidence). It must be *evident* (*required* by the evidence). The claim that *knowledge* is *evident true belief* is very plausible. But there are a few tricky cases where we can have *evident true belief* but not *knowledge*. (See example 34 of Section 8.5B.) So the suggested definition of "knowledge" is flawed. But it's still a useful approximation.

8.4A Exercises

Translate each sentence into a BC wff.

0. You ought to want Jones to sit down. [Use Sx, u, and j.]
ANSWER: O$\underline{u}$:S$\underline{j}$

It might help to paraphrase the sentence as "It's obligatory that you say in your heart 'Would that Jones sit down!' "

1. You ought to believe that Jones is sitting down.
2. It's evident to you that Jones is sitting down.
3. It's reasonable for you to believe that Jones ought to sit down.
4. Belief in God is reasonable (for you). [Use G.]
5. Belief in God is unreasonable for everyone.
6. It isn't reasonable for you to believe that belief in God is unreasonable for everyone.
7. Belief in God is reasonable, only if "There is a God" is logically consistent.
8. You ought not to combine *believing* that there is a God with *not believing* that "There is a God" is logically consistent.
9. You ought not to combine *believing* that you ought to do A with *not acting* to do A.
10. You *know* that everything is self-identical. [Use the flawed definition of *knowledge* given previously.]
11. If agnosticism is reasonable, then theism isn't evident. [Here agnosticism = not believing that G, and not believing that not-G; and theism = believing that G.]
12. You have a true belief that A. [You believe that A, and A.]
13. You mistakenly believe that A.
14. It would be impossible for you mistakenly to believe that A.
15. A is evident to you, if and only if it would be impossible for you mistakenly to believe that A. [Descartes might have liked this principle. But it leads quickly to skepticism.]
16. It's logically possible that you have a belief A which is evident to you and yet false.
17. It's evident to all that, if they doubt, then they exist. [Use Dx and Ex.]
18. If A entails B, and B is unreasonable, then A is unreasonable.
19. It's permissible for you to do A, only if you want everyone to do A.
20. If you want x to do A to you, then you ought to do A to x. [Use Axy. This one and the next are versions of the golden rule.]
21. You ought not to combine *acting* to do A to x with *wanting* x not to do A to you.
22. It's necessary that if you're in pain then it's evident to you that you're in pain. [Use Px. This example claims that "I'm in pain" is a self-justifying belief. Many think that there are two different kinds of self-justifying beliefs: those of experience (as in this example) and those of reason (as in the next example).]

23. It's necessary that, if you believe that (x)x = x, then it's evident to you that (x)x = x. [Perhaps believing "Everything is self-identical" entails understanding it, and understanding it makes it evident.]

24. If you have no reason to doubt your perceptions and it's evident to you that you believe that you see a red object, then it's evident to you that there's an actual red object. [Use Dx for "x has reason to doubt his or her perceptions," Sx for "x sees a red object," and R for "There's an actual red object." Roderick Chisholm thinks we need principles like this (but more complex) to show how beliefs about external objects are based on beliefs about perceptions.]

25. If it's evident to you that Jones shows pain behavior and you have no reason to doubt her sincerity, then it's evident to you that Jones feels pain. [Use Bx, Dx, Fx, and j.]

8.5 DEONTIC BELIEF ARGUMENTS

Deontic belief proofs don't require any new inference rules. But they often use complex world prefixes like "Du" or "Duu." Here's an example:

> **A CONSCIENTIOUSNESS PRINCIPLE**
>
> You ought not to combine *believing* that it's wrong for you to do A with *acting* to do A.
>
> O~(u:O~Au · u:Au)

```
       [∴ O~(u:O~Au · u:Au)
 * 1.  ┌ asm: ~O~(u:O~Au · u:Au)
 * 2.  │ ∴ R(u:O~Au · u:Au) [from 1]
 * 3.  │ D ∴ (u:O~Au · u:Au) [from 2]
   4.  │ D ∴ u:O~Au [from 3]
   5.  │ D ∴ u:Au [from 3]
   6.  │ Du ∴ O~Au [from 4]
   7.  │ Du ∴ Au [from 5]
   8.  └ Du ∴ ~Au [from 6, contradicting 7]
   9.  ∴ O~(u:O~Au · u:Au) [from 1, 7, and 8]
```

We get to line 5 using PC and DC rules. Lines 6 and 7 follow using rule B + . Here we write the belief world prefix "u" after the world prefix ("D") used in lines 4 and 5. (World-Du is a belief world of u that depends on what deontic world-D tells u to accept.) We soon get a contradiction.

Our formula "O~(u:O~Au · u:Au)" is special. It's a *formal ethical principle*—an ethical principle that we can formulate using the abstract notions of our logical systems, but leaving unspecified the meaning of the individual, property, relational, and statement letters. (Here "u" can stand for *any* person and "A" can stand for *any* action.) In the next chapter, we'll prove another formal ethical principle—the golden rule.

8.5A Exercises

Say whether valid (give a proof) or invalid (no refutation necessary).

0. Ru̱:O(A̱ · Ḇ)
∴ Ru̱:OA̱

    ANSWER: * 1.   Ru̱:O(A̱ · Ḇ)   VALID
               [∴ Ru̱:OA̱
      * 2.   ⌈ asm: ~Ru̱:OA̱
        3.   │ ∴ O~u̱:OA̱ [from 2]
        4.   │ D ∴ u̱:O(A̱ · Ḇ) [from 1]
      * 5.   │ D ∴ ~u̱:OA̱ [from 3]
      * 6.   │ Du ∴ ~OA̱ [from 5]
        7.   │ Du ∴ O(A̱ · Ḇ) [from 4]
      * 8.   │ Du ∴ R~A̱ [from 6]
        9.   │ DuD ∴ ~A̱ [from 8]
       10. │ DuD ∴ (A̱ · Ḇ) [from 7]
       11. ⌊ DuD ∴ A̱ [from 10, contradicting 9]
       12. ∴ Ru̱:OA̱ [from 2, 9, and 11]

Note how we get the complex world prefix "DuD." If you can follow this example, you needn't fear proofs involving complex world prefixes.

1. Oa̱:(C · D)
∴ Oḇ:C
3. ∴ O~(u̱:A · ~u̱: ◊ A)
5. □(A ⊃ B)
∴ (R~u̱:B ⊃ Ru̱:~A)
7. O~u̱:A
∴ Ou̱:~A
9. Ru̱:OAu
∴ Ru̱: ◊ Au

2. ~Ru̱:A
∴ (□(Ḇ ⊃ A) ⊃ ~Ru̱:B)
4. ∴ (Ru̱:A ⊃ ◊ A)
6. R(~u̱:A · ~u̱:~A)
∴ ~Ou̱:A
8. Ru̱:~A
∴ R~u̱:A
10. Ou̱:(A ⊃ OBu̱)
∴ ~(u̱:A · ~u̱:Bu̱)

8.5B Exercises

For each argument, first evaluate intuitively and then translate into BC and work it out. Say whether valid (give a proof) or invalid (no refutation necessary). Use G for "There is a God" and u for "you." Some examples say that a belief is "evident" or "reasonable," but without saying *to whom*. When this happens, assume that I mean "evident" or "reasonable" *to you*.

0. It's logically possible that you have a belief A which is evident to you and yet false.
∴ "A is evident to you" doesn't entail "It would be impossible for you mistakenly to believe A."

    ANSWER: * 1.   ◊ (u̱:A · (Ou̱:A · ~A))   VALID
             [∴ ~□(Ou̱:A ⊃ ~ ◊ (u̱:A · ~A))

2.   ⌈ asm: □(Ou:A ⊃ ~◇(u:A · ~A))
* 3. | W ∴ (u:A̲ · (Ou:A · ~A)) [from 1]
* 4. | W ∴ (Ou:A ⊃ ~̲◇(u:A · ~A)) [from 2]
5. | W ∴ u:A̲ [from 3]
* 6. | W ∴ (Ou:A · ~A) [from 3]
7. | W ∴ Ou̲:A [from 6]
8. | W ∴ ~A [from 6]
* 9. | W ∴ ~◇(u:A · ~A) [from 4 and 7]
10. | W ∴ □~(u:A · ~A) [from 9]
*11. | W ∴ ~(u:A · ~A) [from 10]
12.   ⌊ W ∴ A [from 5 and 11, contradicting 8]
13.   ∴ ~□(Ou̲:A ⊃ ~◇(u:A · ~A)) [from 2, 8, and 12]

1. Theism is evident.
∴ Atheism is unreasonable.
[Here theism = believing that there is a God, and atheism = believing that there is no God.]
2. Theism isn't evident.
∴ Atheism is reasonable.
3. ∴ It's reasonable to want A to be done, only if it's reasonable to believe that A would be all right.
4. It's evident that A is true.
∴ A is true.
5. It's reasonable to combine believing that there is a God with believing that T is true.
T entails that there's evil in the world.
∴ It's reasonable to combine believing that there is a God with believing that there's evil in the world.
[Use G, T, and E. Here T is a plausible theodicy (explanation of why God permits evil) which itself entails the existence of evil in the world. T might, for example, be "God's goal in creating the world involves the significant use of human freedom to bring a half-completed world toward its fulfillment; moral evil results from the abuse of human freedom, and physical evil from the half-completed state of the world."]
6. ∴ You ought not to combine *believing* that you ought to do A with *not acting* to do A.
7. ∴ If you believe that you ought to do A, then you ought to act to do A.
8. A entails B.
It wouldn't be reasonable for you to accept B.
∴ Don't accept A.
9. "All men are endowed by their creator with certain unalienable rights" entails "There's a creator."
It would be reasonable not to accept "There's a creator."
∴ "All men are endowed by their creator with certain unalienable rights" isn't evident.
[Use E and C. The opening line of the Declaration of Independence claims E to be self-evident.]
10. You could reasonably believe that A is true.
You could reasonably believe that B is true.
∴ You could reasonably believe that A and B are both true.

11. It's evident to you that, if there are moral obligations, then there's free will.
∴ Don't combine accepting that there are moral obligations with not accepting that there's free will.
[Use M and F.]

12. "If I'm hallucinating, then physical objects aren't as they appear to me" is evident to me. It isn't evident to me that I'm not hallucinating.
∴ It isn't evident to me that physical objects are as they appear to me.
[Use H, P, and i. This powerful argument for skepticism is from Descartes.]

13. "If I'm hallucinating, then physical objects aren't as they appear to me" is evident to me. If I have no special reason to doubt my perceptions, then it's evident to me that physical objects are as they appear to me.
I have no special reason to doubt my perceptions.
∴ It's evident to me that I'm not hallucinating.
[Use H, P, D, and i. This is John Pollock's answer to the previous argument.]

14. Theism is reasonable.
∴ Atheism is unreasonable.

15. Theism is evident.
∴ Agnosticism is unreasonable.
[Here agnosticism = not believing that there is a God, and not believing that there is no God.]

16. ∴ It's reasonable for you to believe that God exists, only if "God exists" isn't self-contradictory.
["Reasonable" in BC is very objective. A belief is "reasonable," only if it *in fact* is consistent. In a more subjective sense, someone could "reasonably" believe a proposition that is reasonably but incorrectly taken to be consistent.]

17. ∴ If A is unreasonable, then don't believe A.

18. You ought not to combine accepting A with not accepting B.
∴ If you accept A, then accept B.

19. It's evident that, if the universe is very much like a machine, then the universe has a creator. It's reasonable to believe that the universe is very much like a machine.
∴ It's reasonable to believe that the universe has a creator.
[Use M and C.]

20. ∴ You ought not to combine wanting A not to be done with believing that A would be all right.

21. It's reasonable not to believe that there's an external world.
∴ It's reasonable to believe that there's no external world.
[Use E.]

22. It's reasonable to believe that A ought to be done.
∴ It's reasonable to want A to be done.

23. ∴ Either theism is reasonable or atheism is reasonable.

24. It's evident to you that if the phone is ringing then you ought to answer it. It's evident to you that the phone is ringing.
∴ Act on the imperative "Answer the phone!"
[Use P and Ax.]

25. A entails B.
Believing A would be reasonable.
∴ Believing B would be reasonable.

26. Atheism isn't evident.
∴ Theism is reasonable.

27. Atheism is unreasonable.
Agnosticism is unreasonable.
∴ Theism is evident.

28. A entails B.
You accept A.
It's unreasonable for you to accept B.
∴ Don't accept A, and don't accept B.

29. It's evident to you that A is true.
It isn't evident to you that B is false.
∴ It would be reasonable for you to believe (A and B).

30. It's reasonable for you to believe that God exists.
If God exists, then there's an afterlife.
∴ You could reasonably believe that there's an afterlife.
[Use G and A.]

31. It's evident to you that taking this means is needed to attain this end.
"Attain this end" entails "If taking this means is needed to attain this end, then take this means."
∴ You ought not to combine *wanting* to attain this end with *not acting* to take this means.
[Use N for "Taking this means is needed to attain this end," E for "You attain this end," M for "You take this means," and u. The conclusion is an ends-means consistency duty.]

32. You ought not to combine *wanting* to attain this end with *not acting* to take this means.
You ought not to act to take this means.
∴ Don't want to attain this end.
[Use E and M.]

38. It would be reasonable for anyone to believe A.
∴ It would be reasonable for everyone to believe A.
[Imagine some controversial issue where everyone has the same evidence. Could it be most reasonable for the community to disagree? If so, the premises of this argument might be true but the conclusion false.]

34. Al believes that Smith owns a Ford.
It's evident to Al that Smith owns a Ford.
Smith doesn't own a Ford.
Smith owns a Chevy.
Al believes that Smith owns a Ford or a Chevy.
Al doesn't *know* that Smith owns a Ford or a Chevy.
∴ Al has an *evident true belief* that Smith owns a Ford or a Chevy, but Al *doesn't know* that Smith owns a Ford or a Chevy.
[Use F for "Smith owns a Ford," C for "Smith owns a Chevy," and K for "Al knows that Smith owns a Ford or a Chevy." This argument from Edmund Gettier attacks the definition of *knowledge* as *evident true belief*.]

35. It's evident to you that if it's all right for you to hit Jones, then it's all right for Jones to hit you.
∴ Don't combine *acting* to hit Jones with *believing* that it would be wrong for Jones to hit you.
[Use Hxy, u, and j. The premise, although generally true, could be false. You might know that Jones needs to be hit to dislodge food he's choking on. The conclusion is a crude form of the version of the golden rule that we'll prove in Chapter 9.]

8.6 SOPHISTICATED BELIEF LOGIC

The system of belief logic that we've developed so far is oversimplified in various ways. We'll now sketch a more sophisticated version of the theory.

First, our "One ought to be consistent" principle requires qualification. For the most part, we *do* have a duty to be consistent. But, since "ought" implies "can," this duty is nullified when we're *unable* to be consistent. Such inability can come from emotional turmoil or from our incapacity to grasp a complex proposition that follows from our beliefs. And the obligation to be consistent can be *overridden* by other factors. If Dr. Evil would destroy the world unless we were inconsistent in some respect, then that would surely override our duty to be consistent. And the duty to be consistent applies, when it does, only to *persons*; yet our principles so far would entail that rocks and trees also have a duty to be consistent.

For these reasons, it would be better to qualify our "One ought to be consistent" principle, as in the following formulation:

> If x is a person who is able to be consistent in
> the relevant ways, and whose being consistent in
> these ways wouldn't have disastrous consequences,
> then x ought to be consistent in these ways.

Here "able to be consistent" includes psychological ability and the ability to grasp the concepts and logical relationships involved without spending an unreasonable amount of time on it.

Let's abbreviate the qualification in the box ("x is a person who is able . . . .") as "Px." Then we can reformulate our inference rules for belief logic by adding "Px":

B+   Px, $\underline{x}$:A → A   The derived line's world prefix must consist of that of the earlier line, plus a string of one or more x's. Also, neither line's world prefix can contain small letters or a "W."

B−   Px, ~$\underline{x}$:A → ~A   The derived line's world prefix must consist of that of the earlier line, plus a *new* string of x's (one not occurring in earlier lines). Also, neither line's world prefix can contain small letters or a "W."

The only change from Section 8.2 is that we now need the premise "Px" to apply either rule. With this revision, the following argument would be provable—but not the corresponding argument without the second premise:

A entails B.

You are a person who is able to be consistent in the relevant ways and whose being consistent in these ways wouldn't have disastrous consequences.

∴ You ought not to combine believing A with not believing B.

□(A ⊃ B)
Pu
∴ O~($\underline{u}$:A · ~$\underline{u}$:B)

If we made these changes, we'd have to qualify almost all of our belief arguments in this chapter with a premise like the second one—or else our arguments would be invalid.

The second problem is that our system can prove a questionable conjunctivity principle:

> You ought not to combine believing A and believing B with not believing (A · B).
>
> O~((u:A · u:B) · ~u:(A · B))

This principle is fine in most cases, but it can lead to questionable results. Consider this "lottery paradox." Suppose that ten people have an equal chance to win a lottery. You know for sure that one of the ten will win, but the probability is against any given person winning. Presumably it could be reasonable for you to accept *each* of the statements from 1 to 10 *without* also accepting statement 11 (which means "No one will win"):

1. Person 1 won't win.
2. Person 2 won't win.
3. Person 3 won't win.
4. Person 4 won't win.
5. Person 5 won't win.
6. Person 6 won't win.
7. Person 7 won't win.
8. Person 8 won't win.
9. Person 9 won't win.
10. Person 10 won't win.

11. Person 1 won't win, and person 2 won't win, and person 3 won't win, and person 4 won't win, and person 5 won't win, and person 6 won't win, and person 7 won't win, and person 8 won't win, and person 9 won't win, and person 10 won't win.

But multiple uses of the conjunctivity principle in the box would entail that one ought not to accept each of the statements from 1 to 10 without also accepting their conjunction 11. So the conjunctivity principle, which is provable using our rules B+ and B−, sometimes leads to questionable results.

I'm not completely convinced that it's *reasonable* to accept each statement from 1 to 10 but not accept 11. But if it *is* reasonable, then we'd have to reject the conjunctivity principle. This would force us to modify our ideas on what sort of consistency is desirable. Let's call the "consistency" defined in Section 8.3 *broad consistency*. Perhaps we should strive, not for broad consistency, but rather for *narrow consistency*:

> Let S be the nonempty set of indicatives and imperatives that x accepts.

> Then x is *broadly consistent*, if and only if:
> 1. set S is logically consistent, and
> 2. x accepts anything that follows logically from set S.

> Then x is *narrowly consistent*, if and only if:
> 1. any pair of items of set S is logically consistent, and
> 2. x accepts any item that follows logically from any single item of set S.

Our lottery person who accepts 10 individual statements but doesn't accept their conjunction could be narrowly, but not broadly, consistent.

To have our rules mirror the ideal of narrow consistency, we'd add one additional proviso to rule B+:

> The world prefix in the derived step cannot have occurred more than once in previous lines.

With this change, only a few of the examples in this chapter that were provable before would cease being provable. And most of these few could still be salvaged by adding an additional conjunctivity premise like the following:

> You ought not to combine believing A and believing B with not believing (A · B).
>
> O~((u:A · u:B) · ~u:(A · B))

Such a conjunctivity premise would be true in most cases. It's only in rare lottery-type cases that conjunctivity fails.

The third problem is that we've been translating the following two statements in the same way (as "Ou:A"), even though they don't mean the same thing:

> You ought to believe A.   A is evident to you.

Imagine a case where you have an obligation to trust another person and give the person the benefit of every reasonable doubt. It could happen that you *ought* to believe what the other person says, even though the evidence for the belief isn't strong enough to make the belief *evident*. So there is a real difference between "ought to believe" and "evident."

Because of such cases, it would be better to use a different symbol (perhaps "O*") for "*evident*":

Ox:A = x ought to believe that A.
      = All things considered, x ought to believe that A.

O*x:A = A is evident to x.
      = Insofar as purely intellectual considerations are concerned (including x's experiences), x ought to believe that A.

Here "O" is an all-things-considered "ought," while "O*" is a *prima facie* "ought" that just considers the intellectual backing for the belief. If we added "O*" to our system, we'd have to add deontic inference rules like D1 to D6 for "O∗." Since "O∗A" is a *prima facie* "ought," it wouldn't entail the corresponding imperative or commit one to action. So we'd have to weaken the "O∗" version of inference rule D4 so that we couldn't derive "u:A" from "O∗u:A."

These refinements overcome some problems, but they make the system harder to use. We seldom need the refinements. So we'll keep the naïve belief logic of earlier sections as our "official system." This is what we'll build on in the next chapter. But we'll be conscious that this system is oversimplified in various ways. If and when the naïve system gives questionable results, we can appeal to the sophisticated system to clear things up.

9

A Formalized Ethical Theory

9.1 INTRODUCTION

Our *golden rule calculus* (GRC) builds on previous systems. It isn't oriented toward analyzing individual arguments. Rather it's meant to give a precise formulation of an ethical theory (one that builds on ideas from R. M. Hare and Immanuel Kant).* GRC shows how we can use logic to construct larger philosophical views. This system grows out of previous chapters, but it represents a shift in emphasis.

Let me sketch what we'll do in this chapter. First we'll consider practical rationality in a general way. Then we'll narrow our focus to one consistency principle: the golden rule. After seeing some problems with the usual phrasing of the rule (''Treat others as you would want to be treated''), we'll work toward a better formulation. Then we'll examine an argument for the rule. To justify part of the argument, we'll need more logical machinery than we have so far. So we'll develop the golden rule calculus (GRC). We'll end the chapter by finishing the proof of the golden rule.

9.2 DIMENSIONS OF PRACTICAL RATIONALITY

Many people stress the nonrational aspects of our ethical thinking. They talk about social and historical influences. And they point to the emotional forces behind our value judgments.

* See my ''Ethics Is Based on Rationality,'' *Journal of Value Inquiry* 20 (1986), 251–64; and ''Ethical Consistency Principles,'' *Philosophical Quarterly* 35 (1985), 157–70. See also R. M. Hare's *Freedom and Reason* (New York: Oxford University Press, 1963); and Immanuel Kant's *Groundwork of the Metaphysics of Morals* (New York: Harper & Row, 1964). Regarding how my view differs from Hare's, see my ''The Prescriptivism Incompleteness Theorem,'' *Mind* 85 (1976), 589–96.

There's much truth in all this. Cultural and emotional influences are strong. But we shouldn't ignore the many ways that reason can contribute to our ethical thinking.

Various elements go into making our actions and ethical judgments "rational." None of these suffices by itself; we need all of them together to be fully rational. Here I'll distinguish five dimensions of practical rationality:

- Ends-means rationality
- Empirical rationality
- Self-knowledge rationality
- Imaginative rationality
- Rationality as consistency

Here's a principle of *ends-means rationality*:

> Don't combine *wanting* to attain this end and *believing* that taking this means is needed to attain this end with *not acting* to take this means.

It's easy to violate this principle. Many people want to lose weight and believe that eating less is needed to attain this end. Yet they *don't* take steps to eat less. Insofar as we violate this principle, we are defective in our rationality. People of all cultures implicitly recognize this principle (or else they wouldn't survive!), but few people know how to formulate it. In Section 8.3C (example 12), we gave the following valid argument (with a clearly true premise) for this principle:

"Attain this end" entails "If taking this means is needed to attain this end, then take this means."

∴ Don't combine *wanting* to attain this end and *believing* that taking this means is needed to attain this end with *not acting* to take this means.

$$\Box(\underline{E} \supset (N \supset \underline{M}))$$
$$\therefore \ \sim((\underline{u{:}E} \cdot \underline{u{:}N}) \cdot \sim\underline{u{:}M})$$

Where: E = You attain this end.

N = Taking this means is needed to attain this end.

M = You take this means.

Ends-means rationality doesn't suffice by itself. Nazis might pursue effective means to their goal of genocide. It doesn't follow that their practical thinking is fully rational. They probably violate other dimensions of rationality (for example, the golden rule).

Empirical rationality demands that we know the facts of the case: circumstances, alternatives, consequences, and so on. To the extent that we are misinformed or ignorant, our practical thinking is flawed. We can never know *all* the facts. So we can satisfy this rationality demand to a greater or lesser degree—but never completely. Again, empirical rationality doesn't suffice by itself. We could know almost all the facts of the case but still be deficient in some other way.

Self-knowledge rationality demands that we understand the origin of our feelings and judgments. Some people have hostility toward a certain group because they were taught this when they were young. They might change their attitudes if they understood how their hostility originated. If so, their attitudes are defective—since they exist because of a lack of self-knowledge.

Imaginative rationality demands that we develop and exercise the ability to imagine ourselves in various situations—especially in the place of another person. This is different from just knowing facts. Empirical rationality demands that we know facts about poor people when deciding how to act toward them. But imaginative rationality demands that we be able to imagine ourselves, vividly and accurately, in their situation. Although the social sciences contribute to empirical rationality, art and literature are more helpful for developing imaginative rationality. Again, imaginative rationality doesn't suffice by itself. We could put ourselves in the place of another to see better how to manipulate the person for our own ends.

Rationality as consistency demands that we be consistent in our thinking and doing. This involves being logical, conscientious, and impartial—and following the golden rule and the formula of universal law. Our systems BC and GRC can help us understand and justify the principles involved. I'll mostly talk about consistency from now on. As a logician, I have more to say about this dimension. I'll focus on the golden rule. But keep in mind that we need all the dimensions working together for our practical thinking to be fully reasonable.

Holistic rationality includes all these kinds of rationality—and probably others that I haven't mentioned. A more traditional term is "practical wisdom." We are "rational" (or "wise") in making our ethical judgments to the extent that we satisfy a variety of requirements. Only God (knowing everything, understanding vividly the inner life of each person, being consistent in every way, and so on) could satisfy them completely. We satisfy the requirements of practical rationality more or less.

9.3 RATIONALITY AS CONSISTENCY

Consistency itself has many dimensions. First there's *logical consistency in beliefs*. We'll call this *logicality*. BC proved many specific instances of this imperative:

> LOGICALITY: Don't believe inconsistent things, and don't believe something without believing its logical consequence.

We often appeal to such things when we argue about ethical principles. Suppose a racist says, "We ought to treat blacks poorly—because they're inferior." We might ask the racist what he means by "inferior." Is "being inferior" a matter of IQ, or education, or wealth, or physical strength, or some combination of these? Let's suppose that he defines "inferior" as "having an IQ of less than 80." Then his argument comes to this:

All blacks have an IQ of less than 80.
Everyone with an IQ of less than 80 ought to be treated poorly.
∴ All blacks ought to be treated poorly.

We could appeal to *empirical rationality* and dispute the racist's ridiculous first premise. Or we could appeal to *logicality* and ask the racist whether he accepts what his second premise logically entails:

> Everyone with an IQ of less than 80 ought to be treated poorly.
> ∴ All *whites* with an IQ of less than 80 ought to be treated poorly.

Of course the racist won't believe that these *whites* ought to be treated poorly. But then he believes a premise but refuses to believe its logical consequence. We needn't be logicians to know that something is wrong with this. But BC lets us formulate the principle involved in a precise way:

> If A logically entails B, then you ought not
> to combine believing A with not believing B.
>
> $(\Box(A \supset B) \supset O{\sim}(\underline{u}{:}A \cdot {\sim}\underline{u}{:}B))$

To be rational, the racist must do one of two things. He could reject his premise. Or he could come to accept its logical consequence about how to treat whites who have a low IQ.

Appealing to consistency in beliefs is a common way to argue about ethical principles. It's often very effective. But it doesn't always suffice by itself. People can be consistent but wrong in their beliefs. Then we need further ammunition.

Our belief calculus also demands *consistency between believing and doing*. We'll call this *conscientiousness*. BC proved many specific instances of this imperative:

> CONSCIENTIOUSNESS: Bring about a harmony
> between your ethical beliefs and your actions,
> resolutions, desires, and so on.

Here's one such instance:

> Don't combine *believing* that it's wrong
> for you to do A with *acting* to do A.
>
> ${\sim}(\underline{u}{:}O{\sim}A\underline{u} \cdot \underline{u}{:}A\underline{u})$

The consistency principles of BC and GRC are *formal*. This means that we can express them using the abstract notions of our logical systems, but leaving unspecified the meaning of the individual, predicate, relational, and statement letters. Our proof of the previous formula works regardless of what person "u" stands for and what action "A" stands for.

Our golden rule calculus extends the methods of BC. It includes three further formal rationality demands:

> IMPARTIALITY: Don't make conflicting ethical
> judgments about cases you recognize to be relevantly
> or exactly similar in their universal properties.

> GOLDEN RULE: Don't act toward another in a way
> that you don't consent to yourself being treated in an
> imagined exactly similar situation.

> UNIVERSAL LAW: Act only as you're willing for anyone
> to act in exactly similar circumstances (regardless
> of where you imagine yourself in the situation).

We'll mostly focus on the golden rule.

9.4 FORMULATING THE GOLDEN RULE

The *golden rule* (GR) says "You ought to treat others as you would want to be treated."
Jesus gave the rule as the summary of the Law and the Prophets (Matthew 7:12). But
the rule isn't uniquely Christian. Rabbi Hillel used it to summarize the Jewish law.
And Confucius used it to sum up his own teachings. All the major religions and many
nonreligious thinkers teach this rule. And GR is important in our own culture.

The golden rule seems clear and simple. But the clarity and simplicity disappear
when we try to explain what the rule means and why we should follow it. We can't
take the usual wording of the rule literally. GR seems to be saying this:

> If you want x to do A to you,
> then you ought to do A to x.
>
> $(u{:}Axu \supset OAux)$

This has absurd implications. Suppose you're a patient who wants Dr. Jones to remove
your inflamed appendix. The rule would tell you to operate on the doctor:

> If you want Jones to remove your appendix,
> then you ought to remove Jones's appendix.

Or suppose you're a little boy who loves to fight with his sister. The rule tells you:

> If you want your sister to fight with you,
> then you ought to fight with her.

Or suppose you want people to hate you:

> If you want others to hate you,
> then you ought to hate them.

Some suggest that we apply GR only to "general" actions (such as treating someone
with kindness) and not to "specific" ones (such as removing someone's appendix).
But our last example uses a general action. So this restriction won't solve the problem.

In previous chapters, we considered the "Follow your conscience" principle.*

---

* See Sections 7.2, 7.2B (problems 10 and 11), and 8.3C (problems 1 and 2). We proposed that
"Follow your conscience" be interpreted as a *don't-combine* imperative.

This seemed to be an *if-then* imperative. But, interpreted this way, it led to absurdities such as, "If you think you ought to blow up the world, then blow up the world!" We tinkered with the wording to get a defensible and provable principle. We'll likewise have to tinker with the wording of GR.

My tinkering leads to this formulation of the golden rule:

> You ought not to combine *acting* to do A to x with *not consenting* to the idea of x doing A to you in an imagined exactly reversed situation.

Notice two key elements of this formulation:

- The rule has you imagine an *exactly reversed situation* in which you are on the receiving end of the action.
- The rule has a *don't-combine* form.

After discussing these elements, we'll try to symbolize the principle. Our symbolization will look like this, but with the dots replaced by further symbols:

> O~(u:Aux · ~u: . . .)

Let's consider what's wrong with our earlier formulation. Recall the case where you're the patient and you want Dr. Jones to remove your appendix. The earlier formulation told you:

> If you want Jones to remove your appendix, then you ought to remove Jones's appendix.

In applying this rule, you'd ask yourself:

| "Do I want Jones to remove my appendix here and now?" | "Sure I do!" |

The silly conclusion would follow that you ought to remove Dr. Jones's appendix!

The problem is that you and Jones are in different situations. *You* need an appendix operation but are ignorant of how to do them. *Jones* knows how to do them but doesn't need one. Our revised GR is sensitive to such differences. Suppose that you were considering whether to remove your doctor's appendix. On our revised rule, the important question would be this:

| "Do I consent to the idea of Jones removing my appendix in an imagined exactly reversed situation?" | "Surely not!" |

To answer this, you'd imagine yourself being a healthy doctor about to be operated on by Jones (a sick patient ignorant of medicine). Surely you don't consent to the idea of Jones operating on you in such a case; the operation would probably kill you! On our revised rule, you'd *violate* the golden rule if you tried to remove your doctor's appendix.

Let me explain "exactly reversed situation" further. Let's call any nonevaluative property describable without proper names (like "Gensler" or "Chicago") or pointer terms (like "I" or "this") a *universal property*. Suppose we list the universal properties of Jones and of myself:

THE ACTUAL SITUATION	
Jones	I
healthy	sick
a doctor	a patient
knows about medicine	ignorant of medicine
brown-eyed	blue-eyed
married	single
. . .	. . .

Imagine that each list contains *all* the universal properties of the person, including complex ones. (We couldn't complete the list; it would be too long.) When I imagine an *exactly reversed situation*, I imagine this:

THE IMAGINED EXACTLY REVERSED SITUATION	
I	Jones
healthy	sick
a doctor	a patient
knows about medicine	ignorant of medicine
brown-eyed	blue-eyed
married	single
. . .	. . .

Here I'm healthy—and Jones is sick. I imagine myself having all the universal properties that Jones now has, and Jones having all those that I now have. And I ask myself:

"Do I consent to the idea of Jones removing my appendix in an imagined situation where I'm a healthy doctor (and so on) and Jones is a sick patient (and so on)?"	"Surely not!"

For the first "(and so on)" we can substitute "and I have all the other universal properties that Jones now has." And for the second "(and so on)" we can substitute "and Jones has all the other universal properties that I now have."

Some might prefer to switch just the properties that are *relevant* to deciding whether the operation ought to be done. It's relevant whether the person operating knows how to operate and whether the person operated upon is sick. It isn't relevant what color eyes each person has. You can switch just the relevant properties if you like. This gives a "*relevantly* reversed situation" instead of an "*exactly* reversed situation." We can still symbolize and prove GR if we make this modification.

Doctors know roughly which factors are relevant in deciding whether to operate. But some factors may be controversial. So I prefer the formulation that uses "*exactly* reversed situation." With this, we needn't worry about which universal properties are *relevant*. We simply switch them *all*.

Understanding our GR involves grasping a subtle distinction. Our GR isn't about *how you'd react if you were in a hypothetical situation*. Rather it's about *your present reaction toward a hypothetical situation*. Compare these two forms:

BAD FORM	You ought not to *act* to do A to x without it being the case that, *if you were* in an exactly reversed situation, *then you'd consent* to the idea of A being done to you.

This has "If R, then you'd consent to A."

GOOD FORM	You ought not to *act* to do A to x without (now) *consenting to the idea that, if you were* in an exactly reversed situation, *then* A would be done to you.

This has "You consent to the idea that if R then A."

The BAD FORM is about *how you'd react if you were in a different situation*. This has little to do with whether you're *now* consistent. The GOOD FORM, on the other hand, forbids an inconsistent combination of *present* activities. The GOOD FORM in this way resembles the BC consistency imperatives of the last chapter.

The BAD FORM has some absurd consequences. Suppose that you're a parent and you need to punish your child. The punishment is just and in the child's own interests. But the child doesn't want to be punished. Using the BAD FORM, you'd ask yourself:

BAD FORM	"If I were in my child's exact place, would I then (as a child) consent to being punished?"

The answer is NO. If you were in your child's *exact* place, you'd have *all* your child's present properties (including that of not wanting to be punished). And so you wouldn't want to be punished. So the BAD FORM wouldn't let you punish your child. This is surely unacceptable.

With the GOOD FORM, you'd ask yourself:

GOOD FORM	"Do I now (as an adult) consent to the idea that, if I were in my child's exact place, then I'd be punished?"

The answer is YES. Since the punishment is just and in the child's own interests, you'd now consent to the idea of yourself being punished in such a hypothetical situation. You might add, "I'm now glad that my parents punished me in such circumstances." So the GOOD FORM seems better than the BAD FORM.

This question invites confusion:

UNCLEAR	"Would I consent to this action being done to me, if I were in the place of the other person?"

It could mean either of these:

| BAD FORM | "If I were in the place of the other person, would I *then* consent to this action being done to me?" |

| GOOD FORM | "Do I *now* consent to the idea that, if I were in the place of the other person, then this action would be done to me?" |

The difference isn't usually important. It becomes important when we deal with someone who isn't very rational—such as a young child, a senile person, or someone in a coma. Then we shouldn't ask *what we'd desire if we were* a young child or senile or in a coma. Rather we should ask *what we now desire* be done in a hypothetical case where we picture ourselves as a young child or senile or in a coma.

Another feature of our GR is that it has a *don't-combine* form—not an *if-then* form. We've seen that *if-then* consistency principles often lead to absurdities (see Sections 7.2 and 8.2). This *if-then* form of the golden rule is defective:

| *IF-THEN*

(~C ⊃ O~<u>A</u>) | If you don't consent to the idea of x doing A to you in an imagined exactly reversed situation, then you ought not to act to do A to x. |

This *if-then* form can prescribe that an agent with perverted desires do evil things:*

| If you demand that x torture you in an imagined exactly reversed situation, then you ought to torture x. |

Imagine some demented person full of self-hatred who demands that others torture him in exactly similar circumstances. The *if-then* form implies that any such person *ought* to torture others. But this is absurd. So the *if-then* form is wrong.

Our GR is a *don't-combine* consistency principle:

| *DON'T-COMBINE*

O~(<u>A</u> · ~<u>C</u>) | You ought not to combine *acting* to do A to x with *not consenting* to the idea of x doing A to you in an imagined exactly reversed situation. |

This form forbids an inconsistent action/desire combination. It tells you not to combine these two:

---

* Take the first box and replace "do(ing) A to" with "refrain(ing) from torturing." This gives us: "If you don't consent to the idea of x refraining from torturing you in an imagined exactly reversed situation, then you ought not to act to refrain from torturing x." Removing the double negatives and rephrasing gives us the statement in the second box.

ACTION
You act to do A to x.

DESIRE
You don't consent to the idea of x doing A to you in an imagined exactly similar situation.

Suppose you *do* combine these two. Then your action/desire combination is inconsistent and has a faulty element. But which element is faulty? Normally the action is faulty. But sometimes the action is fine but the desire is faulty. For example, the demented person might (GOOD ACTION) refrain from torturing others but yet (BAD DESIRE) refuse to consent to the idea of others refraining from torturing him in exactly similar circumstances. The possibility of such perverted desires ruins the *if-then* form of the golden rule and moves us to the *don't-combine* form.

Unlike the *if-then* GR, our *don't-combine* GR never commands or forbids an individual action. It only forbids inconsistent action/desire combinations. So our GR doesn't compete with ordinary ethical principles (like "One ought not to cheat on tests") that *do* command or forbid individual actions. Our golden rule, as a consistency principle, operates on a different level.

So we've had to tinker with the wording of the golden rule to get a defensible principle. Our version of GR involves three key elements:

- an exactly (or relevantly) reversed situation
- one's present consent to a hypothetical case
- a *don't-combine* form

Here are two equivalent formulations of our GR:

> You ought to act toward another only in ways you're willing to be treated in the same exact situation.

> You ought not to combine *acting* to do A to x with *not consenting* to the idea of x doing A to you in an imagined exactly reversed situation.

The first wording is easier to grasp, while the second is more technically precise.

Suppose that you're about to do something to another, and you apply the golden rule. First you'd imagine the exactly reversed situation. This involves imagining yourself having the other person's feelings, desires, background, shoe size, and so on. Then you'd ask yourself, "Do I consent to the idea of this action being done to me in this situation?" If the answer is NO, and yet you act this way toward the other person, then you're inconsistent and violating the golden rule.

As I stressed before, consistency principles like the golden rule aren't sufficient in themselves. To apply GR adequately, we need to know how our actions would influence the lives of others (empirical rationality). We need to understand our own desires and their origin (self-knowledge rationality). We need to develop and exercise the ability to imagine ourselves in the place of another (imaginative rationality). And we may need further elements as well. When combined with other elements, GR can be a powerful tool of ethical thinking. I see the golden rule as the most important principle of life.

But we shouldn't make excessive claims for the rule. It doesn't give all the answers to ethical problems. It doesn't separate concrete actions into "right actions" and "wrong actions." It's only a consistency principle. It prescribes that we not be inconsistent by having our actions (toward another) be out of harmony with our desires (toward the reversed situation action). But, despite its limitations, GR is very useful. The golden rule expresses a formal rational condition that we often violate.

We noted at the end of the last chapter that consistency duties require qualifiers like "insofar as you can follow this principle and no disaster would result from so doing. . . ." This also applies to the golden rule. We'll regard this qualifier as implicit throughout.

The golden rule is a family of related principles. As noted before, we could shift the wording from "exactly similar" to "relevantly similar." And instead of imagining ourselves on the receiving end of the action, we could imagine someone else we care about in this place. We could leave unspecified the agent of this imagined action. We could shift from "You ought not to combine . . . ." to "Don't combine . . . ." We could shift from "acting to . . . ." to "believing that it would be all right to . . . ." Other variations are possible.

Let me give three examples of further members of the GR family. Here's a GR form using "demand" instead of "consent":

> You ought not to *act* to treat another in a given way while also *demanding* that *you not* be treated that way in the exact same situation.

Here's a 3-party version of GR:

> You ought not to (1) *act* in a given way affecting x and y without (2) *consenting* to the idea of this act being done when you imagine yourself in the place of x, and (3) *consenting* to the idea of this act being done when you imagine yourself in the place of y.

Finally, here's a GR covering any number of parties:

> You ought to act only as you're willing for anyone to act in the same situation, regardless of where you imagine yourself in the situation.

We'll call this last one the "formula of universal law," since it resembles Immanuel Kant's principle with that name. Kant's original wording went, "Act only on that maxim through which you can at the same time will that it should be a universal law"; this is interpreted in various ways.

Our golden rule calculus will be able to symbolize and prove all these members of the golden rule family.

9.4A Exercises

Assuming the approach presented here, which examples have correct GR forms? Change those with incorrect forms so that they have correct forms.

0. If you don't want others to keep you awake after midnight by loud radio playing, then you ought not to do this to others.

ANSWER: The form here is incorrect. This particular example isn't absurd, but other examples with this form are. Correct forms talk about similar circumstances and use a *don't-combine* form. Several correct forms are possible. Here's one example:

> Don't act to play your radio loudly after midnight (and thus keep others awake) without consenting to the idea of others doing this same thing to you in relevantly or exactly similar circumstances.

1. If you want Jones to remove your appendix, then remove Jones's appendix.

2. If you don't want your little sister to pull your hair, then don't pull her hair.

3. Don't accept "It's all right for the United States to do nothing to help starving people in other countries," without consenting to the idea of the United States doing nothing to help you if you were a person starving in such circumstances.

4. Don't act to spank Jimmy, without it being the case that, if you were in Jimmy's exact position, then you'd consent to being spanked.

5. Given that if you were in this criminal's place then you wouldn't want to be jailed, it follows that you ought not to jail this criminal.

6. If you believe that you ought to rob another but don't desire that you be robbed in the exactly reversed situation, then your belief is false.

7. If your slave consents to the idea of your beating him, then it's all right for you to beat him.

8. Don't act to drive drunk and hence threaten the lives of others without consenting to the idea of others doing this in parallel circumstances and thus threatening the lives of your loved ones.

9. If, in the exactly reversed situation, you'd want to rob Jones, then you ought to let Jones rob you.

10. Don't think that it would be all right for you to take hostages to further your political goals without consenting to the idea of others taking your loved ones hostage to further their political goals in relevantly similar circumstances.

11. If you demand that in relevantly or exactly similar circumstances your wife not lie to you about your terminal illness, then don't act to lie to your wife about her terminal illness.

12. Don't think that it's all right for you to leave your dirty dishes around, while also demanding that in admittedly similar circumstances others not leave their dirty dishes around.

13. If you don't consent to the idea that, if you were in the exact place of your pet dog, then you'd be tortured just for fun, then don't act to torture your pet dog just for fun.

14. Don't act to let others take advantage of you without consenting to the thought of someone you care about doing this exact same thing in relevantly or exactly similar situations.

15. Don't act to destroy the environment for future generations without consenting to the idea of others having destroyed the environment for your generation in relevantly or exactly similar circumstances.

9.5 STARTING THE PROOF OF THE GOLDEN RULE

Now we'll begin the proof of the golden rule. We'll break the proof up into parts. We'll cover as many parts as we can without introducing further logical machinery. Later we'll construct GRC and complete the proof.

Our main argument for the golden rule goes as follows:

1. You ought not to *act* to do A to x without *believing* that it would be all right for you to do A to x.
2. You ought not to *believe* that it would be all right for you to do A to x without *believing* that, in an imagined exactly reversed situation, it would be all right for x to do A to you.
3. You ought not to *believe* that, in an imagined exactly reversed situation, it would be all right for x to do A to you, without *consenting* to the idea of x doing A to you in an imagined exactly reversed situation.
∴ You ought not to *act* to do A to x without *consenting* to the idea of x doing A to you in an imagined exactly reversed situation.

The conclusion is our GR. First we'll show that the argument is valid. Then we'll prove the premises.

It's easier to show that the argument is valid if we use these abbreviations:

C = You act to do A to x
D = You believe that it would be all right for you to do A to x
E = You believe that, in an imagined exactly reversed situation, it would be all right for x to do A to you
F = You consent to the idea of x doing A to you in an imagined exactly reversed situation

Then our argument goes this way:

$$O{\sim}(\underline{C} \cdot {\sim}\underline{D})$$
$$O{\sim}(\underline{D} \cdot {\sim} \underline{E})$$
$$O{\sim}(\underline{E} \cdot {\sim} \underline{F})$$
$$\therefore O{\sim}(\underline{C} \cdot {\sim} \underline{F})$$

You can probably see that it's valid. But here's a proof anyway (using our deontic calculus):

1. $O{\sim}(\underline{C} \cdot {\sim}\underline{D})$ VALID
2. $O{\sim}(\underline{D} \cdot {\sim}\underline{E})$

3.　　O~(E̲ · ~F̲)
　　　[∴ O~(C̲ · ~F̲)

* 4.　┌ asm: ~O~(C̲ · ~F̲)
* 5.　│　∴ R(C̲ · ~F̲) [from 4]
* 6.　│ D ∴ (C̲ · ~F̲) [from 5]
* 7.　│ D ∴ ~(C̲ · ~D̲) [from 1]
* 8.　│ D ∴ ~(D̲ · ~E̲) [from 2]
* 9.　│ D ∴ ~(E̲ · ~F̲) [from 3]
10.　│ D ∴ C̲ [from 6]
11.　│ D ∴ ~F̲ [from 6]
12.　│ D ∴ D̲ [from 7 and 10]
13.　│ D ∴ E̲ [from 8 and 12]
14.　└ D ∴ F̲ [from 9 and 13, contradicting 11]
15.　∴ O~(C̲ · ~F̲) [from 4, 11, and 14]

Since our argument is valid, we can turn to the premises. Here's the first premise and its formulation:*

> 1. You ought not to *act* to do A to x without *believing* that it would be all right for you to do A to x.
>
> O~(u:Au̲x · ~u:RAu̲x)

We can easily prove this (using our belief calculus):

　　　[∴ O~(u:Au̲x · ~u:RAu̲x)　　　VALID
* 1.　┌ asm: ~O~(u:Au̲x · ~u:RAu̲x)
* 2.　│　∴ R(u:Au̲x · ~u:RAu̲x) [from 1]
* 3.　│ D ∴ (u:Au̲x · ~u:RAu̲x) [from 2]
4.　│ D ∴ u:Au̲x [from 3]
* 5.　│ D ∴ ~u:RAu̲x [from 3]
* 6.　│ Du ∴ ~RAu̲x [from 5]
7.　│ Du ∴ Au̲x [from 4]
8.　│ Du ∴ O~Au̲x [from 6]
9.　└ Du ∴ ~Au̲x [from 8, contradicting 7]
10.　∴ O~(u:Au̲x · ~u:RAu̲x) [from 1, 7, and 9]

Since the argument is valid and the first premise is proved, we can focus on the second and third premises. To symbolize and prove these, we need to expand our system.

---

* For some discussion on this principle, see my "Acting Commits One to Ethical Beliefs," *Analysis* 42 (1983), 40–43.

Consider the second premise:

> 2. You ought not to *believe* that it would be all
> right for you to do A to x without *believing*
> that, in an imagined exactly reversed situation,
> it would be all right for x to do A to you.
>
> O~(u:RAux · ~u: . . .)

To *symbolize* the premise, we need to replace ". . ." with a formula that means this:

> In an exactly reversed situation, it
> would be all right for x to do A to you.

And to *prove* the premise, we need an inference rule to reflect the "universalizability principle."

The *universalizability principle* (U) is one of the few principles whose truth almost all moral philosophers agree on. Here's one formulation of U:

> ### UNIVERSALIZABILITY
>
> If x *ought* to do A, then under the same
> circumstances anyone else *ought* to do A.
>
> If it's *all right* for x to do A, then
> under the same circumstances it would be
> *all right* for anyone else to do A.

This amounts to saying that the morality of an action depends on reasons or principles that apply to similar cases—regardless of the specific persons involved.

We can express U more precisely using the notion of a "universal property." Recall from Section 9.5 that a *universal property* is a nonevaluative property describable without proper names or pointer terms. U says that the morality of an action depends on what kind of action it is—it depends on its universal properties. Two actions with the same universal properties have the same moral status. So we can formulate U this way:

> U   If act A ought to be done, then there's some
> universal property (or set of universal
> properties) Z, such that (1) act A is a Z
> act, and (2) in every actual or hypothetical
> situation, every Z act ought to be done.
>
> (OA ⊃ (∃Z)(ZA · ∎(X)(ZX ⊃ OX)))

The "all right" form is similar, but has "would be all right" ("R") in place of "ought to be done" ("O"). I'll explain the symbolization later.

One corollary of U is U∗:

> U*   If it would be all right for me to do A to
>   x, then, in an exactly reversed situation,
>   it would be all right for x to do A to me.

U* relates closely to our second premise:

> 2. You ought not to *believe* that it would be all
>   right for you to do A to x without *believing*
>   that, in an imagined exactly reversed situation,
>   it would be all right for x to do A to you.

After building U (and hence U*) into our system, we'll be able to prove this premise. Consider our third premise:

> 3. You ought not to *believe* that, in an imagined exactly
>   reversed situation, it would be all right for x to do
>   A to you, without *consenting* to the idea of x doing
>   A to you in an imagined exactly reversed situation.
>
> $$O\sim(\underline{u}\colon \ldots \cdot \sim\underline{u}\colon \ldots)$$

We'll interpret "*consenting* to A being done" as "*accepting* 'A *may* be done.'" Here the permissive "A *may* be done" *isn't* another way to say "A is *all right*." Rather it's a member of the imperative family, but weaker than "Do A," expressing only one's consent to the action. We'll symbolize "A may be done" as "M$\underline{A}$." The third premise says that you ought not to accept this first box without accepting the second:

> In an exactly reversed
> situation, it would be
> *all right* for x to do
> A to you.

> In an exactly reversed
> situation, x *may* do A
> to you.

We need machinery to symbolize these two. Also, we need this new principle to go from one box to the other:

> "It would be *all right*" for x
> to do A" entails "x *may* do A."
>
> $$\Box(R\underline{A} \supset M\underline{A})$$

In Section 7.4 we discussed "Prescriptivity" ("Hare's Law"), which says that an ought-judgment entails the corresponding imperative ["$\Box(O\underline{A} \supset \underline{A})$"]. This new principle is similar. It says that a permissibility-judgment entails the corresponding permissive.*

---

* For some discussion on this principle, see my "How Incomplete is Prescriptivism?" *Mind* 93 (1984), 103–107.

9.6 GRC FORMULAS

Now we start to build our *golden rule calculus* (GRC). Soon things will get complicated and difficult. You may need to read the next few sections a couple of times to follow what's happening.

GRC will build on our previous systems. But now we'll use capital letters in new ways and add the symbols "■" and "∗." Here are examples of some of the new wffs and quantifiers that we'll use:

> M$\underline{A}$ = Act A may be done.
>
> F$\underline{A}$ = Act A has universal property F.
>
> (∃Z) = For some universal property Z . . .
>
> (∃$\underline{X}$) = For some action X . . .
>
> ■O$\underline{A}$ = In every actual or hypothetical case, act A ought to be done.
>
> F∗$\underline{A}$ = F is the complete description of act A in universal terms.

We'll examine each of these items in turn.

When "M" is prefixed to an imperative wff, we'll translate it as "may":

1. The result of prefixing an imperative wff with "M" is a wff.

M$\underline{A}$ = Act A may be done.

MA$\underline{x}$u = x may do A to you.

u:MA$\underline{x}$u = You accept "x may do A to me."
         = You consent to x's doing A to you.
         = You're willing that x do A to you.
         = It's OK by you that x do A to you.

Permissives like "M$\underline{A}$" are weaker members of the imperative family. They express our *consent* to the act, but not necessarily our positive desire that the act take place. We can consistently consent both to the act and to its omission—by saying "You may do A and you may omit A." Here are further wffs:

∼M∼$\underline{A}$ = Act A may not be omitted.

∼M∼A$\underline{x}$u = x may not omit doing A to you.

u:∼M∼A$\underline{x}$u = You accept "x may not omit doing A to me."
         = You demand that x do A to you.

"M$\underline{A}$" is weaker and "∼M∼$\underline{A}$" is stronger than "$\underline{A}$."∗

∗ Capital letters in GRC can have a variety of uses, depending on the context. In "((M · Ma) ⊃ (Mbc · M$\underline{A}$))," for example, "M" is used first for a statement, then for a property of an individual entity, then for a relation between individual entities, and finally to represent "may." The meaning of "M" in each of these four cases may be distinct. For the sake of clarity, we'll generally use different capital letters in such cases.

GRC will be able to use capital letters (including those with primes) as universal property constants and variables:

> Capital letters (except for "M," "O," "R," "W," "X," "Y," and "Z"), when prefixed to an imperative wff (or to "*" followed by an imperative wff), are *universal property constants*. These represent specific universal properties of actions.

> "W," "X," "Y," and "Z," when used in a quantifier or when prefixed to an imperative wff (or to "*" followed by an imperative wff), are *universal property variables*. These range over universal properties of actions.

The following rule and examples show how these letters are used:

2. The result of writing a capital letter (except for "M," "O," and "R") and then an imperative wff is itself a *descriptive* wff.

FA   = Act A has universal property F.
           = Act A is F.

SA   = Act A is an act of stealing.

(SA ⊃ ~RA) = If act A is an act of stealing, then act A is wrong.

BA   = Act A is an act of a blue-eyed philosophy teacher stealing from a poor student with a sick wife.

Note that we translate "FA" as "Act A is F." We don't translate "FA" as "The imperative 'Do A!' is F."

We'll use universal property variables in quantifiers, as in these examples:

(Z)   = For every universal property (or set of properties) Z . . .
(∃Z)   = For some universal property (or set of properties) Z . . .

(∃Z)ZA   = For some universal property Z, act A is Z.
           = Act A has some universal property Z.

We'll also use action variables in quantifiers:

> "W," "X," "Y," and "Z" are *action variables*. These range over specific actions.

(We can also use such letters with primes.) Here are some examples:

(X)   = For every act X . . .
(∃X)   = For some act X . . .

(∃X)FX   = For some act X, X has universal property F.
                = Some act has universal property F.

(X)(FX ⊃ OX)   = For every act X, if act X is F, then act X ought to be done.
                = Every act which is F ought to be done.

(X)(∃Z)ZX   = For every act X there's some universal property Z, such that
                  act X is Z.
                = Every act has some universal property.

"■" is a modal operator somewhat like "□." This rule tells how to construct wffs using "■":

3. The result of prefixing any wff with "■" is a wff.

"■" translates as "in every actual or hypothetical case" or "in every possible world having the same basic moral principles as those true in the actual world." Here's a wff using "■":

■(FA ⊃ OA) = In every actual or hypothetical case, if act A is F, then act A
              ought to be done.
            = In any actual or hypothetical case where act A is F, act A
              ought to be done.
            = If act A *is* or *were* F, then act A ought to be done.

Suppose that, while act A doesn't have property F (for example, act A doesn't maximize pleasure), still, *if it did*, then it would be what ought to be done. We'll use "■(FA ⊃ OA)" to express "If act A *had* property F, then act A ought to be done." "(FA ⊃ OA)" is too weak to express this idea (since this wff is trivially true whenever "FA" is false). "□(FA ⊃ OA)" is too strong (since there's no such entailment). We'll use "■" to formulate claims about what would be right or wrong in hypothetical situations (such as imagined exactly reversed situations).

We can now symbolize the *universalizability principle*:

| U   If act A ought to be done, then there's some |
| universal property (or set of universal |
| properties) Z, such that (1) act A is a Z |
| act, and (2) in every actual or hypothetical |
situation, every Z act ought to be done.
(OA ⊃ (∃Z)(ZA · ■(X)(ZX ⊃ OX)))

This amounts to saying that the morality of an action depends on reasons or principles that apply to similar cases—regardless of the specific persons involved.

We'll use "∗" with universal property constants and variables to represent the complete description of an action in universal terms. Here's the rule for constructing wffs with "∗," together with an example:

4. The result of writing a capital letter (except for "M," "O," and "R"), then "∗," and then an imperative wff is itself a *descriptive* wff.

F*A = F is the *complete description of act A in universal terms.*
     = F is the description of act A in universal terms, which includes all the
        universal properties of act A.

"F*A" means the same as this longer wff:

(FA · (Z)(ZA ⊃ □(X)(FX ⊃ ZX)))
= Act A is F, and every universal property Z that A has is included as part of F.
= Act A is F, and, for every universal property Z that A has, it's logically necessary
   that every act that's F is also Z.

We'll use "*" in symbolizing "exactly reversed situation." Let's take an example. Let "Amx" represent the act of my attacking x. Suppose that this act has
complete description F:

COMPLETE DESCRIPTION	
F*Amx	*My attacking x* has complete universal description F.

Let's flesh this out. Let "G," "G'," "G''," and so on represent my universal
properties. These include properties such as being a logician. Let "H," "H'," "H'',"
and so on represent x's universal properties. Now property F would look something
like this (for simplicity I omit relationships between x and myself):

FA = A is an act of someone who is G, G', G'', and so on, attacking someone
       who is H, H', H'', and so on.

More concretely:

FA = A is the act of someone who is a logician and . . . [adding all of my
       other universal properties] attacking someone who is a poor student
       and . . . [adding all of x's other universal properties].

So then we could describe the ACTUAL SITUATION this way:

ACTUAL SITUATION	
FAmx	*My attacking x* is an act of someone who is G, G', G'', and so on, attacking someone who is H, H', H'', and so on.

Now imagine the situation where *x attacking me* has this same description F:

EXACTLY REVERSED SITUATION	
FAxm	*x's attacking me* is an act of someone who is G, G', G'', and so on, attacking someone who is H, H', H'', and so on.

To imagine this is to imagine an exactly reversed situation. Here x (with properties G, G', G'', and so on) is in my exact place. And I (with properties H, H', H'', and so on) am in x's exact place. All our universal properties are switched.

We'll later need to symbolize this claim:

> In an exactly reversed situation it
> would be all right for x to attack *me*.

We'll analyze this claim as follows:

> For some universal property Z, Z is the
> complete description in universal terms of my
> attacking x, and, in any actual or hypothetical
> situation, if x attacking me is Z, then it
> would be all right for x to attack me.
>
> $(\exists Z)(Z*A\underline{mx} \cdot \blacksquare(ZA\underline{x}m \supset RA\underline{x}m))$

Note how this expresses the idea of an "exactly reversed situation." Given this, it's easy to symbolize corollary U* of universalizability:

> U*   If it would be all right for me to do A to
>   x, then, in an exactly reversed situation,
>   it would be all right for x to do A to me.
>
> =   If it's all right for me to do A to x,
>   then, for some universal property Z, Z is
>   the complete description in universal terms
>   of my doing A to x, and, in any actual or
>   hypothetical situation, if x's doing A to
>   me is Z, then it would be all right for x
>   to do A to me.
>
> $(RA\underline{mx} \supset (\exists Z)(Z*A\underline{mx} \cdot \blacksquare(ZA\underline{x}m \supset RA\underline{x}m)))$

Recall that the last section formulated an argument for the golden rule. We proved the argument valid using DC, and we proved the first premise using BC. But we needed more machinery to symbolize and prove the second and third premises. Now we can symbolize both. Here's how we symbolize the second premise:

> 2. You ought not to *believe* that it would be all
>   right for you to do A to x without *believing*
>   that, in an imagined exactly reversed situation,
>   it would be all right for x to do A to you.
>
> $O{\sim}(u{:}RA\underline{ux} \cdot {\sim}u{:}(\exists Z)(Z*A\underline{ux}$
> $\cdot \blacksquare(ZA\underline{xu} \supset RA\underline{xu})))$

We can paraphrase the wff this way:

> You ought not to combine (1) accepting "It's all
> right for me to do A to x," with (2) not accepting
> "For some universal property Z, Z is the complete
> description in universal terms of my doing A to x,
> and, in any actual or hypothetical situation, if
> x's doing A to me is Z, then it's all right for x
> to do A to me."

Here's how we symbolize the third premise:

> 3. You ought not to *believe* that, in an imagined
>   exactly reversed situation, it would be all
>   right for x to do A to you, without *consenting*
>   to the idea of x doing A to you in an imagined
>   exactly reversed situation.
>
> $O{\sim}(u{:}(\exists Z)(Z{*}Aux \cdot \blacksquare(ZAxu \supset RAxu))$
> $\cdot \sim\underline{u}{:}(\exists Z)(Z{*}Au\underline{x} \cdot \blacksquare(ZA\underline{xu} \supset MA\underline{xu})))$

This says that you ought not to accept this first box without accepting the second:

> In an exactly reversed
> situation, it would be
> *all right* for x to do A
> to you.

> In an exactly reversed
> situation, x *may* do A
> to you.

Our version of the golden rule translates this way:

> You ought not to *act* to do A to x without
> *consenting* to the idea of x doing A to you
> in an imagined exactly reversed situation.
>
> $O{\sim}(u{:}Aux \cdot \sim u{:}(\exists Z)(Z{*}Aux$
> $\cdot \blacksquare(\overline{Z}A\underline{xu} \supset MAxu)))$

Here's a paraphrase:

> You ought not to combine (1) accepting "Do A to
> x," with (2) not accepting "For some universal
> property Z, Z is the complete description in
> universal terms of my doing A to x, and, in any
> actual or hypothetical situation, if x's doing
> A to me is Z, then x may do A to me."

Our version of Kant's formula of universal law translates in the following way:

> You ought to act only as you're willing for
> anyone to act in the same situation, regardless
> of where you imagine yourself in the situation.
>
> $O{\sim}(u{:}Au \cdot \sim u{:}(\exists Z)(Z{*}Au$
> $\cdot \blacksquare(\underline{X})(Z\underline{X} \supset MX)))$

Again, here's a paraphrase:

> You ought not to combine (1) accepting "Do A,"
> with (2) not accepting "For some universal
> property Z, Z is the complete description in
> universal terms of my doing A, and, in any
> actual or hypothetical situation, any act
> that's Z may be done."

9.6A Exercises

Translate each sentence into a GRC wff. Use these equivalences:

j = Jones	V$\underline{X}$ = Act X is a violent act
u = you	H$\underline{X}$ = Act X is a harmless-playful act
Axy = x attacks y	

0. Your attacking Jones is a violent act.   ANSWER: VA$\underline{uj}$

1. F is the complete description in universal terms of your attacking Jones.
2. All violent acts are wrong.
3. If your attacking Jones is a violent act, then you may not attack Jones.
4. In every actual or hypothetical case, all violent acts are wrong.
5. If Jones's attacking you is a violent act, then it's wrong.
6. In the hypothetical case where Jones's attacking you is a violent act, it would be wrong.
7. If Jones's attacking you were a violent act, it would be wrong.
8. Your attacking Jones is a violent act, and, in every actual or hypothetical case, every violent act would be wrong.
9. For some universal property Z, your attacking Jones has property Z, and, in every actual or hypothetical case, every act that has property Z would be wrong.
10. If act A is wrong, then, for some universal property Z, act A has property Z, and, in every actual or hypothetical case, every act that has property Z would be wrong.
11. Don't believe that A is wrong, without believing that, for some universal property Z, act A has property Z, and, in every actual or hypothetical case, every act that has property Z would be wrong.
12. If your attacking Jones is a violent act, don't you attack Jones.
13. Don't combine believing that your attacking Jones is a violent act with acting to attack Jones.
14. No violent act is a harmless-playful act.
15. Jones's attacking you is a harmless-playful act.
16. F is the complete description in universal terms of your attacking Jones; and, in the hypothetical case where Jones's attacking you is F, Jones may not attack you.
17. You believe that Jones's attacking you is a violent act.
18. You don't consent to the idea of Jones's attacking you.
19. You demand that Jones not attack you.
20. You consent to the idea that, in the hypothetical case where Jones's attacking you is a harmless-playful act, Jones may attack you.

21. If you were in the hypothetical case where Jones's attacking you is a harmless-playful act, then you'd consent to the idea of Jones's attacking you.

22. Don't act to do A to x while also demanding that x not do A to you in the exactly reversed situation.

23. If you don't consent to the idea of x doing A to you in an imagined exactly reversed situation, then don't act to do A to x.

24. Don't act to do A to x, without it being the case that, if you were in the imagined exactly reversed situation, then you'd consent to the idea of x doing A to you.

25. Don't act to do A to x without consenting to the idea of x doing A to you in the exactly reversed situation.

26. Don't act to do A to x without consenting to the idea of x doing A to you in any *relevantly similar* actual or hypothetical situation. [In Section 9.4, we mentioned that we could formulate an acceptable golden rule that speaks of "relevantly similar" situations. Let me give you a hint on how to do this problem. Paraphrase "In relevantly similar circumstances, x may do A to me" as "For some universal property Z, my doing A to x is Z, the fact that it's Z makes it morally permissible, and, in every actual or hypothetical situation, if x's doing A to me is Z, then x may do A to me."]

9.7 GRC INFERENCE RULES

To complete our proof of the golden rule, we need to add various inference rules. We need rules for universalizability and for the " 'Right' entails 'may' " part of prescriptivity. We need rules for our new symbols. And we need to expand the quantifier rules to cover property and action quantifiers.

G1 and G2 cover the "ought" and "all right" forms of universalizability. These rules hold regardless of what imperative wff replaces "$\underline{A}$," what universal property variable replaces "$\underline{Z}$," and what action variable replaces "$\underline{X}$":

$$G1 \quad O\underline{A} \rightarrow (\exists Z)(Z\underline{A} \cdot \blacksquare(\underline{X})(Z\underline{X} \supset O\underline{X}))$$
$$G2 \quad R\underline{A} \rightarrow (\exists Z)(Z\underline{A} \cdot \blacksquare(\underline{X})(Z\underline{X} \supset R\underline{X}))$$

In applying G1 and G2, the world prefix of the derived and deriving steps must be identical and must contain no "W."

> Act A ought to be done (would be all right)
>
> $\rightarrow$
>
> There's some universal property (or set of universal properties) Z, such that (1) act A is a Z act, and (2) in every actual or hypothetical situation, every Z act ought to be done (would be all right).

The proviso prevents us from being able to prove that violations of universalizability are logically self-contradictory. I think they *aren't*; but this is too difficult to discuss here.

G3 is the " 'Right' entails 'may' " principle. G3 holds regardless of what imperative wff replaces "A":

$$G3 \quad R\underline{A} \rightarrow M\underline{A}$$

A is all right	$\rightarrow$	A may be done.

Given this and the rules for "M," "O," and "R," we can also prove the reverse entailment from "M$\underline{A}$" to "R$\underline{A}$." Then either of the two logically entails the other; so accepting one commits a person to accepting the other. But the distinction between the two doesn't vanish. "R$\underline{A}$" is true or false. To accept "R$\underline{A}$" is to *believe* that something is true. But "M$\underline{A}$" isn't true or false. To accept "M$\underline{A}$" isn't to believe something but to *will* something—to *consent* to the idea of something being done.*

G4 gives an analysis (or definition) of the notion of a *complete description of an act in universal terms*. G4 holds regardless of what universal property constant or variable replaces F, what imperative wff replaces "A," what universal property variable replaces "Z," and what action variable replaces "X":

$$G4 \quad F*\underline{A} \quad \leftrightarrow \quad (F\underline{A} \cdot (Z)(Z\underline{A} \supset \Box(\underline{X})(F\underline{X} \supset Z\underline{X})))$$

> F is the *complete description*
> *of act A in universal terms*
>
> $\leftrightarrow$
>
> Act A is F and, for every universal property
> Z that A has, it's logically necessary that
> every act that's F is also Z.

A "complete description" would probably be infinitely long, and thus impossible to formulate completely.

G5 says that every act has a complete description in universal terms. G5 is an axiom. It lets us put the wff "$(\underline{X})(\exists Z)Z*\underline{X}$" on any line of a proof:

$$G5 \quad \rightarrow \quad (\underline{X})(\exists Z)Z*\underline{X}$$

> For every act X there's some universal
> property Z, such that Z is the complete
> description of act X in universal terms.

The rules for "■" and "M" involve new kinds of worlds. In GRC, a *world prefix* is any string of zero-or-more instances of letters from the set $\langle$W, D, H, P, a, b, c, . . .$\rangle$—where $\langle$a, b, c, . . .$\rangle$ is the set of small letters. Here "H," "HH," "HHH," and so on represent *hypothetical situation worlds*. These are possible worlds having the same basic moral principles as those true in the actual world (or whatever

---

* In my view, thinking that an act is *all right* commits one to *consenting* to the idea of it being done. In place of "consent" here, we could say "accept," "approve," "allow," "agree to," "condone," or "tolerate"—in one sense of these terms. The sense of "consent" that I have in mind refers to a kind of inner attitude that is incompatible with inwardly *objecting to* (*condemning, disapproving, forbidding, protesting, prohibiting, repudiating*) the act. *Consenting* here is a minimal attitude and need not involve *favoring* (*advocating, endorsing, recommending, supporting, welcoming*) the act. It's consistent to both consent to the idea of A being done and also consent to the idea of A not being done.

world the H-world depends on). G6 and G7 hold regardless of what pair of contradictory wffs replaces "A"/"~A":

$$G6 \quad ■A \quad \rightarrow \quad A$$

In applying G6, the world prefixes in the derived and deriving steps must be identical except perhaps for final H's.

> ■A   →   A is in every H-world.

$$G7 \quad \sim■A \quad \rightarrow \quad \sim A$$

In applying G7, the derived step's world prefix must be like that of the earlier step except that it ends in a *new* string of one or more H's (a string not occurring in earlier steps).

> ~■A   →   not-A is in *some new* H-world.

We won't use G8 to 11 in the proof of the golden rule. But I'll give them anyway. G8 says that "□" and "■" are equivalent when prefixed to descriptive wffs. G8 holds regardless of what *descriptive* wff replaces "A":

$$G8 \quad ■A \quad \leftrightarrow \quad □A$$

> If A is a descriptive wff: ■A   ↔   □A.

G9 to G11 govern "M" and use permission worlds (represented by "P," "PP," "PPP," and so on). These are analogous to deontic worlds. A *permission* world that depends on a given world W1 is a possible world that contains the indicative judgments of W1 and some set of imperatives prescribing actions jointly permitted by the permissives of W1. G9 and G10 hold regardless of what pair of contradictory imperative wffs replaces "A̲"/"~A̲":

$$G9 \quad \sim M\underline{A} \quad \rightarrow \quad \sim\underline{A}$$

In applying G9, the world prefix of the derived step must either be the same as that of the earlier step or else consist of the world prefix of the earlier step, plus a string of one or more P's.

> A may not be done
>
> →
>
> "Don't do A" is in every P-world.

$$G10 \quad M\underline{A} \quad \rightarrow \quad \underline{A}$$

In applying G10, the world prefix of the derived step must consist of that of the earlier step, plus a *new* string of D's (one not occurring in earlier steps).

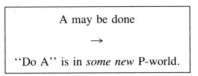

G11 mirrors the deontic indicative transfer rule D5. It holds regardless of what descriptive or deontic wff replaces "A":

$$G11 \quad \underline{A} \quad \leftrightarrow \quad A$$

In applying G11, the world prefixes in the derived and deriving steps must be identical except that one ends in one or more P's.

> If A is an indicative wff and
> W1 is a P-world that depends on world W2:
>
> A is true in W1 $\leftrightarrow$ A is true in W2

Finally, we expand the quantifier dropping rules Q3 and Q4 to cover the new kinds of quantifiers. We have to substitute the right sort of thing for the variable in the quantifier:

FOR	SUBSTITUTE
individual variables w, x, y, z, x', . . .	individual constants a, b, c, d, . . .
universal property variables W, X, Y, Z, W', . . .	universal property constants A, B, C, D, . . .
action variables $\underline{W}, \underline{X}, \underline{Y}, \underline{Z}, \underline{W}', \ldots$	imperative wffs $\underline{A}a, \underline{B}, \underline{A}x\underline{y}, \ldots$

In the last case, we need two further technical provisos. Suppose that we drop a quantifier containing an action variable, and we substitute an imperative wff for this variable:

- This imperative wff can contain no variable that also occurs in a quantifier in the derived wff.
- When dropping an existential quantifier, this substituted imperative wff must be an underlined capital letter that isn't an action variable and that hasn't occurred before in the proof.

9.8 FINISHING THE PROOF OF THE GOLDEN RULE

Before we finish the proof of the golden rule, let's review the larger picture.

When the chapter started, we sketched various dimensions of ethical rationality. Then we focused on rationality as consistency. Then we narrowed our focus further to a single consistency principle—the golden rule. We found that we had to formulate the rule precisely to avoid absurd implications. We arrived at this wording:

> You ought not to *act* to do A to x without *consenting* to the idea of x doing A to you in an imagined exactly reversed situation.

Then we sketched this proof of the golden rule:

1. You ought not to *act* to do A to x without *believing* that it would be all right for you to do A to x.

2. You ought not to *believe* that it would be all right for you to do A to x without *believing* that, in an imagined exactly reversed situation, it would be all right for x to do A to you.

3. You ought not to *believe* that, in an imagined exactly reversed situation, it would be all right for x to do A to you, without *consenting* to the idea of x doing A to you in an imagined exactly reversed situation.

∴ You ought not to *act* to do A to x without *consenting* to the idea of x doing A to you in an imagined exactly reversed situation.

Using the logical machinery that we had developed, we proved that the argument was valid and we proved the first premise. To prove the second and third premises, we had to develop a new system—the golden rule calculus (GRC). Now we'll finish the proof of the golden rule by proving the second and third premises.

Here's how we symbolize the second premise:

> 2. You ought not to *believe* that it would be all right for you to do A to x without *believing* that, in an imagined exactly reversed situation, it would be all right for x to do A to you.
>
> $O{\sim}(u{:}RAux \cdot {\sim}u{:}(\exists Z)(Z{*}Aux \cdot \blacksquare(\bar{Z}A\underline{x}u \supset RA\underline{x}u)))$

We can paraphrase the symbolization in the following way:

> You ought not to combine (1) accepting "It's all right for me to do A to x," with (2) not accepting "For some universal property Z, Z is the complete description in universal terms of my doing A to x, and, in any actual or hypothetical situation, if x's doing A to me is Z, then it's all right for x to do A to me."

The proof of this second premise is the most difficult proof in this book. I'll give the full proof and then sketch the main points in a Proof Chart. The full proof goes as follows (I put "#" next to the steps that use our new GRC inference rules):

[∴ O~(u:RAux · ~u:(∃Z)(Z∗Aux · ■(ZAxu ⊃ RAxu)))

1.	┌ asm: ~O~(u:RAux · ~u:(∃Z)(Z∗Aux · ■(ZAxu ⊃ RAxu)))
2.	∴ R(u:RAux · ~u:(∃Z)(Z∗Aux · ■(ZAxu ⊃ RAxu))) [from 1]
3.	D ∴ (u:RAux · ~u:(∃Z)(Z∗Aux · ■(ZAxu ⊃ RAxu))) [from 2]
4.	D ∴ u:RAux [from 3]
5.	D ∴ ~u:(∃Z)(Z∗Aux · ■(ZAxu ⊃ RAxu)) [from 3]
6.	Du ∴ ~(∃Z)(Z∗Aux · ■(ZAxu ⊃ RAxu)) [from 5]
7.	Du ∴ RAux [from 4]
8.	Du ∴ (Z)~(Z∗Aux · ■(ZAxu ⊃ RAxu)) [from 6]
# **9.**	Du ∴ (∃Z)(ZAux · ■(X)(ZX ⊃ RX)) [from 7 and *G2*]
10.	Du ∴ (FAux · ■(X)(FX ⊃ RX)) [from 9]
11.	Du ∴ FAux [from 10]
12.	Du ∴ ■(X)(FX ⊃ RX) [from 10]
#**13.**	Du ∴ (X)(∃Z)Z∗X [from *G5*]
14.	Du ∴ (∃Z)Z∗Aux [from 13]
15.	Du ∴ G∗Aux [from 14]
#**16.**	Du ∴ (GAux · (Z)(ZAux ⊃ □(X)(GX ⊃ ZX))) [from 15 and *G4*]
17.	Du ∴ GAux [from 16]
18.	Du ∴ (Z)(ZAux ⊃ □(X)(GX ⊃ ZX)) [from 16]
19.	Du ∴ (FAux ⊃ □(X)(GX ⊃ FX)) [from 18]
20.	Du ∴ □(X)(GX ⊃ FX) [from 11 and 19]
21.	Du ∴ ~(G∗Aux · ■(GAxu ⊃ RAxu)) [from 8]
22.	Du ∴ ~■(GAxu ⊃ RAxu) [from 15 and 21]
#**23.**	DuH ∴ ~(Gaxu ⊃ RAxu) [from 22 and *G7*]
#**24.**	DuH ∴ (X)(FX ⊃ RX) [from 12 and *G6*]
25.	DuH ∴ (X)(GX ⊃ FX) [from 20]
26.	DuH ∴ GAxu [from 23]
27.	DuH ∴ ~RAxu [from 23]
28.	DuH ∴ (FAxu ⊃ RAxu) [from 24]
29.	DuH ∴ (GAxu ⊃ FAxu) [from 25]
30.	DuH ∴ FAxu [from 26 and 29]
31.	└ DuH ∴ RAxu [from 28 and 30, contradicting 27]
32.	∴ O~(u:RAux · ~u:(∃Z)(Z∗Aux · ■(ZAxu ⊃ RAxu))) [from 1, 27, 31]

This is a difficult proof. But you should be able to follow the steps and see that everything follows correctly. The proof begins in the normal way. Soon we get these wffs in lines 7 and 6 (using "attack" in place of "do A to"):

7. It would be all right for you to attack x.
6. It's not the case that, in the exactly reversed situation, it would be all right for x to attack you.

Proof Chart. This sketches the key elements in the proof for our second premise.

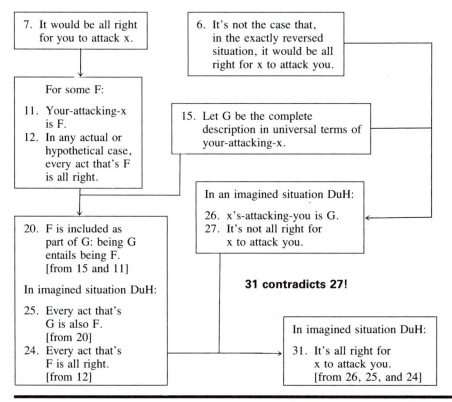

The proof chart sketches how we derive a contradiction from these two. We then apply RAA and derive our conclusion. This finishes our long proof of the second premise of our argument for the golden rule.

Here's how we symbolize the third premise:

> 3. You ought not to *believe* that, in an imagined exactly reversed situation, it would be all right for x to do A to you, without *consenting* to the idea of x doing A to you in an imagined exactly reversed situation.
>
> O~(u:(∃Z)(Z∗Au͟x · ■(ZAxu ⊃ RAxu))
> · ~u:(∃Z)(Z∗Au͟x · ■(ZAxu͟ ⊃ MAxu͟)))

This says that you ought not to accept this first box without accepting the second:

In an exactly reversed situation, it would be *all right* for x to do A to you.	In an exactly reversed situation, x *may* do A to you.

We begin the proof by assuming that it's all right to accept the first box but reject the second (steps 1 and 2). Then we peel off various operators until we get these two wffs in world-DuH (representing the imagined reversed situation):

18. It would be all right for x to do A to you.	17. x may not do A to you.

Then we use the " 'Right' entails 'may' " rule on 18 to get a contradiction:

> 19. x may do A to you.
>
> [from 18, contradicting 17]

The full proof of this third premise is long but not overly difficult:

$[\therefore$ O~(u:(∃Z)(Z∗Au$\underline{x}$ · ■(ZA$\underline{x}$u ⊃ RA$\underline{x}$u)) · ~$\underline{u}$:(∃Z)(Z∗Au$\underline{x}$ · ■(ZA$\underline{x}$u ⊃ MA$\underline{x}$u)))

1. ⌐asm: ~O~(u:(∃Z)(Z∗Au$\underline{x}$ · ■(ZA$\underline{x}$u ⊃ RA$\underline{x}$u)) · ~$\underline{u}$:(∃Z)(Z∗Au$\underline{x}$ · ■(ZA$\underline{x}$u ⊃ MA$\underline{x}$u)))

2. ∴ R(u:(∃Z)(Z∗Au$\underline{x}$ · ■(ZA$\underline{x}$u ⊃ RA$\underline{x}$u)) · ~$\underline{u}$:(∃Z)(Z∗Au$\underline{x}$ · ■(ZA$\underline{x}$u ⊃ MA$\underline{x}$u))) [from $\overline{1}$]

3. D ∴ (u:(∃Z)(Z∗Au$\underline{x}$ · ■(ZA$\underline{x}$u ⊃ RA$\underline{x}$u)) · ~$\underline{u}$:(∃Z)(Z∗Au$\underline{x}$ · ■(ZA$\underline{x}$u ⊃ MA$\underline{x}$u))) [from $\overline{2}$]

4. D ∴ u:(∃Z)(Z∗Au$\underline{x}$ · ■(ZA$\underline{x}$u ⊃ RA$\underline{x}$u)) [from 3]

5. D ∴ ~$\underline{u}$:(∃Z)(Z∗Au$\underline{x}$ · ■(ZA$\underline{x}$u ⊃ MA$\underline{x}$u)) [from 3]

6. Du ∴ ~(∃Z)(Z∗Au$\underline{x}$ · ■(ZA$\underline{x}$u ⊃ MA$\underline{x}$u)) [from 5]

7. Du ∴ (∃Z)(Z∗Au$\underline{x}$ · ■(ZA$\underline{x}$u ⊃ RA$\underline{x}$u)) [from 4]

8. Du ∴ (Z)~(Z∗Au$\underline{x}$ · ■(ZA$\underline{x}$u ⊃ MA$\underline{x}$u)) [from 6]

9. Du ∴ (F∗Au$\underline{x}$ · ■(FA$\underline{x}$u ⊃ RA$\underline{x}$u)) [from 7]

10. Du ∴ ~(F∗Au$\underline{x}$ · ■(FA$\underline{x}$u ⊃ MA$\underline{x}$u)) [from 8]

11. Du ∴ F∗Au$\underline{x}$ [from 9]

12. Du ∴ ■(FA$\underline{x}$u ⊃ RA$\underline{x}$u) [from 9]

13. Du ∴ ~■(FA$\underline{x}$u ⊃ MA$\underline{x}$u) [from 10 and 11]

#14. DuH ∴ ~(FA$\underline{x}$u ⊃ MA$\underline{x}$u) [from 13 and *G7*]

#15. DuH ∴ (FA$\underline{x}$u ⊃ RA$\underline{x}$u) [from 12 and *G6*]

16. DuH ∴ FA$\underline{x}$u [from 14]

17. DuH ∴ ~MA$\underline{x}$u [from 14]

18. DuH ∴ RA$\underline{x}$u [from 15 and 16]

#19. ⌊DuH ∴ MA$\underline{x}$u [from 18 and *G3*, contradicting 17]

20. ∴ O~(u:(∃Z)(Z∗Au$\underline{x}$ · ■(ZA$\underline{x}$u ⊃ RA$\underline{x}$u)) · ~$\underline{u}$:(∃Z)(Z∗Au$\underline{x}$ · ■(ZA$\underline{x}$u ⊃ MA$\underline{x}$u))) [from $\overline{1}$, 17 and 19]

So now we've proved that the golden rule follows from our three premises. And we've proved all three premises. This ends our proof of the golden rule:

THE GOLDEN RULE
You ought not to *act* to do A to x without *consenting* to the idea of x doing A to you in an imagined exactly reversed situation.
$O{\sim}(u{:}Aux \cdot {\sim}u{:}(\exists Z)(Z{*}Aux \cdot \blacksquare(\overline{Z}Axu \supset MAxu)))$
So always treat others as you would want to be treated; that is the summary of the Law and the Prophets. (Mt 7:12)

9.8A Exercise

Using GRC, prove Kant's formula of universal law. Prove the version of the formula that we gave in Section 9.6.

Answers to Selected Problems

(Answers are given for exercise problems 1, 3, 5, 10, 15, 20, 25, 30, and so forth.)

2.2A

1. ~(A · B)
3. (~A ∨ B)
5. (A ∨ (B · C))
10. ((A · B) ∨ C)
15. ((C ⊃ M) ⊃ (C ⊃ ~R))
20. E ["(M ∨ F)" won't work, since the sentence doesn't mean "Everyone is male or everyone is female."]

2.4A

1. ~(1 · 0) = ~0 = 1
3. ~(~1 · ~0) = ~(0 · 1) = ~0 = 1
5. (~0 ≡ 0) = (1 ≡ 0) = 0
10. (~1 ∨ ~(0 ⊃ 0)) = (0 ∨ ~1) = (0 ∨ 0) = 0
15. ((1 · ~0) ⊃ ~1) = ((1 · 1) ⊃ 0) = (1 ⊃ 0) = 0
20. (1 ∨ ~1) = (1 ∨ 0) = 1

2.5A

1. (? ⊃ ~1) = (? ⊃ 0) = ?
3. (? ∨ ~0) = (? ∨ 1) = 1
5. (? · 0) = 0
10. (~0 · ?) = (1 · ?) = ?
15. (? ⊃ 0) = ?
20. (? ⊃ 1) = 1

2.6A

1.

P	Q	(~P ⊃ Q)
0	0	0
0	1	1
1	0	1
1	1	1

3.

P	Q	(~P · Q)
0	0	0
0	1	1
1	0	0
1	1	0

5.

P	Q	$(\sim Q \supset \sim P)$
0	0	1
0	1	1
1	0	0
1	1	1

10.

P	Q	R	$((P \lor \sim Q) \supset R)$
0	0	0	0
0	0	1	1
0	1	0	1
0	1	1	1
1	0	0	0
1	0	1	1
1	1	0	0
1	1	1	1

2.7A

1.

C	I	$(C \supset I)$, I $\therefore$ C		
0	0	1	0	0
0	1	1	1	0
1	0	0	0	1
1	1	1	1	1

INVALID (You might be in Evanston, Illinois.)

***   ← We sometimes get true premises and a false conclusion.

3.

T	A	$(T \supset A)$, $(T \supset \sim A)$ $\therefore$ $\sim T$		
0	0	1	1	1
0	1	1	1	1
1	0	0	1	0
1	1	1	0	0

VALID

5.

R	T	E	$((R \cdot T) \supset E)$, T, $\sim R$ $\therefore$ $\sim E$			
0	0	0	1	0	1	1
0	0	1	1	0	1	0
0	1	0	1	1	1	1
0	1	1	1	1	1	0
1	0	0	1	0	0	1
1	0	1	1	0	0	0
1	1	0	0	1	0	1
1	1	1	1	1	0	0

INVALID

Row 4 marked *** ← We sometimes get true premises and a false conclusion.

(A few years ago, I got a group together but couldn't get Grand Canyon backcountry reservations for spring break. But we explored canyons anyway. We backpacked in Escalante Canyon in Utah. This made R = 0, T = 1, and E = 1.)

10.

M	S	V	$((M \lor \sim S) \supset V)$, $\sim M$, S $\therefore$ $\sim V$			
0	0	0	0	1	0	1
0	0	1	1	1	0	0
0	1	0	1	1	1	1
0	1	1	1	1	1	0
1	0	0	0	0	0	1
1	0	1	1	0	0	0
1	1	0	0	0	1	1
1	1	1	1	0	1	0

INVALID (The warranty might have expired.)

Row 4 marked *** ← We sometimes get true premises and a false conclusion.

2.8A

1. $((T \lor \sim M^0) \supset O^0) \neq 1$ VALID
   $\sim M^0 = 1$
   $\therefore O^0 = 0$

3. $((J^1 \cdot \sim D^1) \supset Z^0) = 1$ INVALID
   $\sim Z^0 = 1$
   $D^1 = 1$
   $\therefore \sim J^1 = 0$

5. $((W^0 \cdot C^1) \supset Z^0) = 1$ INVALID
   $\sim Z^0 = 1$
   $\therefore \sim C^1 = 0$
   (Here we let W be 0 in order for the first premise to be 1.)

10. $K^1 = 1$ INVALID
    $((K^1 \cdot E^0) \supset \sim A^0) = 1$
    $\sim A^0 = 1$
    $\therefore E^0 = 0$

15. $(\sim T^0 \supset (P^1 \supset J^0)) \neq 1$ VALID
    $P^1 = 1$
    $\sim J^0 = 1$
    $\therefore T^0 = 0$

20. $A^1 = 1$ VALID
    $\sim A^1 \neq 1$
    $\therefore B^0 = 0$
    (An argument with inconsistent premises is always *valid*. If the premises can't all be true, we can't have true premises and a false conclusion. So the argument is valid. But it can't be *sound*, since the premises can't all be true.)

2.8B

1. $(\sim F \supset (E \lor S))$ VALID
   $\sim E$
   $\sim S$
   $\therefore F$

3. $((P \cdot A) \supset E)$ VALID
   A
   P
   $\therefore E$

5. $((A \cdot M) \supset G)$ VALID
   $\sim G$
   M
   $\therefore \sim A$

10. $(I \supset (U \lor \sim P))$ INVALID
    $\sim U$
    $\sim P$
    $\therefore I$

15. $((H \cdot \sim D) \supset T)$ VALID
     $\sim D$
     H
∴ T

20. $(T \supset K)$ VALID
     $\sim K$
∴ $\sim T$

2.9A

1. $(A \supset B)$
3. $(A \equiv B)$
5. $(\sim A \supset \sim B)$ or $(B \supset A)$
10. $(F \supset \sim K)$
15. $(S \supset W)$

2.10A

1. Knowledge isn't sensation.
3. Our wide receiver won't be open.
5. You don't get an A.
10. Sam stole the money.
15. There will be a fire.
20. You don't get an A.

2.10B

1. $(K \supset \sim P)$ VALID
     P
∴ $\sim K$

3. $((B \cdot \sim G) \supset O)$ INVALID
     $\sim B$
∴ $\sim O$

5. $(\sim E \supset \sim A)$ VALID
     $\sim E$
∴ $\sim A$

10. $(K \supset S)$ VALID (with a less plausible
     K   premise)
∴ S   or
    $(S \supset K)$ INVALID (with a more
     K   plausible premise)
∴ S

15. $(F \supset O)$ INVALID
     O
∴ F

20. $(E \supset A)$ INVALID
     $\sim E$
∴ $\sim A$

2.11A

1. $\sim I$, V
3. no conclusion
5. $\sim Q$, B
10. A, P
15. no conclusion

2.12A

1. $\sim T$
3. $\sim L$
5. no conclusion
10. J
15. B
20. $\sim O$

2.12B

1. no conclusion
3. ~E
5. ~R, A
10. no conclusion
15. B, ~C

2.13A

1. (A · B), C
3. no conclusion
5. no conclusion
10. no conclusion

3.2A

1.   * 1.   (A ⊃ B)   VALID
          [∴ (~B ⊃ ~A)
    * 2.   ⌈asm: ~(~B ⊃ ~A)
      3.   │ ∴ ~B [from 2]
      4.   │ ∴ A [from 2]
      5.   ⌊∴ B [from 1 and 4, contradicting 3]
      6.   ∴ (~B ⊃ ~A) [from 2, 3, and 5]

3.   * 1.   ~(A · B)   VALID
          [∴ (~A ∨ ~B)
    * 2.   ⌈asm: ~(~A ∨ ~B)
      3.   │ ∴ A [from 2]
      4.   │ ∴ B [from 2]
      5.   ⌊∴ ~B [from 1 and 3, contradicting 4]
      6.   ∴ (~A ∨ ~B) [from 2, 4, and 5]

5.   * 1.   (A · (B · C))   VALID
          [∴ ((A · B) · C)
    * 2.   ⌈asm: ~((A · B) · C)
      3.   │ ∴ A [from 1]
    * 4.   │ ∴ (B · C) [from 1]
      5.   │ ∴ B [from 4]
      6.   │ ∴ C [from 4]
    * 7.   │ ∴ ~(A · B) [from 2 and 6]
      8.   ⌊∴ ~B [from 3 and 7, contradicting 5]
      9.   ∴ ((A · B) · C) [from 2, 5, and 8]

10.   * 1.   (A ∨ B)   VALID
    * 2.   (A ⊃ C)
    * 3.   (B ⊃ D)
          [∴ (C ∨ D)
    * 4.   ⌈asm: ~(C ∨ D)
      5.   │ ∴ ~C [from 4]
      6.   │ ∴ ~D [from 4]
      7.   │ ∴ ~A [from 2 and 5]
      8.   │ ∴ ~B [from 3 and 6]
      9.   ⌊∴ B [from 1 and 7, contradicting 8]
      10.   ∴ (C ∨ D) [from 4, 8, and 9]

3.2B

1.   * 1.   ((S · ~M) ⊃ D)   VALID
      2.   ~D
          [∴ (S ⊃ M)

* 3.   ⌈asm: ~(S ⊃ M)
  4. | ∴ S [from 3]
  5. | ∴ ~M [from 3]
* 6. | ∴ ~(S · ~M) [from 1 and 2]
  7.   ⌊∴ M [from 4 and 6, contradicting 5]
  8.   ∴ (S ⊃ M) [from 3, 5, and 7]

3. * 1.   ((P · U) ⊃ S)   VALID
  2.   P
  3.   ~S
       [∴ ~U
  4.   ⌈asm: U
* 5. | ∴ ~(P · U) [from 1 and 3]
  6.   ⌊∴ ~U [from 2 and 5, contradicting 4]
  7.   ∴ ~U [from 4 and 6]

5. 1.   W   VALID
  2.   M
  3.   U
* 4.   ((M · W) ⊃ P)
* 5.   ((U · P) ⊃ D)
       [∴ D
  6.   ⌈asm: ~D
* 7. | ∴ ~(U · P) [from 5 and 6]
  8. | ∴ ~P [from 3 and 7]
* 9. | ∴ ~(M · W) [from 4 and 8]
  10. ⌊∴ ~M [from 1 and 9, contradicting 2]
  11. ∴ D [from 6, 2, and 10]

10. 1.   G   VALID
* 2.   (P ∨ C)
* 3.   ((P · G) ⊃ ~X)
  4.   A
* 5.   (C ⊃ D)
* 6.   ((D · A) ⊃ H)
* 7.   (H ⊃ ~X)
       [∴ ~X
  8.   ⌈asm: X
  9. | ∴ ~H [from 7 and 8]
* 10. | ∴ ~(D · A) [from 6 and 9]
  11. | ∴ ~D [from 4 and 10]
  12. | ∴ ~C [from 5 and 11]
  13. | ∴ P [from 2 and 12]
* 14. | ∴ ~(P · G) [from 3 and 8]
  15. ⌊∴ ~G [from 13 and 14, contradicting 1]
  16. ∴ ~X [from 8, 1, and 15]

3.3A

1.   1. $(A^0 \supset B^1) = 1$ INVALID
         $[\therefore (B^1 \supset A^0) = 0$

    * 2.   asm: $\sim(B \supset A)$ $\boxed{B, \sim A}$
     3. $\therefore B$ [from 2]
     4. $\therefore \sim A$ [from 2]

3.   * 1. $((A^1 \cdot B^0) \supset C^0) = 1$ INVALID
   * 2. $((C^0 \vee D^0) \supset \sim E^0) = 1$
        $[\therefore \sim(A^1 \cdot E^1) = 0$ $\boxed{A, E, \sim B, \sim C, \sim D}$
   * 3.   asm: $(A \cdot E)$
     4. $\therefore A$ [from 3]
     5. $\therefore E$ [from 3]
   * 6. $\therefore \sim(C \vee D)$ [from 2 and 5]
     7. $\therefore \sim C$ [from 6]
     8. $\therefore \sim D$ [from 6]
   * 9. $\therefore \sim(A \cdot B)$ [from 1 and 7]
    10. $\therefore \sim B$ [from 4 and 9]

5.   1. $((A^1 \supset B^0) \supset (C^1 \supset D^0)) = 1$ INVALID
   * 2. $(B^0 \supset D^0) = 1$
   * 3. $(A^1 \supset C^1) = 1$ $\boxed{A, C, \sim B, \sim D}$
       $[\therefore (A^1 \supset D^0) = 0$
   * 4.   asm: $\sim(A \supset D)$
     5. $\therefore A$ [from 4]
     6. $\therefore \sim D$ [from 4]
     7. $\therefore C$ [from 3 and 5]
     8. $\therefore \sim B$ [from 2 and 6]

10. $[\therefore ((A^0 \vee B^1) \supset A^0) = 0$ INVALID
   * 1.   asm: $\sim((A \vee B) \supset A)$
   * 2. $\therefore (A \vee B)$ [from 1] $\boxed{B, \sim A}$
     3. $\therefore \sim A$ [from 1]
     4. $\therefore B$ [from 2 and 3]

3.3B

1.   1.   T   VALID
   * 2. $(T \supset O)$
   * 3. $(O \supset W)$
       $[\therefore W$
     4.   ⌈ asm: $\sim W$
     5. | $\therefore O$ [from 1 and 2]
     6.   ⌊ $\therefore W$ [from 3 and 5, contradicting 4]
     7. $\therefore W$ [from 4 and 6]

3.   * 1.   (P ⊃ C)   VALID
         * 2.   ((C · D) ⊃ ~G)
                  [∴ (G ⊃ (~P ∨ ~D))
         * 3.   ⌐asm: ~(G ⊃ (~P ∨ ~D))
           4.   │∴ G [from 3]
         * 5.   │∴ ~(~P ∨ ~D) [from 3]
           6.   │∴ P [from 5]
           7.   │∴ D [from 5]
           8.   │∴ C [from 1 and 6]
         * 9.   │∴ ~(C · D) [from 2 and 4]
          10.   └∴ ~D [from 8 and 9, contradicting 7]
          11.   ∴ (G ⊃ (~P ∨ ~D)) [from 3, 7, and 10]

5.   1.   ((M¹ · ~A¹) ⊃ I) = 1   INVALID
           2.   (I ⊃ W¹) = 1
                  [∴ ((M¹ · A¹) ⊃ ~W¹) = 0   ┌─────────┐
         * 3.   asm: ~((M · A) ⊃ ~W)   │ M, A, W │
         * 4.   ∴ (M · A) [from 3]   └─────────┘
           5.   ∴ W [from 3]
           6.   ∴ M [from 4]
           7.   ∴ A [from 4]

The conclusion is false. You might do something wrong and yet be acting on your moral belief that this act is your duty. This more plausible conclusion follows validly from the premises: "If you hold a moral belief and *don't* act on it, then you're doing wrong." (In this case, you're doing wrong by being inconsistent.)

10.   1.   (D⁰ ⊃ ~F⁰) = 1   INVALID
         * 2.   (H¹ ⊃ E¹) = 1
         * 3.   (E¹ ⊃ ~D⁰) = 1   ┌──────────────┐
                  [∴ (H¹ ⊃ F⁰) = 0   │ H, E, ~D, ~F │
         * 4.   asm: ~(H ⊃ F)   └──────────────┘
           5.   ∴ H [from 4]
           6.   ∴ ~F [from 4]
           7.   ∴ E [from 2 and 5]
           8.   ∴ ~D [from 3 and 7]

We'd have a valid argument if the first premise read "If determinism isn't true then we have free will."

15. * 1.   (W ≡ B)   VALID
                  [∴ ((B · L) ⊃ W)
         * 2.   ⌐asm: ~((B · L) ⊃ W)
         * 3.   │∴ (B · L) [from 2]
           4.   │∴ ~W [from 2]
           5.   │∴ B [from 3]
           6.   │∴ L [from 3]
           7.   │∴ (W ⊃ B) [from 1]
         * 8.   │∴ (B ⊃ W) [from 1]
           9.   └∴ W [from 5 and 8, contradicting 4]
          10.   ∴ ((B · L) ⊃ W) [from 2, 4, and 9]

20.   1. $(M^0 \supset \sim I^0) = 1$ INVALID
      2. $(I^0 \supset \sim M^0) = 1$
      * 3. $(E^1 \supset \sim M^0) = 1$ $\boxed{\text{T, E, } \sim\text{I, } \sim\text{M}}$
      * 4. $(T^1 \supset E^1) = 1$
         [∴ $(T^1 \supset I^0) = 0$
      * 5.   asm: $\sim(T \supset I)$
      6.   ∴ T [from 5]
      7.   ∴ $\sim$I [from 5]
      8.   ∴ E [from 4 and 6]
      9.   ∴ $\sim$M [from 3 and 8]

We'd have a valid argument if the first premise read "If materialism is false, then idealism is true" or "Either materialism is true or idealism is true."

25.   1.   S   VALID
      * 2. $(Q \supset A)$
      * 3. $(A \supset B)$
      * 4. $(B \supset X)$
      * 5. $((X \cdot B) \supset Y)$
      * 6. $((X \cdot Y) \supset \sim S)$
         [∴ $\sim$Q
      7.   ⌈asm: Q
      8.   │∴ A [from 2 and 7]
      9.   │∴ B [from 3 and 8]
      10.   │∴ X [from 4 and 9]
      * 11.   │∴ $\sim(X \cdot Y)$ [from 1 and 6]
      12.   │∴ $\sim$Y [from 10 and 11]
      * 13.   │∴ $\sim(X \cdot B)$ [from 5 and 12]
      14.   ⌊∴ $\sim$B [from 10 and 13, contradicting 9]
      15.   ∴ Q [from 7, 9, and 14]

3.5A

1.   * 1. $(B \lor A)$ VALID
      2. $(B \supset A)$
        [∴ $\sim(A \supset \sim A)$
      * 3.   ⌈asm: $(A \supset \sim A)$
      4.   │⌈asm: B [nice to have "$\sim$B" for 1]
      5.   ││∴ A [from 2 and 4]
      6.   │⌊∴ $\sim$A [from 3 and 5, contradicting 5]
      7.   │∴ $\sim$B [from 4, 5, and 6]
      8.   │∴ A [from 1 and 7]
      9.   ⌊∴ $\sim$A [from 3 and 8, contradicting 8]
      10.   ∴ $\sim(A \supset \sim A)$ [from 3, 8, and 9]
3.   * 1. $(((A \cdot B) \supset C) \supset (D \supset E))$ VALID
      2.   D
        [∴ $(C \supset E)$

    * 3.   ⌈asm: ~(C ⊃ E)
     4. | ∴ C [from 3]
     5. | ∴ ~E [from 3]
     6. | ⌈asm: ~((A · B) ⊃ C) [nice to have "((A · B) ⊃ C)" for 1]
     7. | | ∴ (A · B) [from 6]
     8. | ⌊∴ ~C [from 6, contradicting 4]
     9. | ∴ ((A · B) ⊃ C) [from 6, 4, and 8]
   * 10. | ∴ (D ⊃ E) [from 1 and 9]
    11.   ⌊∴ E [from 2 and 10, contradicting 5]
    12. ∴ (C ⊃ E) [from 3, 5, and 11]

5.   * 1.   ((A · B) ∨ (A · C))   VALID
         [∴ (A · (B ∨ C))
   * 2.   ⌈asm: ~(A · (B ∨ C))
     3. | ⌈asm: (A · B) [nice to have "~(A · B)" for 1]
     4. | | ∴ A [from 3]
     5. | | ∴ B [from 3]
     6. | | ∴ ~(B ∨ C) [from 2 and 4]
     7. | ⌊∴ ~B [from 6, contradicting 5]
     8. | ∴ ~(A · B) [from 3, 5, and 7]
   * 9. | ∴ (A · C) [from 1 and 8]
    10. | ∴ A [from 9]
    11. | ∴ C [from 9]
  * 12. | ∴ ~(B ∨ C) [from 2 and 10]
    13. | ∴ ~B [from 12]
    14.   ⌊∴ ~C [from 12, contradicting 11]
    15. ∴ (A · (B ∨ C)) [from 2, 11, and 14]

10.   * 1.   (A ⊃ (B · C))   VALID
        [∴ ((A ⊃ B) · (A ⊃ C))
   * 2.   ⌈asm: ~((A ⊃ B) · (A ⊃ C))
     3. | ⌈asm: ~(A ⊃ B) [nice to have "(A ⊃ B)" for 2]
     4. | | ∴ A [from 3]
     5. | | ∴ ~B [from 3]
     6. | | ∴ (B · C) [from 1 and 4]
     7. | ⌊∴ B [from 6, contradicting 5]
     8. | ∴ (A ⊃ B) [from 3, 5, and 7]
   * 9. | ∴ ~(A ⊃ C) [from 2 and 8]
    10. | ∴ A [from 9]
    11. | ∴ ~C [from 9]
  * 12. | ∴ (B · C) [from 1 and 10]
    13. | ∴ B [from 12]
    14.   ⌊∴ C [from 12, contradicting 11]
    15. ∴ ((A ⊃ B) · (A ⊃ C)) [from 2, 11, and 14]

This one gets difficult (but still possible) if you assume "~A" for step 3. Another strategy is to first prove "(A ⊃ B)" (by assuming the opposite and then getting a contradiction), and then prove "(A ⊃ C)" in a similar way, and then derive their conjunction.

3.5B

1.   * 1.   ((H · S) ⊃ (P · ~N))   VALID
     2.   S

      3.   N
           [∴ ~H
      4.   ⌐asm: H
      5.   │ ∴ (H · S) [from 4 and 2, nice to have for 1]
 * 6.   │ ∴ (P · ~N) [from 1 and 5]
      7.   │ ∴ P [from 6]
      8.   └ ∴ ~N [from 6, contradicting 3]
      9. ∴ ~H [from 4, 3, and 8]

3. * 1.   (A ≡ (H ∨ B))   VALID
     2.   ~H
         [∴ (A ≡ B)
 * 3.   ⌐asm: ~(A ≡ B)
 * 4.   │ ∴ (A ∨ B) [from 3]
    5.   │ ∴ ~(A · B) [from 3]
    6.   │ ∴ (A ⊃ (H ∨ B)) [from 1]
 * 7.   │ ∴ ((H ∨ B) ⊃ A) [from 1]
    8.   │ ⌐asm: A [nice to have "~A" for 4]
    9.   │ │ ∴ ~B [from 5 and 8]
   10. │ │ ∴ (H ∨ B) [from 6 and 8]
   11. │ └ ∴ H [from 9 and 10, contradicting 2]
   12. │ ∴ ~A [from 8, 2, and 11]
   13. │ ∴ B [from 4 and 12]
   14. │ ∴ ~(H ∨ B) [from 7 and 12]
   15. └ ∴ ~B [from 14, contradicting 13]
   16. ∴ (A ≡ H) [from 3, 13, and 15]

5. 1.   M   VALID
    2.   ~G
 * 3.   ((M · E) ⊃ (B ∨ G))
        [∴ (~E ∨ B)
 * 4.   ⌐asm: ~(~E ∨ B)
    5.   │ ∴ E [from 4]
    6.   │ ∴ ~B [from 4]
    7.   │ ∴ (M · E) [from 1 and 5, nice to have for 3]
 * 8.   │ ∴ (B ∨ G) [from 3 and 7]
    9.   └ ∴ G [from 6 and 8, contradicting 2]
   10. ∴ (~E ∨ B) [from 4, 2, and 9]

10. * 1.   ((D · L) ⊃ (F · A))   VALID
    2.   ((D · ~L) ⊃ (G · A))
        [∴ (D ⊃ A)
 * 3. ⌐ asm: ~(D ⊃ A)
    4. │ ∴ D [from 3]
    5. │ ∴ ~A [from 3]
    6. │ ⌐asm: ~(D · L) [nice to have "(D · L)" for 1]
    7. │ │ ∴ ~L [from 4 and 6]
    8. │ │ ∴ (D · ~L) [from 4 and 7, nice to have for 2]
    9. │ │ ∴ (G · A) [from 2 and 8]
   10. │ │ ∴ G [from 9]
   11. │ └ ∴ A [from 9, contradicting 5]
   12. │ ∴ (D · L) [from 6, 5, and 11]

* 13. $\quad\Big|\ \therefore\ (F \cdot A)$ [from 1 and 12]
14. $\quad\Big|\ \therefore\ F$ [from 13]
15. $\quad\Big\lfloor\ \therefore\ A$ [from 13, contradicting 5]
16. $\quad \therefore\ (D \supset A)$ [from 3, 5, and 15]

3.6A

1. ** 1. $\quad \sim(A^0 \cdot B^1) = 1$ $\qquad\qquad$ INVALID

   $\quad [\therefore\ (\sim A^0 \cdot \sim B^1) = 0$

   ** 2. $\quad$ asm: $\sim(\sim A \cdot \sim B)$ $\qquad$ $\boxed{B, \sim A}$

   3. $\qquad$ asm: $\sim A$ [nice to have "A" for 1]

   4. $\qquad \therefore\ B$ [from 2 and 3]

   Various refutations are often possible. Here this
   one would also work: $\qquad\qquad$ $\boxed{A, \sim B}$

3. ** 1. $\quad (A^0 \supset (B \cdot C^0)) = 1$ $\qquad\qquad$ INVALID

   ** 2. $\quad ((D^1 \supset E^0) \supset A^0) = 1$

   $\quad [\therefore\ (E^0 \vee C^0) = 0$ $\qquad$ $\boxed{D, \sim A, \sim C, \sim E}$

   * 3. $\quad$ asm: $\sim(E \vee C)$

   4. $\quad \therefore\ \sim E$ [from 3]

   5. $\quad \therefore\ \sim C$ [from 3]

   6. $\qquad$ asm: $\sim A$ [nice to have "A" for 1]

   ** 7. $\qquad \therefore\ \sim(D \supset E)$ [from 2 and 6]

   8. $\qquad \therefore\ D$ [from 7]

   9. $\qquad \therefore\ \sim E$ [from 7]

5. ** 1. $\quad (\sim A^1 \supset B^1) = 1$ $\qquad\qquad$ INVALID

   $\quad [\therefore\ \sim(A^1 \supset B^1) = 0$

   ** 2. $\quad$ asm: $(A \supset B)$ $\qquad\qquad$ $\boxed{A, B}$

   3. $\qquad$ asm: A [nice to have "$\sim A$" for 1]

   4. $\qquad \therefore\ B$ [from 2 and 3]

10. * 1. $\quad (\sim(A \cdot B) \supset \sim(C \cdot D))$ $\qquad$ VALID

    2. $\quad C$

    $\quad [\therefore\ (\sim A \supset \sim D)$

    * 3. $\quad \big\lceil$ asm: $\sim(\sim A \supset \sim D)$

    4. $\quad \big|\ \therefore\ \sim A$ [from 3]

    5. $\quad \big|\ \therefore\ D$ [from 3]

    6. $\quad \big|\ \big\lceil$ asm: $(A \cdot B)$ [nice to have "$\sim(A \cdot B)$" for 1]

    7. $\quad \big|\ \big\lfloor\ \therefore\ A$ [from 6, contradicting 4]

    8. $\quad \big|\ \therefore\ \sim(A \cdot B)$ [from 6, 4, and 7]

    * 9. $\quad \big|\ \therefore\ \sim(C \cdot D)$ [from 1 and 8]

    10. $\quad \big\lfloor\ \therefore\ \sim D$ [from 2 and 9, contradicting 5]

    11. $\quad \therefore\ (\sim A \supset \sim D)$ [from 3, 5, and 10]

15. ** 1. $\quad \sim(\sim A^1 \cdot \sim B^1) = 1$ $\qquad\qquad$ INVALID

    ** 2. $\quad (C^1 \vee D^0) = 1$

    ** 3. $\quad \sim(D^0 \cdot A^1) = 1$ $\qquad$ $\boxed{A, B, C, \sim D}$

    $\quad [\therefore\ \sim(C^1 \equiv B^1) = 0$

```
 *  4.      asm: (C ≡ B)
**  5.      ∴ (C ⊃ B) [from 4]
**  6.      ∴ (B ⊃ C) [from 4]
    7.         asm: A [nice to have "~A" for 1]
    8.         ∴ ~D [from 3 and 7]
    9.         ∴ C [from 2 and 8]
   10.         ∴ B [from 5 and 9]
   11.         ∴ C [from 6 and 10 (redundant)]
```

3.6B

1. $*$ 1. $(M^0 \supset (F \cdot S^0)) = 1$ INVALID
   $**$ 2. $(S^0 \supset (W \cdot C^0)) = 1$
            $[\therefore (\sim M^0 \supset C^0) = 0$ ~M, ~C, ~S
   $*$ 3.   asm: $\sim(\sim M \supset C)$
     4.   ∴ ~M [from 3]
     5.   ∴ ~C [from 3]
     6.   asm: ~S [nice to have "S" for 2]

Here we could assume "~M" for line 1. But this would be redundant (since line 4 already contains "~M"). So we can skip this if we like but still star line 1. We'd have a valid argument if the conclusion read "If the world contains moral goodness then the creator is imperfect."

3. $**$ 1. $((C^0 \lor P^0) \supset (S \cdot G^1)) = 1$ INVALID
            $[\therefore (\sim C^0 \supset \sim G^1) = 0$
   $*$ 2.   asm: $\sim(\sim C \supset \sim G)$ G, ~C, ~P
     3.   ∴ ~C [from 2]
     4.   ∴ G [from 2]
   $**$ 5.   asm: $\sim(C \lor P)$ [nice to have "$(C \lor P)$" for 1]
     6.   ∴ ~C [from 5]
     7.   ∴ ~P [from 5]

Maybe you took biology to satisfy the requirement.   G, P, S, ~C
Or maybe you took physics:

5. $**$ 1. $((R^0 \lor S^0) \supset M^0) = 1$ INVALID
   $***$ 2. $((M^0 \cdot W^0) \supset F^1) = 1$
            $[\therefore (F^1 \supset W^0) = 0$ F, ~W, ~R, ~S, ~M
   $*$ 3.   asm: $\sim(F \supset W)$
     4.   ∴ F [from 3]
     5.   ∴ ~W [from 3]
   $**$ 6.   asm: $\sim(R \lor S)$ [nice to have "$(R \lor S)$" for 1]
     7.   ∴ ~R [from 6]
     8.   ∴ ~S [from 6]
   $****$ 9.   asm: $\sim(M \cdot W)$ [nice to have "$(M \cdot W)$" for 2]
     10.   asm: ~M [nice to have "M" for 9]

10. $**$ 1. $(A^0 \supset (H^1 \cdot L)) = 1$ INVALID
   $*$ 2. $(C^0 \supset A^0) = 1$
            $[\therefore (\sim H^1 \lor C^0) = 0$ H, ~C, ~A

```
 *  3.      asm: ~(~H ∨ C)
    4.      ∴ H [from 3]
    5.      ∴ ~C [from 3]
    6.         asm: ~A [nice to have "A" for 1]
```

Here we could conclude "~C" from 2 and 6. But this would be redundant (since line 5 already has "~C"). So we can skip this if we like but still star line 2.

```
15.   *  1.      (P ⊃ (D ∨ V))                           VALID
      *  2.      (D ⊃ (T · ~M))
      *  3.      ((T · ~M) ⊃ S)
         4.      (V ⊃ T)
      *  5.      ((V · M) ⊃ Q)
         6.      ~Q
              [∴ (P ⊃ (S · ~M))
      *  7.    ┌ asm: ~(P ⊃ (S · ~M))
         8.    │ ∴ P [from 7]
      *  9.    │ ∴ ~(S · ~M) [from 7]
        10.    │ ∴ ~(V · M) [from 5 and 6]
        11.    │ ∴ (D ∨ V) [from 1 and 8]
        12.    │ ┌ asm: ~D [nice to have "D" for 2]
        13.    │ │ ∴ V [from 11 and 12]
        14.    │ │ ∴ ~M [from 10 and 13]
        15.    │ │ ∴ ~S [from 9 and 14]
        16.    │ │ ∴ ~(T · ~M) [from 3 and 15]
        17.    │ │ ∴ T [from 4 and 13]
        18.    │ └ ∴ M [from 16 and 17, contradicting 14]
        19.    │ ∴ D [from 12, 14, and 18]
      * 20.    │ ∴ (T · ~M) [from 2 and 19]
        21.    │ ∴ T [from 20]
        22.    │ ∴ ~M [from 20]
        23.    │ ∴ ~S [from 9 and 22]
        24.    └ ∴ ~(T · ~M) [from 3 and 23, contradicting 20]
        25.    ∴ (P ⊃ (S · ~M)) [from 7, 20, and 24]
```

4.2A

1. (∃x)~Ex
3. ~(∃x)Ex
5. (x)(Lx ⊃ Ex)
10. ~(x)Ex
15. (x)(~Lx ⊃ Ex)
20. (x)(Ex · Lx) [NOT "(x)(Ex ⊃ Lx)"]

4.3A

```
1.    *  1.      (∃x)(Fx · Gx)                           VALID
              [∴ (∃x)Fx
```

* 2. ⌈asm: ~(∃x) Fx
  3. │ ∴ (x)~Fx [from 2]
  4. │ ∴ (Fa · Ga) [from 1]
  5. │ ∴ ~Fa [from 3]
  6. └ ∴ Fa [from 4, contradicting 5]
  7. ∴ (x)Fx [from 2, 5, and 6]

3. 1. (x)(Fx ⊃ Gx)   VALID
      [∴ ~(∃x)(Fx · ~Gx)
* 2. ⌈asm: (∃x)(Fx · ~Gx)
* 3. │ ∴ (Fa · ~Ga) [from 2]
  4. │ ∴ Fa [from 3]
  5. │ ∴ ~Ga [from 3]
* 6. │ ∴ (Fa ⊃ Ga) [from 1]
  7. └ ∴ Ga [from 4 and 6, contradicting 5]
  8. ∴ ~(∃x)(Fx · ~Gx) [from 2, 5, and 7]

5. 1. (x)((Fx ∨ Gx) ⊃ Hx)   VALID
      [∴ (x)(Fx ⊃ Hx)
* 2. ⌈asm: ~(x)(Fx ⊃ Hx)
* 3. │ ∴ (∃x)~(Fx ⊃ Hx) [from 2]
* 4. │ ∴ ~(Fa ⊃ Ha) [from 3]
  5. │ ∴ Fa [from 4]
  6. │ ∴ ~Ha [from 4]
* 7. │ ∴ ((Fa ∨ Ga) ⊃ Ha) [from 1]
  8. │ ∴ ~(Fa ∨ Ga) [from 6 and 7]
  9. └ ∴ ~Fa [from 8, contradicting 5]
 10. ∴ (x)(Fx ⊃ Hx) [from 2, 5, and 9]

10. 1. (x)(Fx ⊃ Gx)   VALID
       2. (x)(~Fx ⊃ Gx)
      [∴ (x)Gx
* 3. ⌈asm: ~(x)Gx
* 4. │ ∴ (∃x)~Gx [from 3]
  5. │ ∴ ~Ga [from 4]
* 6. │ ∴ (Fa ⊃ Ga) [from 1]
* 7. │ ∴ (~Fa ⊃ Ga) [from 2]
  8. │ ∴ ~Fa [from 5 and 6]
  9. └ ∴ Ga [from 7 and 8, contradicting 5]
 10. ∴ (x)Gx [from 3, 5, and 9]

4.3B

1. 1. (x)(Dx ⊃ Bx)   VALID
       2. (x)Dx
      [∴ (x)Bx
* 3. ⌈asm: ~(x)Bx
* 4. │ ∴ (∃x)~Bx [from 3]
  5. │ ∴ ~Ba [from 4]
* 6. │ ∴ (Da ⊃ Ba) [from 1]
  7. │ ∴ Da [from 2]
  8. └ ∴ Ba [from 6 and 7, contradicting 5]
  9. ∴ (x)Bx [from 3, 5, and 8]

3.   1.   (x)Mx   VALID
          [∴ (x)(Lx ⊃ Mx)
   * 2.   ⌈asm: ~(x)(Lx ⊃ Mx)
   * 3.   │∴ (∃x)~(Lx ⊃ Mx) [from 2]
   * 4.   │∴ ~(La ⊃ Ma) [from 3]
     5.   │∴ Ma [from 1]
     6.   │∴ La [from 4]
     7.   ⌊∴ ~Ma [from 4, contradicting 5]
     8. ∴ (x)(Lx ⊃ Mx) [from 2, 5, and 7]

5. * 1.   ~(∃x)(Fx · Ox)   VALID
     2.   (x)(Cx ⊃ Ox)
         [∴ ~(∃x)(Fx · Cx)
   * 3.   ⌈asm: (∃x)(Fx · Cx)
     4.   │∴ (x)~(Fx · Ox) [from 1]
   * 5.   │∴ (Fa · Ca) [from 3]
   * 6.   │∴ (Ca ⊃ Oa) [from 2]
   * 7.   │∴ ~(Fa · Oa) [from 4]
     8.   │∴ Fa [from 5]
     9.   │∴ Ca [from 5]
    10.   │∴ Oa [from 6 and 9]
    11.   ⌊∴ ~Oa [from 7 and 8, contradicting 10]
    12. ∴ ~(∃x)(Fx · Cx) [from 3, 10, and 11]

10. * 1.   ~(∃x)(Bx · Cx)   VALID
   * 2.   (∃x)(Lx · Cx)
     3.   (x)((Cx · ~Bx) ⊃ Rx)
        [∴ (∃x)(Rx · Cx)
   * 4.   ⌈asm: ~(∃x)(Rx · Cx)
     5.   │∴ (x)~(Rx · Cx) [from 4]
     6.   │∴ (x)~(Bx · Cx) [from 1]
   * 7.   │∴ (La · Ca) [from 2]
   * 8.   │∴ ((Ca · ~Ba) ⊃ Ra) [from 3]
   * 9.   │∴ ~(Ra · Ca) [from 5]
  * 10.   │∴ ~(Ba · Ca) [from 6]
    11.   │∴ La [from 7]
    12.   │∴ Ca [from 7]
    13.   │∴ ~Ra [from 9 and 12]
    14.   │∴ ~Ba [from 10 and 12]
  * 15.   │∴ ~(Ca · ~Ba) [from 8 and 13]
    16.   ⌊∴ Ba [from 12 and 15, contradicting 14]
    17. ∴ (∃x)(Rx · Cx) [from 4, 14, and 16]

4.4A

1. * 1.   (∃x)Fx   INVALID
         [∴ (x)Fx   a, b
   * 2.   asm: ~(x)Fx
   * 3.   ∴ (∃x)~Fx [from 2]   ┌─────────┐
     4.   ∴ Fa [from 1]   │ Fa, ~Fb │
     5.   ∴ ~Fb [from 3]   └─────────┘

3. * 1.   (∃x)Fx   INVALID
     * 2.   ~(x)Gx   a, b
           [∴ ~(x)(Fx ⊃ Gx)   ┌─────────────┐
        3.   asm: (x)(Fx ⊃ Gx)   │ Fa, Ga   │
     * 4.   ∴ (∃x)~Gx [from 2]   │ ~Fb, ~Gb   │
        5.   ∴ Fa [from 1]   └─────────────┘
        6.   ∴ ~Gb [from 4]
     * 7.   ∴ (Fa ⊃ Ga) [from 3]
     * 8.   ∴ (Fb ⊃ Gb) [from 3]
        9.   ∴ Ga [from 5 and 7]
       10.   ∴ ~Fb [from 6 and 8]

5. * 1.   ~(x)Fx   INVALID
        2.   (∃x)~Gx   a, b
           [∴ ~(x)(Fx ∨ Gx)   ┌─────────────┐
        3.   asm: (x)(Fx ∨ Gx)   │ Fa, ~Ga   │
     * 4.   ∴ (∃x)~Fx [from 1]   │ ~Fb, Gb   │
        5.   ∴ ~Ga [from 2]   └─────────────┘
        6.   ∴ ~Fb [from 4]
     * 7.   ∴ (Fa ∨ Ga) [from 3]
     * 8.   ∴ (Fb ∨ Gb) [from 3]
        9.   ∴ Fa [from 5 and 7]
       10.   ∴ Gb [from 6 and 8]

10. * 1.   (∃x)(Fx ∨ ~Gx)   INVALID
        2.   (x)(~Gx ⊃ Hx)   a, b
     * 3.   (∃x)(Fx ⊃ Hx)   ┌──────────────────┐
           [∴ (∃x) Hx   │ Fa, Ga, ~Ha   │
     * 4.   asm: ~(∃x)Hx   │ ~Fb, Gb, ~Hb   │
        5.   ∴ (x)~Hx [from 4]   └──────────────────┘
     * 6.   ∴ (Fa ∨ ~Ga) [from 1]
     * 7.   ∴ (Fb ⊃ Hb) [from 3]
     * 8.   ∴ (~Ga ⊃ Ha) [from 2]
     * 9.   ∴ (~Gb ⊃ Hb) [from 2]
       10.   ∴ ~Ha [from 5]
       11.   ∴ ~Hb [from 5]
       12.   ∴ Ga [from 8 and 10]
       13.   ∴ Gb [from 9 and 11]
       14.   ∴ ~Fb [from 7 and 11]
       15.   ∴ Fa [from 6 and 12]

4.4B

1. * 1.   ~(∃x)(Mx · Ix)   INVALID
     * 2.   ~(x)Mx   a
           [∴ (∃x)Ix   ┌─────────────┐
     * 3.   asm: ~(∃x)Ix   │ ~Ma, ~Ia   │
        4.   ∴ (x)~Ix [from 3]   └─────────────┘
        5.   ∴ (x)~(Mx · Ix) [from 1]

* 6.   ∴ (∃x)~Mx [from 2]
  7.   ∴ ~Ma [from 6]
  8.   ∴ ~Ia [from 4]
* 9.   ∴ ~(Ma · Ia) [from 5]

Here we could assume "~Ma" for line 9. But this would be redundant (since line 7 already has "~Ma"). So we can skip this but still star line 9.

3. * 1.   (∃x)Sx   INVALID
    * 2.   ~(x)Cx   a, b
            [∴ (∃x)(Sx · ~Cx)]
    * 3.   asm: ~(∃x)(Sx · ~Cx)   ┌─────────────┐
      4.   ∴ (x)~(Sx · ~Cx) [from 3]   │ Sa, Ca   │
    * 5.   ∴ (∃x)~Cx [from 2]   │ ~Sb, ~Cb   │
      6.   ∴ Sa [from 1]   └─────────────┘
      7.   ∴ ~Cb [from 5]
    * 8.   ∴ ~(Sa · ~Ca) [from 4]
    * 9.   ∴ ~(Sb · ~Cb) [from 4]
      10.   ∴ Ca [from 6 and 8]
      11.   ∴ ~Sb [from 7 and 9]

5.   1.   (x)(~Cx ⊃ Ax)   INVALID
            [∴ ~(∃x)(Cx · Ax)]   a
    * 2.   asm: (∃x)(Cx · Ax)
    * 3.   ∴ (Ca · Aa) [from 2]   ┌─────────┐
    * 4.   ∴ (~Ca ⊃ Aa) [from 1]   │ Ca, Aa │
      5.   ∴ Ca [from 3]   └─────────┘
      6.   ∴ Aa [from 3]

Here we could assume "Ca" for line 4. But this would be redundant (since line 5 already has "Ca"). So we can skip this but still star line 4.

10.   1.   (x)(Tx ⊃ Ex)   INVALID
            [∴ (∃x)(Tx · Ex)]   a
    * 2.   asm: ~(∃x)(Tx · Ex)
      3.   ∴ (x)~(Tx · Ex) [from 2]   ┌──────┐
   ** 4.   ∴ (Ta ⊃ Ea) [from 1]   │ ~Ta │
   ** 5.   ∴ ~(Ta · Ea) [from 3]   └──────┘
      6.   asm: ~Ta [nice to have "Ta" for 4]

15.   1.   (x)(Tx ⊃ (Dx ∨ Ex))   VALID
    * 2.   ~(∃x)(Mx · Dx)
    * 3.   ~(∃x)(Mx · Ex)
            [∴ ~(∃x)(Mx · Tx)]
    * 4.   ┌ asm: (∃x)(Mx · Tx)
      5.   │ ∴ (x)~(Mx · Dx) [from 2]
      6.   │ ∴ (x)~(Mx · Ex) [from 3]
    * 7.   │ ∴ (Ma · Ta) [from 4]
    * 8.   │ ∴ (Ta ⊃ (Da ∨ Ea)) [from 1]
    * 9.   │ ∴ ~(Ma · Da) [from 5]
   * 10.   │ ∴ ~(Ma · Ea) [from 6]

11. | ∴ Ma [from 7]
12. | ∴ Ta [from 7]
* 13. | ∴ (Da ∨ Ea) [from 8 and 12]
14. | ∴ ~Da [from 9 and 11]
15. | ∴ ~Ea [from 10 and 11]
16. |_∴ Ea [from 13 and 14, contradicting 15]
17. ∴ ~(∃x)(Mx · Tx) [from 4, 15, and 16]

4.5A

1. Eg
3. (Cg ∨ Eg)
5. (x)~Ex *or, equivalently*, ~(∃x)Ex
10. (x)(Ex ⊃ R) *or, equivalently*, ((∃x)Ex ⊃ R)
15. (x)~(Ex · Lx) *or, equivalently*, ~(∃x)(Ex · Lx)

4.6A

1. 1.   (x)(Fx ⊃ P)   VALID
       [∴ ((∃x)Fx ⊃ P)
* 2.   ⌈asm: ~((∃x)Fx ⊃ P)
* 3. | ∴ (∃x)Fx [from 2]
  4. | ∴ ~P [from 2]
  5. | ∴ Fa [from 3]
* 6. | ∴ (Fa ⊃ P) [from 1]
  7. |_∴ P [from 5 and 6, contradicting 4]
  8. ∴ ((∃x)Fx ⊃ P) [from 2, 4, and 7]
3. 1.   (x)((Fx ∨ Gx) ⊃ Hx)   VALID
  2.   Fm
       [∴ Hm
  3.   ⌈asm: ~Hm
* 4. | ∴ ((Fm ∨ Gm) ⊃ Hm) [from 1]
  5. | ∴ ~(Fm ∨ Gm) [from 3 and 4]
  6. |_∴ ~Fm [from 5, contradicting 2]
  7. ∴ Hm [from 3, 2, and 6]
5. * 1.   ((∃x)Fx ⊃ (x)Gx)   VALID
  2.   ~Gp
       [∴ ~Fp
  3.   ⌈asm: Fp
  4. | ⌈asm: ~(∃x)Fx [nice to have "(∃x)Fx" for 1]
  5. | | ∴ (x)~Fx [from 4]
  6. | |_∴ ~Fp [from 5, contradicting 3]
* 7. | ∴ (∃x)Fx [from 4, 3, and 6]
  8. | ∴ (x)Gx [from 1 and 7]
  9. | ∴ Fa [from 7 (unnecessary)]
  10. |_∴ Gp [from 8, contradicting 2]
  11. ∴ ~Fp [from 3, 2, and 10]
10. 1.   (x)(Fx ⊃ P)   VALID
       [∴ ((x)Fx ⊃ P)

```
   * 2.  ┌asm: ~((x)Fx ⊃ P)
     3.  │ ∴ (x)Fx [from 2]
     4.  │ ∴ ~P [from 2]
     5.  │ ∴ Fa [from 3]
   * 6.  │ ∴ (Fa ⊃ P) [from 1]
     7.  └ ∴ P [from 5 and 6, contradicting 4]
     8.  ∴ ((x)Fx ⊃ P) [from 2, 4, and 7]
```

15.   1.   (x)((Fx · Gx) ⊃ P)   VALID
           2.   ~P
           3.   Fo
              [∴ ~(x)(Fx ⊃ Gx)
```
     4.  ┌asm: (x)(Fx ⊃ Gx)
   * 5.  │ ∴ ((Fo · Go) ⊃ P) [from 1]
   * 6.  │ ∴ (Fo ⊃ Go) [from 4]
   * 7.  │ ∴ ~(Fo · Go) [from 2 and 5]
     8.  │ ∴ Go [from 3 and 6]
     9.  └ ∴ ~Go [from 3 and 7, contradicting 8]
    10.  ∴ ~(x)(Fx ⊃ Gx) [from 4, 8, and 9]
```

20. * 1.   (∃x)(Fx · (Gx ∨ Hx))   INVALID
              [∴ (~Hk ⊃ (∃x)(Fx · Gx))   k, a
   * 2.   asm: ~(~Hk ⊃ (∃x)(Fx · Gx))
     3.   ∴ ~Hk [from 2]   ┌─────────────┐
   * 4.   ∴ ~(∃x)(Fx · Gx) [from 2]   │ ~Fk, ~Hk   │
     5.   ∴ (x)~(Fx · Gx) [from 4]   │ Fa, ~Ga, Ha │
   * 6.   ∴ (Fa · (Ga ∨ Ha)) [from 1]   └─────────────┘
     7.   ∴ Fa [from 6]
   * 8.   ∴ (Ga ∨ Ha) [from 6]
  ** 9.   ∴ ~(Fk · Gk) [from 5]
  * 10.   ∴ ~(Fa · Ga) [from 5]
    11.   ∴ ~Ga [from 7 and 10]
    12.   ∴ Ha [from 8 and 11]
    13.   asm: ~Fk [nice to have "Fk" for 9]

4.6B

1.   1.   (x)Cx   VALID
        * 2.   (Cw ⊃ G)
              [∴ G
```
     3.  ┌asm: ~G
     4.  │ ∴ ~Cw [from 2 and 3]
     5.  └ ∴ Cw [from 1, contradicting 4]
     6.  ∴ G [from 3, 4, and 5]
```

3. * 1.   ((x)Lx ⊃ D)   INVALID
              [∴ (Lu ⊃ D)   u, a
        * 2.   asm: ~(Lu ⊃ D)
          3.   ∴ Lu [from 2]   ┌─────────────────┐
          4.   ∴ ~D [from 2]   │ Lu, ~La, ~D   │
        * 5.   ∴ ~(x)Lx [from 1 and 4]   └─────────────────┘
        * 6.   ∴ (∃x)~Lx [from 5]
          7.   ∴ ~La [from 6]

5. 1. (x)(Ex ⊃ (Sx ∨ Fx))   VALID
  2. ~St
    [∴ (~Et ∨ Ft)
 * 3. ⌈ asm: ~(~Et ∨ Ft)
  4. │ ∴ Et [from 3]
  5. │ ∴ ~Ft [from 3]
 * 6. │ ∴ (Et ⊃ (St ∨ Ft)) [from 1]
 * 7. │ ∴ (St ∨ Ft) [from 4 and 6]
  8. └ ∴ Ft [from 2 and 7, contradicting 5]
  9. ∴ (~Et ∨ Ft) [from 3, 5, and 8]

10. 1. (x)(Lx ⊃ Lu)   VALID
  2. (x)(Px ⊃ Lx)
  3. Pe
    [∴ Lu
  4. ⌈asm: ~Lu
 * 5. │ ∴ (Pe ⊃ Le) [from 2]
 * 6. │ ∴ (Le ⊃ Lu) [from 1]
  7. │ ∴ Le [from 3 and 5]
  8. └ ∴ Lu [from 6 and 7, contradicting 4]
  9. ∴ Lu [from 4 and 8]

15. * 1. ((∃x)Tx ⊃ (x)Jx)   VALID
 * 2. (Ps ⊃ Ts)
    [∴ (Ps ⊃ Ji)
 * 3. ⌈asm: ~(Ps ⊃ Ji)
  4. │ ∴ Ps [from 3]
  5. │ ∴ ~Ji [from 3]
  6. │ ∴ Ts [from 2 and 4]
  7. │ ⌈ asm: ~(∃x)Tx [nice to have "(∃x)Tx" for 1]
  8. │ │ ∴ (x)~Tx [from 7]
  9. │ └ ∴ ~Ts [from 8, contradicting 6]
  10. │ ∴ (∃x)Tx [from 7, 6, and 9]
  11. │ ∴ (x)Jx [from 1 and 10]
  12. └ ∴ Ji [from 11, contradicting 5]
  13. ∴ (Ps ⊃ Ji) [from 3, 5, and 12]

20. 1. (x)(Ax ⊃ Jx)   VALID
 * 2. (Ae · Ad)
    [∴ (Je · Jd)
 * 3. ⌈asm: ~(Je · Jd)
  4. │ ∴ Ae [from 2]
  5. │ ∴ Ad [from 2]
 * 6. │ ∴ (Ae ⊃ Je) [from 1]
 * 7. │ ∴ (Ad ⊃ Jd) [from 1]
  8. │ ∴ Je [from 4 and 6]
  9. │ ∴ Jd [from 5 and 7]
  10. └ ∴ ~Jd [from 3 and 8, contradicting 9]
  11. ∴ (Je · Jd) [from 3, 9, and 10]

25. * 1. (~G ⊃ (x)(Ex ⊃ Fx))   VALID
  2. T
 * 3. (((x)(Ex ⊃ Fx) · T) ⊃ (∃x)(Ex · Ax))

4.   ~(∃x)(Fx · Ax)
      [∴ G
5.   ┌asm: ~G
6.   │ ∴ (x)(Ex ⊃ Fx) [from 1 and 5]
7.   │ ∴ ((x)(Ex ⊃ Fx) · T) [from 6 and 2, nice to have for 3]
* 8.   │ ∴ (∃x)(Ex · Ax) [from 3 and 7]
9.   │ ∴ (x)~(Fx · Ax) [from 4]
* 10. │ ∴ (Ea · Aa) [from 8]
11.   │ ∴ Ea [from 10]
12.   │ ∴ Aa [from 10]
.* 13. │ ∴ (Ea ⊃ Fa) [from 6]
* 14. │ ∴ ~(Fa · Aa) [from 9]
15.   │ ∴ Fa [from 11 and 13]
16.   └ ∴ ~Fa [from 12 and 14, contradicting 15]
17. ∴ G [from 5, 15, and 16]

5.1A

1. a = t
3. ~a = p
5. (La · ~(∃x)(~x = a · Lx))
10. (∃x)((Ex · Lx) · ~(∃y)(~y = x · (Ey · Ly)))
15. c = s

5.2A

1.   1.   Fa   VALID
        [∴ (~Fb ⊃ ~b = a)
* 2.   ┌asm: ~(~Fb ⊃ ~b = a)
3.   │ ∴ ~Fb [from 2]
4.   │ ∴ b = a [from 2]
5.   └ ∴ Fb [from 1 and 4, contradicting 3]
6.   ∴ (~Fb ⊃ ~b = a) [from 2, 3, and 5]

3.   1.   a = b   VALID
2.   b = c
        [∴ a = c
3.   ┌asm: ~a = c
4.   └ ∴ ~a = b [from 2 and 3, contradicting 1]
5.   ∴ a = c [from 3, 1, and 4]

5.   1.   a = b   VALID
2.   (x)(Fx ⊃ Gx)
3.   ~Ga
        [∴ ~Fb
4.   ┌asm: Fb
* 5.   │ ∴ (Fb ⊃ Gb) [from 2]
6.   │ ∴ Gb [from 4 and 5]
7.   └ ∴ Ga [from 1 and 6, contradicting 3]
8.   ∴ ~Fb [from 4, 3, and 7]

10.   [∴ (∃x)(∃y)~x = y   INVALID
  * 1.   asm: ~(∃x)(∃y)~x = y   a
    2.   ∴ (x)~(∃y)~x = y [from 1]
  * 3.   ∴ ~(∃y)~a = y [from 2]   ┌─────────┐
    4.   ∴ (y)a = y [from 3]   │ a = a │
    5.   ∴ a = a [from 4]   └─────────┘

5.2B

1.   1.   k = m   VALID
          2.   Bm
          3.   Tk
                [∴ (∃x)(Tx · Bx)
  * 4.   ┌asm: ~(∃x)(Tx · Bx)
    5.   │ ∴ (x)~(Tx · Bx) [from 4]
  * 6.   │ ∴ ~(Tk · Bk) [from 5]
    7.   │ ∴ ~Bk [from 3 and 6]
    8.   └ ∴ ~Bm [from 1 and 7, contradicting 2]
    9.   ∴ (∃x)(Tx · Bx) [from 4, 2, and 8]
3.   1.   Oc   VALID
          2.   ~Op
                [∴ ~p = c
          3.   ┌asm: p = c
          4.   └ ∴ ~Oc [from 2 and 3, contradicting 1]
          5.   ∴ ~p = c [from 3, 1, and 4]
5. ** 1.   (Rm ∨ Hm)   INVALID
          2.   ~Ru   u, m
                [∴ (Hu ⊃ u = m)   ┌──────────────────┐
  * 3.   asm: ~(Hu ⊃ u = m)   │ Hu, ~Ru   │
    4.   ∴ Hu [from 3]   │ Rm, ~u = m   │
    5.   ∴ ~u = m [from 3]   └──────────────────┘
    6.   asm: Rm [nice to have "~Rm" for 1]

10. * 1.   (∃x)(Nx · ~(∃y)(~y = x · Ny))   VALID
          2.   Np
          3.   Fp
                [∴ (x)(Nx ⊃ Fx)
  * 4.   ┌asm: ~(x)(Nx ⊃ Fx)
  * 5.   │ ∴ (∃x)~(Nx ⊃ Fx) [from 4]
  * 6.   │ ∴ ~(Na ⊃ Fa) [from 5]
    7.   │ ∴ Na [from 6]
    8.   │ ∴ ~Fa [from 6]
  * 9.   │ ∴ (Nb · ~(∃y)(~y = b · Ny)) [from 1]
    10.   │ ∴ Nb [from 9]
  * 11.   │ ∴ ~(∃y)(~y = b · Ny) [from 9]
    12.   │ ∴ (y)~(~y = b · Ny) [from 11]
  * 13.   │ ∴ ~(~a = b · Na) [from 12]
  * 14.   │ ∴ ~(~p = b · Np) [from 12]
    15.   │ ∴ a = b [from 7 and 13]

16. | ∴ p = b [from 2 and 14]
17. | ∴ Fb [from 3 and 16]
18. |_∴ Fa [from 15 and 17, contradicting 8]
19. ∴ (x)(Nx ⊃ Fx) [from 4, 8, and 18]

5.3A

1. Cgw
3. (∃x)(Ex · Gax)
5. (x)(y)((Ex · ~Ey) ⊃ Gxy)
10. (∃y)Cyx
15. (x)(∃y)Gxy
20. (∃x)(Ex · (y)(Ey ⊃ Cxy))
25. (∃x)((∃y)Cxy · ~(∃y)Cyx)

5.4A

1. 1.   (x)(y)Lxy   VALID
          [∴ (∃x)Lax
   * 2. ┌ asm: ~(∃x)Lax
     3. │ ∴ (x)~Lax [from 2]
     4. │ ∴ ~Laa [from 3]
     5. │ ∴ (y)Lay [from 1]
     6. └ ∴ Laa [from 5, contradicting 4]
     7. ∴ (∃x)Lax [from 2, 4, and 6]
3. 1.   (x)Lxa   INVALID
          [∴ (x)Lax   a, b
   * 2.   asm: ~(x)Lax
   * 3.   ∴ (∃x)~Lax [from 2]   ┌─────────────────┐
     4.   ∴ ~Lab [from 3]   │ Laa, Lba, ~Lab │
     5.   ∴ Laa [from 1]   └─────────────────┘
     6.   ∴ Lba [from 1]
5. 1.   (x)Lxx   INVALID
          [∴ (∃x)(y)Lxy   a, b
   * 2.   asm: ~(∃x)(y)Lxy
     3.   ∴ (x)~(y)Lxy [from 2]   ┌─────────────┐
   * 4.   ∴ ~(y)Lay [from 3]   │ Laa, ~Lab   │
   * 5.   ∴ (∃y)~Lay [from 4]   │ Lbb, ~Lba   │
     6.   ∴ ~Lab [from 5]   └─────────────┘
     7.   ∴ Laa [from 1]
     8.   ∴ Lbb [from 1]
             ∴ . . . .

The normal strategy leads to an endless loop. We'd drop the universal quantifier in 3 using
"b" and then get a new constant two steps later. We'd drop the universal quantifier in 3
again using this new constant, and then get another new constant two steps later, and so
on. It's easy to expand the simple wffs we have so far to get a refutation. We only have
to add "~Lba."

10.   1.   (x)(y)(Uxy ⊃ Lxy)   VALID
            2.   (x)(∃y)Uxy
              [∴   (x)(∃y)Lxy

```
    * 3.      ┌ asm: ~(x)(∃y)Lxy
    * 4.      │ ∴ (∃x)~(∃y)Lxy [from 3]
    * 5.      │ ∴ ~(∃y)Lay [from 4]
      6.      │ ∴ (y)~Lay [from 5]
    * 7.      │ ∴ (∃y)Uay [from 2]
      8.      │ ∴ Uab [from 7]
      9.      │ ∴ ~Lab [from 6]
      10.     │ ∴ (y)(Uay ⊃ Lay) [from 1]
    * 11.     │ ∴ (Uab ⊃ Lab) [from 10]
      12.     └ ∴ Lab [from 8 and 11, contradicting 9]
      13.   ∴  (x)(∃y)Lxy [from 3, 9, and 12]
```

15.   1.   (x)(y)(z)((Lxy · Lyz) ⊃ Lxz)   INVALID
       2.   (x)(y)(Lxy ⊃ Lyx)   a
           [∴ (x)Lxx
    * 3.   asm: ~(x)Lxx   ┌─────────┐
    * 4.   ∴ (∃x)~Lxx [from 3]   │ ~Laa   │
      5.   ∴ ~Laa [from 4]   └─────────┘
      6.   ∴ (y)(Lay ⊃ Lya) [from 2]
    * 7.   ∴ (Laa ⊃ Laa) [from 6]
      8.   ∴ (y)(z)((Lay · Lyz) ⊃ Laz) [from 1]
      9.   ∴ (z)((Laa · Laz) ⊃ Laz) [from 8]
    * 10.   ∴ ((Laa · Laa) ⊃ Laa) [from 9]
      11.   ∴ ~Laa [from 5 and 7 (redundant)]
    * 12.   ∴ ~(Laa · Laa) [from 10 and 5]
```

### 5.4B

1.  1.    (x)Ljx                                   VALID
        [∴  (∃x)Lxu
    * 2.    ┌ asm: ~(∃x)Lxu
      3.    │ ∴ (x)~Lxu [from 2]
      4.    │ ∴ Lju [from 1]
      5.    └ ∴ ~Lju [from 3, contradicting 4]
      6.    ∴ (∃x)Lxu [from 2, 4, and 5]

3.  1.    (x)(∃y)Lxy                               INVALID
        [∴  (∃x)Lxx                                 a, b
    * 2.    asm: ~(∃x)Lxx                        ┌──────────────┐
      3.    ∴ (x)~Lxx [from 2]                   │ Lab, ~Laa    │
      4.    ∴ ~Laa [from 3]                      │ Lba, ~Lbb    │
    * 5.    ∴ (∃y)Lay [from 1]                   └──────────────┘
      6.    ∴ Lab [from 5]
      7.    ∴ ~Lbb [from 3]
          ∴ . . .

The normal strategy leads to an infinite loop. We'd drop the universal quantifier in 1 using "b" and then get a new constant a step later. We'd drop the universal quantifier in 1 again using this new constant, and then get another new constant a step later, and so on. It's easy to expand the simple wffs we have so far to get a refutation. We only have to add "Lba."
```

5.   1.   (x)Lxb   VALID
  * 2.   ~(∃x)(~x = m · Lbx)
          [∴ b = m
    3.   ⌈asm: ~b = m
    4.   │ ∴ (x)~(~x = m · Lbx) [from 2]
    5.   │ ∴ Lmb [from 1]
    6.   │ ∴ Lbb [from 1]
    7.   │ ∴ ~(~m = m · Lbm) [from 4]
  * 8.   │ ∴ ~(~b = m · Lbb) [from 4]
    9.   ⌊∴ ~Lbb [from 3 and 8, contradicting 6]
    10.   ∴ b = m [from 3, 6, and 9]

10.   1.   (x)(∃y)Dxy   INVALID
          [∴ (∃y)(x)Dxy   a, b
  * 2.   asm: ~(∃y)(x)Dxy
    3.   ∴ (y)~(x)Dxy [from 2] | Daa, Dbb, ~Dab, ~Dba |
  * 4.   ∴ ~(x)Dxa [from 3]
  * 5.   ∴ (∃x)~Dxa [from 4] | Dab, Dba, ~Daa, ~Dbb |
    6.   ∴ ~Dba [from 5]
          ∴ . . . .

The normal strategy leads to an infinite loop. We'd drop the universal quantifier in 3 using "b" and then get a new constant two steps later. We'd drop the universal quantifier in 3 again using this new constant, and then get another new constant two steps later, and so on. (Line 1 leads to an infinite loop of its own.) We have to think up the refutation on our own. Either of the two refutations given does the job.

15.   1.   (x)((∃y)Lxy ⊃ (y)Lyx)   VALID
    2.   Lrj
          [∴ Liu
    3.   ⌈asm: ~Liu
  * 4.   │ ∴ ((∃y)Lry ⊃ (y)Lyr) [from 1]
    5.   │ ⌈asm: ~(∃y)Lry [nice to have "(∃y)Lry" for 4]
    6.   │ │ ∴ (y)~Lry [from 5]
    7.   │ ⌊∴ ~Lrj [from 6, contradicting 2]
    8.   │ ∴ (∃y)Lry [from 5, 2, and 7]
    9.   │ ∴ (y)Lyr [from 4 and 8]
    10.   │ ∴ Lur [from 9]
  * 11.   │ ∴ ((∃y)Luy ⊃ (y)Lyu) [from 1]
    12.   │ ⌈asm: ~(∃y)Luy [nice to have "(∃y)Luy" for 11]
    13.   │ │ ∴ (y)~Luy [from 12]
    14.   │ ⌊∴ ~Lur [from 13, contradicting 10]
    15.   │ ∴ (∃y)Luy [from 12, 14, and 10]
    16.   │ ∴ (y)Lyu [from 11 and 15]
    17.   ⌊∴ Liu [from 16, contradicting 3]
    18.   ∴ Liu [from 3 and 17]

20.   1.   (x)(Cx ⊃ (∃t)~Ext)   INVALID
                 [∴ ((x)Cx ⊃ (∃t)(x)~Ext)   a, b, u, u′
     * 2.   asm: ~((x)Cx ⊃ (∃t)(x)~Ext)
        3.   ∴ (x)Cx [from 2]   ┌─────────────────┐
     * 4.   ∴ ~(∃t)(x)~Ext [from 2]   │   Ca, Cb   │
        5.   ∴ (t)~(x)~Ext [from 4]   │ Eau, ~Eau′   │
     * 6.   ∴ (Ca ⊃ (∃t)~Eat) [from 1]   │ Ebu′, ~Ebu   │
        7.   ∴ Ca [from 3]   └─────────────────┘
     * 8.   ∴ (∃t)~Eat [from 6 and 7]
        9.   ∴ ~Eau [from 8]
     * 10.   ∴ ~(x)~Exu [from 5]
     * 11.   ∴ (∃x)Exu [from 10]
       12.   ∴ Ebu [from 11]
                    ∴ . . .

Our possible universe here contains two contingent beings (a and b) and two moments of
time (u and u′). Each contingent being exists at one moment in time but not the other. Here
the premise "For any contingent being, there is some time at which it fails to exist" is
true. But the conclusion is false. Its antecedent, "Everything is contingent," is true. But
its consequent, "There is some time at which everything fails to exist," is false. The normal
strategy leads to an infinite loop. We have to think up the refutation on our own.

5.5A

1. On Russell's theory, the first premise means "There's exactly one round square and 'It's
   false that there's exactly one round square' is a true statement about this thing." The first
   premise is false or implausible if it means this.
3. On Russell's theory, the first premise means "If it's false that there's exactly one most
   perfect being conceivable, then there's exactly one most perfect being conceivable and this
   book is greater than it." The first premise is implausible if it means this.

6.3A

1. □G
3. ~□M
5. □(R ⊃ P)
10. G
15. ◇(M · E)
20. *Ambiguous:* □(G ⊃ ~E) *or* (G ⊃ □~E) [The latter form could also be put as
    "(G ⊃ ~◇E).")
25. (R · ◇~R)
30. □(G ⊃ □G)

6.4A

1.   * 1.   ◇(A · B)   VALID
                 [∴ ◇A

   * 2.   ⌈asm: ~ ◇ A
    3. | ∴ □~A [from 2]
    4. | W ∴ (A · B) [from 1]
    5. | W ∴ ~A [from 3]
    6.   ⌊W ∴ A [from 4, contradicting 5]
    7.   ∴ ◇A [from 2, 5, and 6]

3.   1.   □(A ⊃ B)   VALID
   * 2.   ◇A
       [∴ ◇B
   * 3.   ⌈asm: ~ ◇B
    4. | ∴ □~B [from 3]
    5. | W ∴ A [from 2]
   * 6. | W ∴ (A ⊃ B) [from 1]
    7. | W ∴ ~B [from 4]
    8.   ⌊W ∴ B [from 5 and 6, contradicting 7]
    9.   ∴ ◇B [from 3, 7, and 8]

5. * 1.   ~◇(A · ~B)   VALID
      [∴ □(A ⊃ B)
   * 2.   ⌈asm: ~□(A ⊃ B)
   * 3. | ∴ ◇~(A ⊃ B) [from 2]
    4. | ∴ □~(A · ~B) [from 1]
   * 5. | W ∴ ~(A ⊃ B) [from 3]
   * 6. | W ∴ ~(A · ~B) [from 4]
    7. | W ∴ A [from 5]
    8. | W ∴ ~B [from 5]
    9.   ⌊W ∴ B [from 6 and 7, contradicting 8]
    10.   ∴ □(A ⊃ B) [from 2, 8, and 9]

10.   1.   □(A ⊃ B)   VALID
      [∴ (□A ⊃ □B)
   * 2.   ⌈asm: ~(□A ⊃ □B)
    3. | ∴ □A [from 2]
   * 4. | ∴ ~□B [from 2]
   * 5. | ∴ ◇~B [from 4]
    6. | W ∴ ~B [from 5]
   * 7. | W ∴ (A ⊃ B) [from 1]
    8. | W ∴ A [from 3]
    9.   ⌊W ∴ B [from 7 and 8, contradicting 6]
    10.   ∴ (□A ⊃ □B) [from 2, 6, and 9]

6.4B

1.   1.   □(T ⊃ L)   VALID
   * 2.   ◇(T · ~I)
      [∴ ~□(L ⊃ I)
    3.   ⌈asm: □(L ⊃ I)
   * 4. | W ∴ (T · ~I) [from 2]
   * 5. | W ∴ (T ⊃ L) [from 1]

* 6.   W ∴ (L ⊃ I) [from 3]

7.   W ∴ T [from 4]

8.   W ∴ ~I [from 4]

9.   W ∴ L [from 5 and 7]

10.   W ∴ I [from 6 and 9, contradicting 8]

11.   ∴ ~□(L ⊃ I) [from 3, 8, and 10]

3.   * 1.   ◇(W · D)   VALID

2.   □(W ⊃ F)

[∴ ◇(F · D)

* 3.   ⌈asm: ~◇(F · D)

4.   ∴ □~(F · D) [from 3]

* 5.   W ∴ (W · D) [from 1]

* 6.   W ∴ (W ⊃ F) [from 2]

* 7.   W ∴ ~(F · D) [from 4]

8.   W ∴ W [from 5]

9.   W ∴ D [from 5]

10.   W ∴ F [from 6 and 8]

11.   W ∴ ~D [from 7 and 10, contradicting 9]

12.   ∴ ◇(F · D) [from 3, 9, and 11]

5.   * 1.   ◇(G · T)   VALID

2.   □(T ⊃ E)

[∴ ◇(G · E)

* 3.   ⌈asm: ~◇(G · E)

4.   ∴ □~(G · E) [from 3]

* 5.   W ∴ (G · T) [from 1]

* 6.   W ∴ (T ⊃ E) [from 2]

* 7.   W ∴ ~(G · E) [from 4]

8.   W ∴ G [from 5]

9.   W ∴ T [from 5]

10.   W ∴ E [from 6 and 9]

11.   W ∴ ~E [from 7 and 8, contradicting 10]

12.   ∴ ◇(G · E) [from 3, 10, and 11]

10.   * 1.   (C ⊃ □(T ⊃ M))   VALID

2.   □T

* 3.   ~□M

[∴ ~C

4.   ⌈asm: C

5.   ∴ □(T ⊃ M) [from 1 and 4]

* 6.   ∴ ◇~M [from 3]

7.   W ∴ ~M [from 6]

8.   W ∴ T [from 2]

* 9.   W ∴ (T ⊃ M) [from 5]

10.   W ∴ M [from 8 and 9, contradicting 7]

11.   ∴ ~C [from 4, 7, and 10]

6.5A

1.   * 1.   ◇ A   INVALID

               [∴ □A

      * 2.   asm: ~□A

      * 3.   ∴ ◇~A [from 2]

       4.   W ∴ A [from 1]

       5.   WW ∴ ~A [from 3]

W	A
WW	~A

3.   * 1.   ◇ A   INVALID

      * 2.   ~□B

             [∴ ~□(A ⊃ B)

       3.   asm: □(A ⊃ B)

      * 4.   ∴ ◇~B [from 2]

       5.   W ∴ A [from 1]

       6.   WW ∴ ~B [from 4]

      * 7.   W ∴ (A ⊃ B) [from 3]

      * 8.   WW ∴ (A ⊃ B) [from 3]

       9.   W ∴ B [from 5 and 7]

     10.   WW ∴ ~A [from 6 and 8]

W	A, B
WW	~A, ~B

5.   * 1.   □(A ∨ B)   INVALID

             [∴ (□A ∨ □B)

      * 2.   asm: ~(□A ∨ □B)

      * 3.   ∴ ~□A [from 2]

      * 4.   ∴ ~□B [from 2]

      * 5.   ∴ ◇~A [from 3]

      * 6.   ∴ ◇~B [from 4]

       7.   W ∴ ~A [from 5]

       8.   WW ∴ ~B [from 6]

      * 9.   W ∴ (A ∨ B) [from 1]

    * 10.   WW ∴ (A ∨ B) [from 1]

     11.   W ∴ B [from 7 and 9]

     12.   WW ∴ A [from 8 and 10]

W	B, ~A
WW	A, ~B

10. ** 1.   (□A ⊃ □B)   INVALID

            [∴ □(A ⊃ B)

      * 2.   asm: ~□(A ⊃ B)

      * 3.   ∴ ◇~(A ⊃ B) [from 2]

      * 4.   W ∴ ~(A ⊃ B) [from 3]

       5.   W ∴ A [from 4]

       6.   W ∴ ~B [from 4]

     ** 7.   asm: ~□A [nice to have "□A" for 1]

     ** 8.   ∴ ◇~A [from 7]

       9.   WW ∴ ~A [from 8]

W	A, ~B
WW	~A

6.5B

1.   * 1.   (U ⊃ □(T ⊃ B))   VALID

      * 2.   ◇(T · ~B)

             [∴ ~U

3.   ⌈asm: U
4.   │ ∴ □(T ⊃ B) [from 1 and 3]
* 5.   │ W ∴ (T · ~B) [from 2]
* 6.   │ W ∴ (T ⊃ B) [from 4]
7.   │ W ∴ T [from 5]
8.   │ W ∴ ~B [from 5]
9.   ⌊W ∴ B [from 6 and 7, contradicting 8]
10.   ∴ ~U [from 3, 8, and 9]

3. 1.   □(B ⊃ B)   INVALID
         [∴ (B ⊃ □B)

  * 2.   asm: ~(B ⊃ □B)
   3.   ∴ B [from 2]
  * 4.   ∴ ~□B [from 2]
  * 5.   ∴ ◇~B [from 4]
   6.   W ∴ ~B [from 5]

	B
W	~B

We could use the first premise to drop "(B ⊃ B)" into the actual world and world-W. But these steps won't help us.

5. 1.   □(B ⊃ A)   VALID
  * 2.   ~◇A
      [∴ ~◇B

  * 3.   ⌈asm: ◇B
   4.   │ ∴ □~A [from 2]
   5.   │ W ∴ B [from 3]
  * 6.   │ W ∴ (B ⊃ A) [from 1]
   7.   │ W ∴ ~A [from 4]
   8.   ⌊W ∴ A [from 5 and 6, contradicting 7]
   9.   ∴ ~◇B [from 3, 7, and 8]

10. * 1.   (□(~C ⊃ ~M) ⊃ □(C ⊃ A))   VALID
   2.   □(A ⊃ ~M)
      [∴ (◇M ⊃ ~□(~C ⊃ ~M))

  * 3.   ⌈asm: ~(◇M ⊃ ~□(~C ⊃ ~M))
  * 4.   │ ∴ ◇M [from 3]
   5.   │ ∴ □(~C ⊃ ~M) [from 3]
   6.   │ ∴ □(C ⊃ A) [from 1 and 5]
   7.   │ W ∴ M [from 4]
  * 8.   │ W ∴ (A ⊃ ~M) [from 2]
  * 9.   │ W ∴ (~C ⊃ ~M) [from 5]
 * 10.   │ W ∴ (C ⊃ A) [from 6]
  11.   │ W ∴ ~A [from 7 and 8]
  12.   │ W ∴ C [from 7 and 9]
  13.   ⌊W ∴ A [from 10 and 12, contradicting 11]
  14.   ∴ (◇M ⊃ ~□(~C ⊃ ~M)) [from 3, 11, and 13]

15. The first premise here is ambiguous. It might mean either

  • It's logically necessary that if you have knowledge then you aren't mistaken: □(H ⊃ ~A)

                                   or

  • If you have knowledge, then "You're mistaken" (taken by itself) is self-contradictory: (H ⊃ □~A)

Taken the first way, the argument is invalid:

```
  1.    □(H ⊃ ~A)                                    INVALID
* 2.      ◇A
       [∴ ~H                                      ┌─────────┐
  3.    asm: H                                    │  H, ~A  │
  4.    W ∴ A [from 2]                      W     ├─────────┤
* 5.    ∴ (H ⊃ ~A) [from 1]                       │  ~H, A  │
* 6.    W ∴ (H ⊃ ~A) [from 1]                     └─────────┘
  7.    ∴ ~A [from 3 and 5]
  8.    W ∴ ~H [from 4 and 6]
```

Our refutation has an actual world (where you have knowledge and aren't mistaken) and a possible world-W (where you are mistaken and don't have knowledge). Here the first premise is true, since in every world in which you have knowledge you aren't mistaken. The second premise is true, since in world-W you are mistaken. But the conclusion is false, since in the actual world you have knowledge. Taken the second way, the argument is valid:

```
* 1.    (H ⊃ □~A)                                   VALID
* 2.      ◇A
       [∴ ~H
  3.    ⌈asm: H
  4.    │∴ □~A [from 1 and 3]
  5.    │W ∴ A [from 2]
  6.    ⌊W ∴ ~A [from 4, contradicting 5]
  7.    ∴ ~H [from 3, 5, and 6]
```

On this reading, the first premise is *false*. Note that the premise is false in the galaxy of possible worlds used to refute the first form of the argument, since then "H" is true and "□~A" is false.

20. The second premise here is ambiguous. It could mean either

- It's logically necessary that if it was always true that you'd do it then you'd do it: □(T ⊃ D)

or

- If it was always true that you'd do it, then "You'll do it" (taken by itself) is logically necessary: (T ⊃ □D)

Taken the first way, the argument is invalid:

```
  1.    T                                          INVALID
  2.    □(T ⊃ D)
* 3.    (□D ⊃ ~F)                                  ┌─────────┐
       [∴ ~F                                       │ T, D, F │
  4.    asm: F                                W    ├─────────┤
* 5.    ∴ ~□D [from 3 and 4]                       │ ~T, ~D  │
* 6.    ∴ ◇~D [from 5]                             └─────────┘
  7.    W ∴ ~D [from 6]
* 8.    ∴ (T ⊃ D) [from 2]
```

* 9.　　W ∴ (T ⊃ D) [from 2]
10.　　∴ D [from 1 and 8]
11.　　W ∴ ~T [from 7 and 9]

Taken the second way, the argument is valid but has a false second premise:

1.　　　T　　　　　　　　　　　　　　　　　　　　　　VALID
* 2.　　　(T ⊃ □D)
* 3.　　　(□D ⊃ ~F)
　　　[∴ ~F
4.　　⌈asm: F
5.　　│∴ □D [from 1 and 2]
6.　　└∴ ~F [from 3 and 5, contradicting 4]
7.　　∴ ~H [from 4 and 6]

6.6A

1.　* 1.　　◇□A　　　　　　　　　　　　　　　　　　　VALID
　　　　　[∴ A
2.　　⌈asm: ~A
3.　　│W ∴ □A [from 1]　　　　　　　　　　　　　# → W
4.　　└∴ A [from 3, contradicting 2]　　　　　　*NEED B OR S5*
5.　　∴ A [from 2 and 4]

3.　* 1.　　◇◇A　　　　　　　　　　　　　　　　　　　VALID
　　　　　[∴ ◇A
* 2.　　⌈asm: ~◇A
3.　　│∴ □~A [from 2]
* 4.　　│W ∴ ◇A [from 1]　　　　　　　　　　　　　# → W
5.　　│WW ∴ A [from 4]　　　　　　　　　　　　　W → WW
6.　　└WW ∴ ~A [from 3, contradicting 5]　　　*NEED S4 OR S5*
7.　　∴ ◇A [from 2, 5, and 6]

5.　* 1.　　(□A ⊃ □B)　　　　　　　　　　　　　　　　VALID
　　　　　[∴ □(□A ⊃ □B)　　　　　　　　　　　　　[NEED S5]
* 2.　　⌈asm: ~□(□A ⊃ □B)
* 3.　　│∴ ◇~(□A ⊃ □B) [from 2]
* 4.　　│W ∴ ~(□A ⊃ □B) [from 3]　　　　　　　　# → W
5.　　│W ∴ □A [from 4]
* 6.　　│W ∴ ~□B [from 4]
* 7.　　│W ∴ ◇~B [from 6]
8.　　│WW ∴ ~B [from 7]　　　　　　　　　　　　W → WW
9.　　│　⌈asm: ~□A [nice to have "□A" for 1]
10.　　│　│∴ ◇~A [from 9]
11.　　│　│WWW ∴ ~A [from 10]　　　　　　　　# → WWW
12.　　│　└WWW ∴ A [from 5, contradicting 11]　*NEED S5*
13.　　│∴ □A [from 9, 11, and 12]
14.　　│∴ □B [from 1 and 13]
15.　　└WW ∴ B [from 14, contradicting 8]　　　*NEED S4 OR S5*
16.　　∴ □(□A ⊃ □B) [from 2, 8, and 15]

10.　* 1.　　◇A　　　　　　　　　　　　　　　　　　　VALID
　　　　　[∴ ◇□◇A　　　　　　　　　　　　　　　　[NEED B OR S5]

*	2.	⌈asm: ~◇□◇A	
	3.	│∴ □~□◇A [from 2]	
	4.	│W ∴ A [from 1]	# → W
*	5.	│W ∴ ~□◇A [from 3]	*ANY SYSTEM OK*
*	6.	│W ∴ ◇~◇A [from 5]	
*	7.	│WW ∴ ~◇A [from 6]	W → WW
	8.	│WW ∴ □~A [from 7]	
	9.	⌊W ∴ ~A [from 8, contradicting 4]	*NEED B OR S5*
	10.	∴ ◇□◇A [from 2, 4, and 9]	

15.

	1.	□A	VALID
		[∴ □□□A	
*	2.	⌈asm: ~□□□A	
*	3.	│∴ ◇~□□A [from 2]	
*	4.	│W ∴ ~□□A [from 3]	# → W
*	5.	│W ∴ ◇~□A [from 4]	
*	6.	│WW ∴ ~□A [from 5]	W → WW
*	7.	│WW ∴ ◇~A [from 6]	
	8.	│WWW ∴ ~A [from 7]	WW → WWW
	9.	⌊WWW ∴ A [from 1, contradicting 8]	*NEED S4 OR S5*
	10.	∴ □□□A [from 2, 8, and 9]	

6.6B

1.

*	1.	~(◇N · ◇~N)	VALID
	2.	◇N	
		[∴ N	
	3.	⌈asm: ~N	
*	4.	│∴ ~◇~N [from 1 and 2]	
	5.	│∴ □N [from 4]	
	6.	⌊∴ N [from 5, contradicting 3]	*ANY SYSTEM OK*
	7.	∴ N [from 3 and 6]	

Any of the four systems suffices: T, B, S4, or S5.

3.

	1.	□(N ⊃ □N)	VALID
*	2.	◇~N	
		[∴ ~N	
	3.	⌈asm: N	
	4.	│W ∴ ~N [from 2]	# → W
*	5.	│∴ (N ⊃ □N) [from 1]	*ANY SYSTEM OK*
	6.	│∴ □N [from 3 and 5]	
	7.	⌊W ∴ N [from 6, contradicting 4]	*ANY SYSTEM OK*
	8.	∴ ~N [from 1, 4, and 7]	

Any of the four systems suffices: T, B, S4, or S5. We could also derive "~□N" from 2 to contradict 6.

5.

	1.	□((M · ◇R) ⊃ I)	VALID
*	2.	~□I	[NEED S5]
*	3.	◇R	
		[∴ ~□M	

4. ⌐asm: □M
* 5. │ ∴ ◇~I [from 2]
6. │ W ∴ ~I [from 5]   # → W
7. │ WW ∴ R [from 3]   # → WW
8. │ W ∴ M [from 4]   *ANY SYSTEM OK*
* 9. │ W ∴ ((M · ◇R) ⊃ I) [from 1]   *ANY SYSTEM OK*
* 10. │ W ∴ ~(M · ◇R) [from 6 and 9]
* 11. │ W ∴ ~◇R [from 8 and 10]
12. │ W ∴ □~R [from 11]
13. └WW ∴ ~R [from 12, contradicting 7]   *NEED S5*
14. ∴ ~□M [from 4, 7, and 13]

6.7A

1. *Ambiguous*: □(x)(Bx ⊃ Ux) *or* (x)(Bx ⊃ □Ux)
3. (x)(Bx ⊃ □Ux)
5. □s = s
10. (x)(Nx ⊃ □Ax)
15. (x)(Lx ⊃ (Px · ◇~Px))
20. □On
25. ◇(x)Ux
30. *Ambiguous*: ◇(Ex)(~Sx · Sx) *or* (Ex)(~Sx · ◇Sx)

6.8A

1. * 1. (∃x)□Fx   VALID
[∴ □(∃x)Fx
* 2. ⌐asm: ~□(∃x)Fx
* 3. │ ∴ ◇~(∃x)Fx [from 2]
* 4. │ W ∴ ~(∃x)Fx [from 3]
5. │ W ∴ (x)~Fx [from 4]
6. │ ∴ □Fa [from 1]
7. │ W ∴ Fa [from 6]
8. └W ∴ ~Fa [from 5, contradicting 7]
9. ∴ □(∃x)Fx [from 2, 7, and 8]

3.   [∴ □(∃x)x = a   VALID
* 1. ⌐asm: ~□(∃x)x = a
* 2. │ ∴ ◇~(∃x)x = a [from 1]
* 3. │ W ∴ ~(∃x)x = a [from 2]
4. │ W ∴ (x)~x = a [from 3]
5. │ W ∴ ~a = a [from 4]
6. └W ∴ a = a [from Q5, contradicting 5]
7. ∴ □(∃x)x = a [from 1, 5, and 6]

5. 1. □(x)Fx   VALID
[∴ (x)□Fx
* 2. ⌐asm: ~(x)□Fx
* 3. │ ∴ (∃x)~□Fx [from 2]
* 4. │ ∴ ~□Fa [from 3]
* 5. │ ∴ ◇~Fa [from 4]
6. │ W ∴ ~Fa [from 5]
7. │ W ∴ (x)Fx [from 1]
8. └W ∴ Fa [from 7, contradicting 6]

     9.   ∴ (x)□Fx [from 2, 6, and 8]

10. * 1.   (∃x)◊Fx   VALID

           [∴ ◊(∃x)Fx

    * 2.   ⎡asm: ~◊(∃x)Fx

     3.   │∴ □~(∃x)Fx [from 2]

    * 4.   │∴ ◊Fa [from 1]

     5.   │W ∴ Fa [from 4]

    * 6.   │W ∴ ~(∃x)Fx [from 3]

     7.   │W ∴ (x)~Fx [from 6]

     8.   ⎣W ∴ ~Fa [from 7, contradicting 5]

     9.   ∴ ◊(∃x)Fx [from 2, 5, and 8]

15.   1.   ~a=b   VALID

           [∴ □~a=b

    * 2.   ⎡asm: ~□~a=b

    * 3.   │∴ ◊a=b [from 2]

     4.   │W ∴ a=b [from 3]

     5.   │⎡W asm: ~□a=b

     6.   ││W ∴ ◊~a=b [from 5]

     7.   ││W ∴ ◊~a=a [from 4 and 6]

     8.   ││WW ∴ ~a=a [from 7]

     9.   │⎣WW ∴ a=a [from Q5, contradicting 8]

     10.   │W ∴ □a=b [from 5, 8, and 9]

     11.   ⎣∴ a=b [from 10, contradicting 1]

     12. ∴ □~a=b [from 2, 1, and 11]

This proof is fairly devious—especially the second assumption. Step 11 requires system B or S5 (see Section 6.6).

6.8B

1.   1.   (x)□(~Bx ⊃ ~x=i)   VALID

           [∴ ~◊~Bi

    * 2.   ⎡asm: ◊~Bi

     3.   │W ∴ ~Bi [from 2]

     4.   │∴ □(~Bi ⊃ ~i=i) [from 1]

    * 5.   │W ∴ (~Bi ⊃ ~i=i) [from 4]

     6.   │W ∴ ~i=i [from 3 and 5]

     7.   ⎣W ∴ i=i [from Q5, contradicting 6]

     8.   ∴ ~◊~Bi [from 2, 6, and 7]

We could also translate the premise as "□(x)(~Bx ⊃ ~x=i)." This is provably equivalent in the present system.

3.   1.   ~a=p   VALID

    * 2.   (((∃x)□x=p · ~(x)□x=p) ⊃ S)

    * 3.   (S ⊃ E)

         [∴ E

     4.   ⎡asm: ~E

     5.   │∴ ~S [from 3 and 4]

    * 6.   │∴ ~((∃x)□x=p · ~(x)□x=p) [from 2 and 5]

7.   ⌐ asm: ~(∃x)□x = p [nice to have "(∃x)□x = p" for 6]

8.   ∴ (x)~□x = p [from 7]

9.   ∴ ~□p = p [from 8]

10.   ∴ ◇~p = p [from 9]

11.   W ∴ ~p = p [from 10]

12.   ⌐ W ∴ p = p [from Q5, contradicting 11]

13.   ∴ (∃x)□x = p [from 7, 11, and 12]

14.   ∴ (x)□x = p [from 6 and 13]

15.   ∴ □a = p [from 14]

16.   ∴ a = p [from 15, contradicting 1]

17. ∴ E [from 4, 1, and 16]

5. 1.   □(∃x)Ux   **INVALID**

    [∴ (∃x)□Ux   a, b

* 2.   asm: ~(∃x)□Ux

3.   ∴ (x)~□Ux [from 2] | Ua, ~Ub |

* 4.   ∴ ~□Ua [from 3]

* 5.   ∴ ◇~Ua [from 4]   W | Ub, ~Ua |

6.   W ∴ ~Ua [from 5]

* 7.   W ∴ (∃x)Ux [from 1]

8.   W ∴ Ub [from 7]

          ....

The normal strategy leads into an infinite loop. We'd next drop the universal quantifier in 3 using "b" and then get a new world WW two lines later. We'd drop the box in 1 using this new world WW and then get a new letter "c" in the next line. And the process would repeat again—and again. We have to get the refutation on our own.

10. * 1.   ◇(Ts · ~Bs)   **VALID**

2.   □(x)(Tx ⊃ Px)

    [∴ ~□(x)(Px ⊃ Bx)

3.   ⌐asm: □(x)(Px ⊃ Bx)

* 4.   W ∴ (Ts · ~Bs) [from 1]

5.   W ∴ (x)(Tx ⊃ Px) [from 2]

6.   W ∴ (x)(Px ⊃ Bx) [from 3]

7.   W ∴ Ts [from 4]

8.   W ∴ ~Bs [from 4]

* 9.   W ∴ (Ts ⊃ Ps) [from 5]

* 10.   W ∴ (Ps ⊃ Bs) [from 6]

11.   W ∴ Ps [from 7 and 9]

12.   W ∴ Bs [from 10 and 11, contradicting 8]

13.   ∴ ~□(x)(Px ⊃ Bx) [from 3, 8, and 12]

15. The first premise is ambiguous. If we take it in the *de dicto* sense (to mean "It's necessary that all (well-formed) cyclists are two-legged"), then the argument is invalid:

1.   □(x)(Cx ⊃ Tx)   **INVALID**

2.   Cp   p

    [∴ □Tp

* 3.   asm: ~□Tp | Cp, Tp |

* 4.   ∴ ◇~Tp [from 3]

5.   W ∴ ~Tp [from 4]   W | ~Cp, ~Tp |

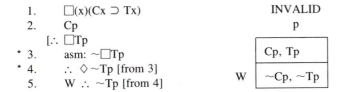

6.   ∴ (x)(Cx ⊃ Tx) [from 1]
7.   W ∴ (x)(Cx ⊃ Tx) [from 1]
* 8.   ∴ (Cp ⊃ Tp) [from 6]
* 9.   W ∴ (Cp ⊃ Tp) [from 7]
10.   ∴ Tp [from 2 and 8]
11.   W ∴ ~Cp [from 5 and 9]

The argument is valid if we take the first premise *de re* (to mean "Each (well-formed) cyclist has the property of being necessarily two-legged"):

1.   (x)(Cx ⊃ □Tx)   VALID
2.   Cp
          [∴ □Tp
3.   ⎡asm: ~□Tp
* 4.   │ ∴ (Cp ⊃ □Tp) [from 1]
5.   ⎣∴ □Tp [from 2 and 4, contradicting 3]
6.   ∴ □Tp [from 3 and 5]

6.9A

1.   [∴ (∃x)x = a   INVALID
* 1.   asm: ~(∃x)x = a   ┌─────────────────┐
2.   ∴ (x)~x = a   │ Nothing exists! │
                                 └─────────────────┘

If an actual world where nothing exists is too nihilistic for you, you could use an actual world where b exists but a doesn't exist. The conclusion would be false in this world. (Here we can't do anything with line 2, since we have no premises asserting the existence of anything.)

3. * 1.   (∃x)□Fx   VALID
             [∴ □(∃x)Fx
* 2.   ⎡asm: ~□(∃x)Fx
* 3.   │ ∴ ◊~(∃x)Fx [from 2]
4.   │ ∴ □Fa [from 1]
5.   │ ∴ (∃x)x = a [from 1]
* 6.   │ W ∴ ~(∃x)Fx [from 3]
7.   │ W ∴ Fa [from 4]
8.   │ W ∴ (x)~Fx [from 6]
9.   │ W ∴ (∃x)x = a [from 7 *using Q7*]
10.   ⎣W ∴ ~Fa [from 8 and 9, contradicting 7]
11.   ∴ □(∃x)Fx [from 2, 7, and 10]

5. * 1.   (∃x)◊Fx   VALID
             [∴ ◊(∃x)Fx
* 2.   ⎡asm: ~◊(∃x)Fx
3.   │ ∴ □~(∃x)Fx [from 2]
* 4.   │ ∴ ◊Fa [from 1]
5.   │ ∴ (∃x)x = a [from 1]
6.   │ W ∴ Fa [from 4]
* 7.   │ W ∴ ~(∃x)Fx [from 3]
8.   │ W ∴ (x)~Fx [from 7]
9.   │ W ∴ (∃x)x = a [from 6 *using Q7*]
10.   ⎣W ∴ ~Fa [from 8 and 9, contradicting 6]
11.   ∴ ◊(∃x)Fx [from 2, 6, and 10]

10.   * 1. $(\exists x)\Diamond((\exists y)y = x \cdot \sim Fx)$ VALID
         [∴ $\Diamond(\exists x)\sim Fx$
    * 2.   ⌈asm: $\sim\Diamond(\exists x)\sim Fx$
      3.   │ ∴ $\Box\sim(\exists x)\sim Fx$ [from 2]
    * 4.   │ ∴ $\Diamond((\exists y)y = a \cdot \sim Fa)$ [from 1]
      5.   │ ∴ $(\exists x)x = a$ [from 1]
    * 6.   │ W ∴ $((\exists y)y = a \cdot \sim Fa)$ [from 4]
    * 7.   │ W ∴ $\sim(\exists x)\sim Fx$ [from 3]
      8.   │ W ∴ $(\exists y)y = a$ [from 6]
      9.   │ W ∴ $\sim Fa$ [from 6]
     10.   │ W ∴ $(x)Fx$ [from 7]
     11.   ⌊W ∴ Fa [from 8 and 10, contradicting 9]
     12.   ∴ $\Diamond(\exists x)\sim Fx$ [from 2, 9, and 11]

15.   1. $\Box(x)Fx$ VALID
         [∴ $(x)\Box((\exists y)y = x \supset Fx)$
    * 2.   ⌈asm: $\sim(x)\Box((\exists y)y = x \supset Fx)$
    * 3.   │ ∴ $(\exists x)\sim\Box((\exists y)y = x \supset Fx)$ [from 2]
    * 4.   │ ∴ $\sim\Box((\exists y)y = a \supset Fa)$ [from 3]
    * 5.   │ ∴ $\Diamond\sim((\exists y)y = a \supset Fa)$ [from 4]
    * 6.   │ W ∴ $\sim((\exists y)y = a \supset Fa)$ [from 5]
      7.   │ W ∴ $(\exists y)y = a$ [from 6]
      8.   │ W ∴ $\sim Fa$ [from 6]
      9.   │ W ∴ $(x)Fx$ [from 1]
     10.   ⌊W ∴ Fa [from 7 and 9, contradicting 8]
     11.   ∴ $(x)\Box((\exists y)y = x \supset Fx)$ [from 2, 8, and 10]

20. "Descartes doesn't think" is ambiguous. It could mean "Descartes is an existing being who doesn't think." [This is the internal negation sense: "$(\exists x)(x = d \cdot \sim Td)$."] Then the first premise would mean "It's necessary that either Descartes is an existing being who thinks or Descartes is an existing being who doesn't think." The premise is false if it means this (since Descartes might not have existed). Or "Descartes doesn't think" could mean "It isn't the case that Descartes is an existing being who thinks." [This is the external negation sense: "$\sim Td$."] Then the third premise would mean "It's necessary that if it isn't the case that Descartes is an existing being who thinks then Descartes exists." The premise is false if it means this. So on either reading we get a false premise.

7.1A

1. $(\underline{L} \lor \underline{S})$
3. $(\underline{A} \supset \underline{W})$
5. $\sim(\underline{A} \cdot \underline{B})$
10. $\sim(\underline{B} \cdot \sim\underline{A})$
15. $(H\underline{ju} \supset H\underline{uj})$
20. $\Box(\underline{B} \supset (\underline{N} \supset \underline{S}))$
25. $(\exists x)(W\underline{ux} \cdot L\underline{ux})$

7.2A

1. ** 1. $(\underline{A} \supset \underline{B})$ INVALID
         [∴ $(\sim\underline{B} \supset \sim\underline{A})$
    * 2.   asm: $\sim(\sim\underline{B} \supset \sim\underline{A})$ | $\sim A,\ \sim B,\ \underline{A}$ |

    3.   ∴ ~B [from 2]
    4.   ∴ A̱ [from 2]
    5.   asm: ~A [nice to have "A" for 1]

3. * 1.   (A ⊃ B)   INVALID
        [∴ ~(A̱ · ~B)
    * 2.   asm: (A̱ · ~B)   ~A, A̱, ~Ḇ
    3.   ∴ A̱ [from 2]
    4.   ∴ ~Ḇ [from 2]
    5.   ∴ ~A̱ [from 1 and 4]

5. * 1.   ~◇(A · B)   VALID
    * 2.   ~(C̱ · ~A̱)
        [∴ ~(C̱ · B)
    * 3.   ⌈ asm: (C̱ · B)
    4.   │ ∴ C̱ [from 3]
    5.   │ ∴ Ḇ [from 3]
    6.   │ ∴ A̱ [from 2 and 4]
    7.   │ ∴ □~(A · B) [from 1]
    * 8.   │ ∴ ~(A̱ · B) [from 7]
    9.   ⌊ ∴ ~Ḇ [from 6 and 8, contradicting 5]
    10.   ∴ ~(C̱ · B) [from 3, 5, and 9]

10. * 1.   ~(A · ~B)   VALID
        [∴ (~A̱ ∨ Ḇ)
    * 2.   ⌈ asm: ~(~A̱ ∨ B)
    3.   │ ∴ A̱ [from 2]
    4.   │ ∴ ~Ḇ [from 2]
    5.   ⌊ ∴ Ḇ [from 1 and 3, contradicting 4]
    6.   ∴ (~A̱ ∨ Ḇ) [from 2, 4, and 5]

    7.2B

1. 1.   (B ∨ H)   VALID
    2.   ~Ḇ
        ∴ H̱ [from 1 and 2]

This argument is already a formal proof. The conclusion follows from the premises using an I-rule.

3. 1.   ~E̱   INVALID
    * 2.   (~E̱ ⊃ G̱)
        [∴ G̱
    3.   asm: ~G̱   E, ~E̱, ~G̱
    4.   ∴ E [from 2 and 3]

5. * 1.   ~(S̱ · ~P̱)   INVALID
    2.   S
        [∴ P̱
    3.   asm: ~P̱   S, ~S̱, ~P̱
    4.   ∴ ~S̱ [from 1 and 3]

10.   1.   (B ⊃ C̲)   VALID
           2.   B
                 ∴ C̲ [from 1 and 2]
This argument is already a formal proof. The conclusion follows from the premises using an I-rule.

15.   * 1.   ~(B̲ · A̲)   VALID
                [∴ (~B̲ ∨ ~A̲)]
          * 2.   ┌asm: ~(~B̲ ∨ ~A̲)
            3.   │ ∴ B̲ [from 2]
            4.   │ ∴ A̲ [from 2]
            5.   └∴ ~A̲ [from 1 and 3, contradicting 4]
            6.   ∴ (~B̲ ∨ ~A̲) [from 2, 4, and 5]

20.   1.   N   VALID
           2.   □(B̲ ⊃ D̲)
           3.   □(D̲ ⊃ (N ⊃ S̲))
                [∴ (S̲ ∨ ~B̲)]
          * 4.   ┌asm: ~(S̲ ∨ ~B̲)
            5.   │ ∴ ~S̲ [from 4]
            6.   │ ∴ B̲ [from 4]
          * 7.   │ ∴ (B̲ ⊃ D̲) [from 2]
          * 8.   │ ∴ (D̲ ⊃ (N ⊃ S̲)) [from 3]
            9.   │ ∴ D̲ [from 6 and 7]
          * 10. │ ∴ (N ⊃ S̲) [from 8 and 9]
            11. └ ∴ S̲ [from 1 and 10, contradicting 5]
            12. ∴ (S̲ ∨ ~B̲) [from 4, 5, and 11]

7.3A

1. (A ⊃ O~B̲)   **15.** (OA̲ ⊃ ◇A)
3. O~(A̲ · B̲)   **20.** (∃x)OAx̲
5. (A ⊃ RA̲)   **25.** (x)~RAx̲
10. (~OA̲ · O~A̲)   **30.** (x)~OAx̲

7.4A

1.   1.   O~A   VALID
                [∴ O~(A̲ · B̲)]
          * 2.   ┌asm: ~O~(A̲ · B̲)
          * 3.   │ ∴ R(A̲ · B̲) [from 2]
            4.   │D ∴ (A̲ · B̲) [from 3]
            5.   │D ∴ ~A̲ [from 1]
            6.   └D ∴ A̲ [from 4, contradicting 5]
            7.   ∴ O~(A̲ · B̲) [from 2, 5, and 6]

3.   1.   b = c   VALID
                [∴ (OFab ⊃ OFac)]
          * 2.   ┌asm: ~(OFab ⊃ OFac)
            3.   │ ∴ OFab [from 2]
            4.   │ ∴ ~OFac [from 2]
            5.   └∴ OFac [from 1 and 3, contradicting 4]

    6.   ∴ (OF<u>ab</u> ⊃ OF<u>ac</u>) [from 2, 4, and 5]

5.   1.   O(<u>A</u> ∨ <u>B</u>)   VALID
    * 2.   (<u>B</u> ⊃ <u>C</u>)
        [∴ (~<u>C</u> ⊃ <u>A</u>)
    * 3.   ⌈asm: ~(~<u>C</u> ⊃ <u>A</u>)
      4. | ∴ ~<u>C</u> [from 3]
      5. | ∴ ~<u>A</u> [from 3]
    * 6. | ∴ (<u>A</u> ∨ <u>B</u>) [from 1]
      7. | ∴ <u>B</u> [from 5 and 6]
      8.   ⌊∴ <u>C</u> [from 2 and 7, contradicting 4]
      9.   ∴ (~<u>C</u> ⊃ <u>A</u>) [from 3, 4, and 8]

10.   1.   O<u>A</u>   VALID
     2.   O<u>B</u>
        [∴ O(<u>A</u> · <u>B</u>)
    * 3.   ⌈asm: ~O(<u>A</u> · <u>B</u>)
    * 4. | ∴ R~(<u>A</u> · <u>B</u>) [from 3]
    * 5. | D ∴ ~(<u>A</u> · <u>B</u>) [from 4]
      6. | D ∴ <u>A</u> [from 1]
      7. | D ∴ <u>B</u> [from 2]
      8.   ⌊D ∴ ~<u>B</u> [from 5 and 6, contradicting 7]
      9.   ∴ O(<u>A</u> · <u>B</u>) [from 3, 7, and 8]

15.   1.   □(<u>A</u> ⊃ <u>B</u>)   VALID
     2.   O<u>A</u>
        [∴ O<u>B</u>
    * 3.   ⌈asm: ~O<u>B</u>
    * 4. | ∴ R~<u>B</u> [from 3]
      5. | D ∴ ~<u>B</u> [from 4]
      6. | D ∴ <u>A</u> [from 2]
    * 7. | D ∴ (<u>A</u> ⊃ <u>B</u>) [from 1]
      8.   ⌊D ∴ <u>B</u> [from 6 and 7, contradicting 5]
      9.   ∴ O<u>B</u> [from 3, 5, and 8]

20.   [∴ (R<u>A</u> ∨ R~<u>A</u>)   VALID
    * 1.   ⌈asm: ~(R<u>A</u> ∨ R~<u>A</u>)
    * 2. | ∴ ~R<u>A</u> [from 1]
    * 3. | ∴ ~R~<u>A</u> [from 1]
      4. | ∴ O~<u>A</u> [from 2]
      5. | ∴ O<u>A</u> [from 3]
      6. | ∴ ~<u>A</u> [from 4]
      7.   ⌊∴ <u>A</u> [from 5, contradicting 6]
      8.   ∴ (R<u>A</u> ∨ R~<u>A</u>) [from 1, 6, and 7]

25.   1.   O(<u>A</u> ⊃ <u>B</u>)   VALID
        [∴ (<u>A</u> ⊃ O<u>B</u>)
    * 2.   ⌈asm: ~(<u>A</u> ⊃ O<u>B</u>)
      3. | ∴ <u>A</u> [from 2]
    * 4. | ∴ ~O<u>B</u> [from 2]
    * 5. | ∴ R~<u>B</u> [from 4]
      6. | D ∴ ~<u>B</u> [from 5]

* 7. |D ∴ (A ⊃ B̄) [from 1]
  8. |D ∴ A [from 3, *using D5*]
  9. |_D ∴ B̄ [from 7 and 8, contradicting 6]
  10.   ∴ (A ⊃ ŌB̄) [from 2, 6, and 9]

30.   1.   (A ∨ ŌB̄)   VALID
  2.   ~A
       ∴ OB̄ [from 1 and 2]

This argument is already a formal proof. The conclusion follows from the premises using an I-rule.

7.4B

1. * 1.   ~R(K̄ · V̄)   VALID
  2.   OV̄
       [∴ ~K̄
  3.   ⌈asm: K̄
  4. |∴ O~(K̄ · V̄) [from 1]
  * 5. |∴ ~(K̄ · V̄) [from 4]
  6. |∴ V̄ [from 2]
  7. |_∴ ~V̄ [from 3 and 5, contradicting 6]
  8.   ∴ ~K̄ [from 3, 6, and 7]

3. 1.   A   VALID
  2.   O~A
  * 3.   ((A · ◇~A) ⊃ F)
       [∴ F
  4.   ⌈asm: ~F
  * 5. |∴ ~(A · ◇~A) [from 3 and 4]
  6. |∴ ~◇~A [from 1 and 5]
  7. |_∴ ◇~A [from 2 *using D6*, contradicting 6]
  8.   ∴ F [from 4, 6, and 7]

5.   [∴ (OA ⊃ A)   INVALID
  * 1.   asm; ~(OA ⊃ A)
  2.   ∴ OA [from 1]
  3.   ∴ ~A [from 1]
  4.   ∴ A [from 2]

10.   [∴ □(A ⊃ RA)   VALID
  * 1.   ⌈asm: ~□(A ⊃ RA)
  * 2. |∴ ◇~(A ⊃ RA) [from 1]
  * 3. |W ∴ ~(A ⊃ RA) [from 2]
  4. |W ∴ A [from 3]
  * 5. |W ∴ ~RA [from 3]
  6. |W ∴ O~A [from 5]
  7. |_W ∴ ~A [from 6, contradicting 4]
  8.   ∴ □(A ⊃ RA) [from 1, 4, and 7]

15. * 1.   (RC̄ ⊃ OT̄)   VALID
       [∴ O(T̄ ∨ ~C̄)
  * 2.   ⌈asm: ~O(T̄ ∨ ~C̄)
  * 3. |∴ R~(T̄ ∨ ~C̄) [from 2]
  * 4. |D ∴ ~(T̄ ∨ ~C̄) [from 3]

5.   D ∴ ~T [from 4]
6.   D ∴ C [from 4]
7.   ⌈ asm: ~RC [nice to have "RC" for 1]
8. | ∴ O~C [from 7]
9.   ⌊ D ∴ ~C [from 8, contradicting 6]
10.   ∴ RC [from 7, 6, and 9]
11.   ∴ OT [from 1 and 10]
12.   D ∴ T [from 11, contradicting 5]
13. ∴ O(T ∨ ~C) [from 2, 5, and 12]

20.   * 1.   RA   **INVALID**
        * 2.   RB
                [∴ R(A · B)
        * 3.   asm: ~R(A · B)
          4.   ∴ O~(A · B) [from 3]
          5.   D ∴ A [from 1]
          6.   DD ∴ B [from 2]
        * 7.   D ∴ ~(A · B) [from 4]
        * 8.   DD ∴ ~(A · B) [from 4]
          9.   D ∴ ~B [from 5 and 7]
       10.   DD ∴ ~A [from 6 and 8]

Suppose that it's all right to promise to work tomorrow—and also all right to take tomorrow off. It doesn't follow that it's all right to do both together.

25.   * 1.   ◇(A · R~A)   **VALID**
               [∴ ~□(OA ≡ A)
        * 2.   ⌈ asm: □(OA ≡ A)
        * 3. | W ∴ (A · R~A) [from 1]
        * 4. | W ∴ (OA ≡ A) [from 2]
          5. | W ∴ A [from 3]
        * 6. | W ∴ R~A [from 3]
          7. | W ∴ (OA ⊃ A) [from 4]
        * 8. | W ∴ (A ⊃ OA) [from 4]
          9. | W ∴ OA [from 5 and 8]
       10. | WD ∴ ~A [from 6]
       11.   ⌊ WD ∴ A [from 9, contradicting 10]
       12.   ∴ ~□(OA ≡ A) [from 2, 10, and 11]

Alternatively, we could derive "W ∴ ~OA" (from 6) for step 10. This would contradict 9.

30.   1.   O(∃x)(Hsx · Bjx)   **INVALID**
              [∴ O(∃x)Bjx
        * 2.   asm: ~O(∃x)Bjx
        * 3.   ∴ R~(∃x)Bjx [from 2]
        * 4.   D ∴ ~(∃x)Bjx [from 3]
        * 5.   D ∴ (∃x)(Hsx · Bjx) [from 1]
          6.   D ∴ (x)~Bjx [from 4]
        * 7.   D ∴ (Hsa · Bja) [from 5]
          8.   D ∴ ~Bja [from 6]
          9.   D ∴ Hsa [from 7]
       10.   D ∴ Bja [from 7]

35.   * 1.   (~RK ⊃ O~I)   VALID
     * 2.   (RN ⊃ RI)
        [∴ (~RK ⊃ ~RN)
     * 3.   ⌈asm: ~(~RK ⊃ ~RN)
      4.   │ ∴ ~RK [from 3]
      5.   │ ∴ RN [from 3]
      6.   │ ∴ O~I [from 1 and 4]
      7.   │ ∴ RI [from 2 and 5]
      8.   ⌊ ∴ ~RI [from 6, contradicting 7]
      9.   ∴ (~RK ⊃ ~RN) [from 3, 7, and 8]

8.1A

1. u:~G
3. ~u:G
5. □(u:G ⊃ ~u:~G)
10. ~(u:A · u:~A)

8.2A

1.   * 1.   ~◇(A · B)   VALID
        [∴ ~(u:A · u:B)
     * 2.   ⌈asm: (u:A · u:B)
      3.   │ ∴ u:A [from 2]
      4.   │ ∴ u:B [from 2]
      5.   │ ∴ □~(A · B) [from 1]
      6.   │ u ∴ A [from 3]
      7.   │ u ∴ B [from 4]
     * 8.   │ u ∴ ~(A · B) [from 5]
      9.   ⌊ u ∴ ~B [from 6 and 8, contradicting 7]
      10.   ∴ ~(u:A · u:B) [from 2, 7, and 9]

3.   * 1.   ~◇(A · B)   INVALID
        [∴ (u:B ⊃ ~u:A)
     * 2.   asm: ~(u:B ⊃ ~u:A)
      3.   ∴ u:B [from 2]
      4.   ∴ u:A [from 2]
      5.   ∴ □~(A · B) [from 1]
      6.   u ∴ A [from 4]
     * 7.   u ∴ ~(A · B) [from 5]
      8.   u ∴ ~B [from 6 and 7]

5.   1.   □(A ⊃ B)   INVALID
      2.   u:A
        [∴ u:B
      3.   asm: ~u:B

10.   * 1.   ~◇(A · B)   INVALID
        [∴ ~(~u:~A · ~u:~B)
     * 2.   asm: (~u:~A · ~u:~B)
     * 3.   ∴ ~u:~A [from 2]
     * 4.   ∴ ~u:~B [from 2]
      5.   ∴ □~(A · B) [from 1]

   6.   u ∴ A [from 3]
   7.   uu ∴ B [from 4]
* 8.   u ∴ ~(A · B) [from 5]
* 9.   uu ∴ ~(A · B) [from 5]
  10.   u ∴ ~B [from 6 and 8]
  11.   uu ∴ ~A [from 7 and 9]

   8.2B

1. * 1.   ~◇(A · B)   INVALID
       2.   u:A
               [∴ ~u:B
       3.   asm: u:B
       4.   ∴ □~(A · B) [from 1]
       5.   u ∴ B [from 3]
    * 6.   u ∴ ~(A · B) [from 4]
       7.   u ∴ ~A [from 5 and 6]
3. * 1.   ~◇A   VALID
               [∴ ~u:A
       2.   ⎡asm: u:A
       3.   ⎮ ∴ □~A [from 1]
       4.   ⎮ u ∴ A [from 2]
       5.   ⎣ u ∴ ~A [from 3, contradicting 4]
       6.   ∴ ~u:A [from 2, 4, and 5]
5.   [∴ (u:A ⊃ ~u:~A)   INVALID
    * 1.   asm: ~(u:A ⊃ ~u:~A)
       2.   ∴ u:A [from 1]
       3.   ∴ u:~A [from 1]
10.   [∴ (u:A ⊃ ◇A)   VALID
    * 1.   ⎡asm: ~(u:A ⊃ ◇A)
       2.   ⎮ ∴ u:A [from 1]
    * 3.   ⎮ ∴ ~◇A [from 1]
       4.   ⎮ ∴ □~A [from 3]
       5.   ⎮ u ∴ A [from 2]
       6.   ⎣ u ∴ ~A [from 4]
       7.   ∴ (u:A ⊃ ◇A) [from 1, 5, and 6]

   8.3A

1. u:Sj
3. (x)~Eux
5. (Fu · ~u:Fu)
10. ~(u:(x)~Eux · u:Eut)
15. (u:Axu ⊃ Aux)

   8.3B

1.   [∴ ~(u:A · u:~A)   VALID
    * 1.   ⎡asm: (u:A · u:~A)
       2.   ⎮ ∴ u:A [from 1]
       3.   ⎮ ∴ u:~A [from 1]
       4.   ⎮ u ∴ A [from 2]
       5.   ⎣ u ∴ ~A [from 3, contradicting 4]

6. ∴ ∴ ~(u:A · u:~A) [from 1, 4, and 5]

3. [∴ (u:Ba ∨ u:~Ba)   INVALID
* 1.   asm: ~(u:Ba ∨ u:~Ba)
* 2.   ∴ ~u:Ba [from 1]
* 3.   ∴ ~u:~Ba [from 1]
  4.   u ∴ ~Ba [from 2]
  5.   uu ∴ Ba [from 3]

5. 1.   u:(x)OAx   INVALID
    [∴ u:Au
* 2.   asm: ~u:Au
  3.   u ∴ ~Au [from 2]

10. 1.   □(A ⊃ B)   VALID
    [∴ ~(u:OA · ~u:B)
* 2.   ⌈asm: (u:OA · ~u:B)
  3.   │ ∴ u:OA [from 2]
* 4.   │ ∴ ~u:B [from 2]
  5.   │ u ∴ ~B [from 4]
  6.   │ u ∴ OA [from 3]
* 7.   │ u ∴ (A ⊃ B) [from 1]
  8.   │ u ∴ A [from 6]
  9.   ⌊u ∴ B [from 7 and 8, contradicting 5]
  10.   ∴ ~(u:OA · ~u:B) [from 2, 5, and 9]

8.3C

1.   [∴ (u:OAu ⊃ Au)   INVALID
* 1.   asm: ~(u:OAu ⊃ Au)
  2.   ∴ u:OAu [from 1]
  3.   ∴ ~Au [from 1]

3.   [∴ ~(u:~RAj · u:Aj)   VALID
* 1.   ⌈asm: (u:~RAj · u:Aj)
  2.   │ ∴ u:~RAj [from 1]
  3.   │ ∴ u:Aj [from 1]
* 4.   │ u ∴ ~RAj [from 2]
  5.   │ u ∴ Aj [from 3]
  6.   │ u ∴ O~Aj [from 4]
  7.   ⌊u ∴ ~Aj [from 6, contradicting 5]
  8.   ∴ ~(u:~RAj · u:Aj) [from 1, 5, and 7]

5.   [∴ ~(u:Au · ~u:RAu)   VALID
* 1.   ⌈asm: (u:Au · ~u:RAu)
  2.   │ ∴ u:Au [from 1]
* 3.   │ ∴ ~u:RAu [from 1]
* 4.   │ u ∴ ~RAu [from 3]
  5.   │ u ∴ Au [from 2]
  6.   │ u ∴ O~Au [from 4]
  7.   ⌊u ∴ ~Au [from 6, contradicting 5]
  8.   ∴ ~(u:Au · ~u:RAu) [from 1, 5, and 7]

10.   1.   u:OAu   VALID
           [∴ u:Au
    * 2.   ┌ asm: ~u:Au
      3.   │ u ∴ ~Au [from 2]
      4.   │ u ∴ OAu [from 1]
      5.   └ u ∴ Au [from 4, contradicting 3]
      6.   ∴ u:Au [from 2, 3, and 5]
15.   [∴ ~(u:(x)~RKx · ~u:(N ⊃ ~Ku))   VALID
    * 1.   ┌ asm: (u:(x)~RKx · ~u:(N ⊃ ~Ku))
      2.   │ ∴ u:(x)~RKx [from 1]
    * 3.   │ ∴ ~u:(N ⊃ ~Ku) [from 1]
    * 4.   │ u ∴ ~(N ⊃ ~Ku) [from 3]
      5.   │ u ∴ (x)~RKx [from 2]
      6.   │ u ∴ N [from 4]
      7.   │ u ∴ Ku [from 4]
    * 8.   │ u ∴ ~RKu [from 5]
      9.   │ u ∴ O~Ku [from 8]
      10.   └ u ∴ ~Ku [from 9, contradicting 7]
      11.   ∴ ~(u:(x)~RKx · ~u:(N ⊃ ~Ku)) [from 1, 7, and 10]

8.4A

1. Ou:Sj   **15.** (Ou:A ≡ ~◇(u:A · ~A))
3. Ru:OSj   **20.** (u:Axu ⊃ OAux)
5. (x)~Rx:G   **25.** ((Ou:Bj · ~Du) ⊃ Ou:Pj)
10. (Ou:(x)x = x · ((x)x = x · u:(x)x = x))

8.5A

1.   1.   Oa:(C · D)   INVALID
           [∴ Ob:C
    * 2.   asm: ~Ob:C
    * 3.   ∴ R~b:C [from 2]
    * 4.   D ∴ ~b:C [from 3]
      5.   D ∴ a:(C · D) [from 1]
      6.   Db ∴ ~C [from 4]
    * 7.   Da ∴ (C · D) [from 5]
      8.   Da ∴ C [from 7]
      9.   Da ∴ D [from 7]
3.   [∴ O~(u:A · ~u:◇A)   VALID
    * 1.   ┌ asm: ~O~(u:A · ~u:◇A)
    * 2.   │ ∴ R(u:A · ~u:◇A) [from 1]
    * 3.   │ D ∴ (u:A · ~u:◇A) [from 2]
      4.   │ D ∴ u:A [from 3]
    * 5.   │ D ∴ ~u:◇A [from 3]
    * 6.   │ Du ∴ ~◇A [from 5]
      7.   │ Du ∴ A [from 4]
      8.   │ Du ∴ □~A [from 6]
      9.   └ Du ∴ ~A [from 8, contradicting 7]
      10.   ∴ O~(u:A · ~u:◇A) [from 1, 7, and 9]

5. 1. □(A ⊃ B) INVALID
  [∴ (R~u:B ⊃ Ru:~A)
* 2. asm: ~(R~u:B ⊃ Ru:~A)
* 3. ∴ R~u:B [from 2]
* 4. ∴ ~Ru:~A [from 2]
  5. ∴ O~u:~A [from 4]
* 6. D ∴ ~u:B [from 3]
* 7. D ∴ ~u:~A [from 5]
  8. Du ∴ ~B [from 6]
  9. Duu ∴ A [from 7]
* 10. Du ∴ (A ⊃ B) [from 1]
* 11. Duu ∴ (A ⊃ B) [from 1]
  12. Du ∴ ~A [from 8 and 10]
  13. Duu ∴ B [from 9 and 11]

10. 1. Ou:(A ⊃ OBu) VALID
  [∴ ~(u:A · ~u:Bu)
* 2. ⌈asm: (u:A · ~u:Bu)
  3. │ ∴ u:A [from 2]
* 4. │ ∴ ~u:Bu [from 2]
  5. │ ∴ u:(A ⊃ OBu) [from 1]
  6. │ u ∴ ~Bu [from 4]
  7. │ u ∴ A [from 3]
* 8. │ u ∴ (A ⊃ OBu) [from 5]
  9. │ u ∴ OBu [from 7 and 8]
  10. ⌊ u ∴ Bu [from 9, contradicting 6]
  11. ∴ ~(u:A · ~u:Bu) [from 2, 6, and 10]

8.5B

1. 1. Ou:G VALID
  [∴ ~Ru:~G
* 2. ⌈asm: Ru:~G
  3. │ D ∴ u:~G [from 2]
  4. │ D ∴ u:G [from 1]
  5. │ Du ∴ ~G [from 3]
  6. ⌊ Du ∴ G [from 4, contradicting 5]
  7. ∴ ~Ru:~G [from 2, 5, and 6]

3. [∴ (Ru:A ⊃ Ru:RA) VALID
* 1. ⌈asm: ~(Ru:A ⊃ Ru:RA)
* 2. │ ∴ Ru:A [from 1]
* 3. │ ∴ ~Ru:RA [from 1]
  4. │ ∴ O~u:RA [from 3]
  5. │ D ∴ u:A [from 2]
* 6. │ D ∴ ~u:RA [from 4]
* 7. │ Du ∴ ~RA [from 6]
  8. │ Du ∴ A [from 5]
  9. │ Du ∴ O~A [from 7]
  10. ⌊ Du ∴ ~A [from 9, contradicting 8]
  11. ∴ (Ru:A ⊃ Ru:RA) [from 1, 8, and 10]

5.   * 1.   R(u:G · u:T)   VALID
     2.   □(T ⊃ E̅)
           [∴ R(u:G · u:E)
   * 3.   ⌈asm: ~R(u:G · u:E)
     4.   │ ∴ O~(u:G̅ · u:E̅) [from 3]
   * 5.   │ D ∴ (u:G · u:T) [from 1]
   * 6.   │ D ∴ ~(u:G̅ · u:E) [from 4]
     7.   │ D ∴ u:G̅ [from 5]
     8.   │ D ∴ u:T [from 5]
   * 9.   │ D ∴ ~u:E [from 6 and 7]
    10.   │ Du ∴ ~E [from 9]
    11.   │ Du ∴ G [from 7]
    12.   │ Du ∴ T [from 8]
 * 13.   │ Du ∴ (T ⊃ E) [from 2]
    14.   ⌊Du ∴ E [from 12 and 13, contradicting 10]
    15.   ∴ R(u:G · u:E) [from 3, 10, and 14]

10. * 1.   Ru:A   INVALID
   * 2.   Ru:B
           [∴ Ru:(A · B)
   * 3.   asm: ~Ru:(A · B)
     4.   ∴ O~u:(A̅ · B̅) [from 3]
     5.   D ∴ u:A [from 1]
     6.   DD ∴ u:B [from 2]
   * 7.   D ∴ ~u:(A · B) [from 4]
   * 8.   DD ∴ ~u:(A · B) [from 4]
   * 9.   Du ∴ ~(A · B) [from 7]
 * 10.   DDuu ∴ ~(A · B) [from 8]
   11.   Du ∴ A [from 5]
   12.   DDuu ∴ B [from 6]
   13.   Du ∴ ~B [from 9 and 11]
   14.   DDuu ∴ ~A [from 10 and 12]

To see the invalidity here, consider this example. If the evidence for rain today is mixed, you might be able to reasonably believe ''It will rain'' and you might be able to reasonably believe ''It won't rain.'' Either view might be reasonable. It doesn't follow that you could reasonably believe ''It will rain and it won't rain.''

15.   1.   Ou:G   VALID
           [∴ ~R̅(~u:G · ~u:~G)
   * 2.   ⌈asm: R(~u:G · ~u:~G)
     3.   │ D ∴ (~u:G · ~u:~G) [from 2]
     4.   │ D ∴ u:G [from 1]
     5.   ⌊D ∴ ~u:G [from 3, contradicting 4]
     6.   ∴ ~R(~u:G · ~u:~G) [from 2, 4, and 5]

20.   [∴ O~(u:~A · u:R̅A)   INVALID
   * 1.   asm: ~O~(u:~A · u:RA)
   * 2.   ∴ R(u:~A · u:R̅A) [from 1]
   * 3.   D ∴ (u:~A · u:R̅A) [from 2]
     4.   D ∴ u:~A [from 3]
     5.   D ∴ u:RA̅ [from 3]

      6.   Du ∴ ~A̱ [from 4]
* 7.   Du ∴ RA̱ [from 5]
      8.   DuD ∴ A̱ [from 7]

25. 1.   □(A ⊃ Ḇ)   VALID
* 2.   Ru:A̱
        [∴ Ru:Ḇ
* 3.   ⌈asm: ~Ru:Ḇ
      4.   │∴ O~u:Ḇ [from 3]
      5.   │D ∴ u̱:A [from 2]
* 6.   │D ∴ ~u̱:B [from 4]
      7.   │Du ∴ ~Ḇ [from 6]
      8.   │Du ∴ A [from 5]
* 9.   │Du ∴ (A ⊃ B) [from 1]
    10.   ⌊Du ∴ B [from 8 and 9, contradicting 7]
    11.   ∴ Ru:Ḇ [from 3, 7, and 10]

30. * 1.   Ru̱:G   INVALID
      2.   (G̱ ⊃ A)
        [∴ Ru̱:A
* 3.   asm: ~Ru̱:A
      4.   ∴ O~u:A̱ [from 3]
      5.   D ∴ u̱:G [from 1]
* 6.   D ∴ ~u̱:A [from 4]
      7.   Du ∴ ~A̱ [from 6]
      8.   Du ∴ G [from 5]

35. 1.   Ou:(RHuj̱ ⊃ RHju)   VALID
        [∴ ~(u:Huj̱ · u:~RHju̱)
* 2.   ⌈asm: (u:Huj̱ · u:~RHju̱)
      3.   │∴ u:Huj̱ [from 2]
      4.   │∴ u:~RHju̱ [from 2]
      5.   │∴ u:(RHuj̱ ⊃ RHju̱) [from 1]
      6.   │u ∴ Huj̱ [from 3]
      7.   │u ∴ ~RHju̱ [from 4]
* 8.   │u ∴ (RHuj̱ ⊃ RHju̱) [from 5]
* 9.   │u ∴ ~RHuj̱ [from 7 and 8]
    10.   │u ∴ O~Huj̱ [from 9]
    11.   ⌊u ∴ ~Huj̱ [from 10, contradicting 6]
    12.   ∴ ~(u:Huj̱ · u:~RHju̱) [from 2, 6, and 11]

9.4A

1. This form is incorrect. Correct forms talk about similar circumstances and use a *don't-combine* wording. Here's a correct formulation: "Don't act to remove Jones's appendix without consenting to the idea of Jones removing your appendix in an exactly reversed situation."

3. This form is correct.

5. This form is incorrect. Correct forms use a *don't-combine* wording and talk about one's present consent to a hypothetical situation. Here's a correct formulation: "Don't act to jail this criminal without consenting to the idea that if you were in this criminal's exact place then you'd be jailed."

10. This form is correct.
15. This form is correct.

9.6A

1. $F*A\underline{u}j$
3. $(VA\underline{u}j \supset {\sim}MA\underline{u}j)$
5. $(VA\underline{j}u \supset {\sim}RA\underline{j}u)$
10. $({\sim}R\underline{A} \supset (\exists Z)(Z\underline{A} \cdot \blacksquare(\underline{X})(Z\underline{X} \supset {\sim}R\underline{X})))$
15. $HA\underline{j}u$
20. $u{:}\blacksquare(HA\underline{j}u \supset MA\underline{j}u)$
25. ${\sim}(u{:}A\underline{u}x \cdot {\sim}\underline{u}{:}(\exists Z)(Z*A\underline{u}x \cdot \blacksquare(ZA\underline{x}u \supset MA\underline{x}u)))$

INDEX

DEONTIC LOGIC

"A" or "Au" = "You do A" (indicative)

"$\underline{A}$" or "A$\underline{u}$" = "Do A!" (imperative)

O$\underline{A}$ = A is obligatory ($\underline{A}$ is in all deontic worlds)
R$\underline{A}$ = A is all right ($\underline{A}$ is in some deontic world)

Reverse squiggle rules
$\dfrac{\sim R\underline{A}}{\therefore\ O\sim\underline{A}} \qquad \dfrac{\sim O\underline{A}}{\therefore\ R\sim\underline{A}}$

Dropping "R"		Drop R's before O's!	Dropping "O"	
$\dfrac{R\underline{A}}{D\ \therefore\ \underline{A}}$	(use a new string of D's)		$\dfrac{O\underline{A}}{D\ \therefore\ \underline{A}}$	(Use any string of D's — or no D's at all)

Indicative transfer rule			Kant's Law
$\dfrac{A}{D\ \therefore\ A}$	$\dfrac{D\ \therefore\ A}{\therefore\ A}$	(Use an indicative for "A"; use any string of D's)	$\dfrac{O\underline{A}}{\therefore\ \Diamond A}$

BELIEF LOGIC

u:A = You believe A. u:$\underline{A}$ = You will A.	$\underline{u}$:A = Believe A! $\underline{u}$:$\underline{A}$ = Will A!

Dropping "$\sim\underline{u}$:"		Drop "$\sim\underline{u}$:" before "$\underline{u}$:"!	Dropping "$\underline{u}$:"	
$\dfrac{\sim\underline{u}:A}{u\ \therefore\ \sim A}$	(Use a new string of u's)		$\dfrac{\underline{u}:A}{u\ \therefore\ A}$	(Use any string of one or more u's)